Hydrosystems Engineering
Uncertainty Analysis

Hydrosystems Engineering Uncertainty Analysis

Yeou-Koung Tung, Ph.D.
Department of Civil Engineering
Hong Kong University of Science & Technology

Ben-Chie Yen, Ph.D.
Late Professor
Department of Civil and Environmental Engineering
University of Illinois at Urbana–Champaign

McGraw-Hill

New York   Chicago San Francisco Lisbon   London   Madrid
Mexico City   Milan   New Delhi   San Juan   Seoul
Singapore   Sydney   Toronto

Cataloging-in-Publication Data is on file with the Library of Congress.

The sponsoring editor for this book was Larry Hager and the production supervisor was Pamela Pelton. The art director for the cover was Anthony Landi. It was set in Century Schoolbook by International Typesetting and Composition.

Printed and bound by RR Donnelley.

 This book was printed on recycled, acid-free paper containing 10% postconsumer waste.

McGraw-Hill books are available at special quantity discounts to use as premiums and sales promotions, or for use in corporate training programs. For more information, please write to the Director of Special Sales, Professional Publishing, McGraw-Hill, Two Penn Plaza, New York, NY 10121-2298. Or contact your local bookstore.

To humanity and human welfare

As far as laws of mathematics refer to reality,
They are not certain;
And
As far as they are certain,
They do not refer to reality.

—Albert Einstein

Contents

Preface

Over the past two decades or so, there has been a steady growth in the development and application of uncertainty analysis techniques in hydrosystems engineering and other disciplines. More and more hydrosystem engineering designs and analyses have been found to go beyond the conventional frequency analysis of rainstorms and floods to consider other aspects of uncertainties. A recent U.S. National Research Council study report in 2000 on the U.S. Army Corps of Engineers' risk analysis approach for flood damage reduction highlights the ideological change in designing flood defense systems in the States, which could be extended to other hydrosystem engineering systems. Engineers dealing with hydrosystem designs and management in the future may be expected to explicitly consider uncertainties involved and to make assessment of the performance reliability of the systems they designed. The aim of this book is to bring together these uncertainty analysis techniques in one book and to demonstrate their applications and limitations for a wide variety of hydrosystem engineering problems. A complementary book, *Hydrosystem Engineering Reliability Assessment and Risk Analysis*, will be devoted to the reliability-related issues in hydrosystem engineering infrastructural designs.

The main areas of concern and applications in the proposed book are hydrosystems and related environmental engineering systems. The term "hydrosystems" was first coined by the late Professor Ven-Te Chow and is now being used widely to encompass various water resource systems including, but not limited to, surface water storage, groundwater, water distribution, flood control, and drainage. As many hydrosystem engineering and management problems address both quantity and quality aspects of water and other environmental issues, it is almost mandatory for an engineer involved in major hydrosystem designs or hazardous waste management problems to quantify the potential risk of failure due to uncertainties and to assess the associated consequences. Uncertainty analysis serves as the basis for the reliability assessment of hydrosystems engineering, which includes designing the geometry and dimension of hydraulic facilities, planning hydraulic projects, developing operation procedures and management strategies, and conducting risk-cost analysis or risk-based decision-making.

This book integrates uncertainty analysis with knowledge in hydrosystems engineering. Many of the examples and problems in the book bring together the use of probability and statistics, along with the knowledge of hydrology, hydraulics, and water resources for the uncertainty analysis of various water-related problems. Hence the book is primarily for the upper-level undergraduate and graduate students for uncertainty analysis of hydrosystem engineering problems. Most of the principles and methodologies presented in the book can equally be applied to other civil engineering disciplines. It presents relevant theories of uncertainty analysis in a systematic fashion to illustrate applications to various hydrosystem engineering problems. Although more advanced statistical and mathematical skills are occasionally required, a great majority of the uncertainty analysis methods can be understood with a basic knowledge of probability and statistics.

The book consists of six chapters. Chapter 1 provides a general introduction of various uncertainties present in hydrosystem engineering design and management. Issues related to engineering design and management due to uncertainties are discussed. In Chapter 2, the fundamentals of probability and statistics pertinent to uncertainty analysis are summarized. Regression analysis is described in Chapter 3 as it is widely used by engineers to develop empirical models. However, uncertainties embedded in a regression model are often overlooked in its application during engineering design. Emphasis, therefore, is placed on the quantification of the uncertainty associated with model outputs to be used in hydrosystems engineering and management. The book does not dwell on data analysis and related subjects (e.g., hypothesis tests) since many excellent textbooks are available on the subject matter. Chapters 4 to 6 provide detailed descriptions of various techniques applicable to uncertainty analysis. Chapter 4 focuses on the different analytical techniques that allow the direct derivation of the exact uncertainty features of hydrosystems. The advantages and weaknesses of the techniques are elaborated. For most real-life problems, the complexity of the systems often prohibits attainability of exact solutions. Hence, Chapter 5 covers several approximate techniques for estimating uncertainty features of complex hydrosystem models in the form of figures, tables, and computer programs. As Monte-Carlo simulation is commonly applied to uncertainty analysis, it is elaborated in Chapter 6. In particular, emphasis is placed on the multivariate settings since most real-life systems involve several correlated, nonnormal variables. For each technique described in the book, ample examples are given to illustrate the methodology for better understanding of the materials. In addition, a large number of end-of-chapter problems are provided for practice.

The intended uses and audiences for the book are: (1) as a textbook for an introductory course at the undergraduate senior level or graduate master level in hydrosystems engineering on the uncertainty and reliability related subjects; (2) as a textbook for a second course at master and Ph.D. levels in uncertainty and reliability analysis of hydrosystems engineering covering more advanced topics; (3) as a reference book for researchers and practicing engineers dealing with hydrosystems engineering, planning, management, and decision-making.

The expected background knowledge for the readers of this book is a minimum of 12 credits of mathematics including calculus, matrix algebra, probability, and statistics, a one-semester course in elementary fluid mechanics, and a one-semester course in elementary water resources covering the basic principles in hydrology and hydraulics. Additional knowledge of engineering economics, water-quality models, and optimization would be desirable.

Two possible one-semester courses could be taught from this book depending on the background of the students and the type of course designed by the instructor.

Course outline 1.   (For students having the first exposure to uncertainty and reliability analysis of hydrosystem infrastructures.) It is suitable for upper-level undergraduate and first-year graduate students. The subject materials could include Chapter 1; Chapter 2 (2.1 to 2.6); Chapter 3; Chapter 4 (4.1 to 4.3); Chapter 5 (5.1 to 5.3); Chapter 6 (6.1 to 6.5)

Course outline 2.   (For students having taken a one-semester probability and statistics course.) The objective aims at achieving a higher level of capability to conduct uncertainty analysis. The course contents could include Chapters 1, 3, 4, 5, and 6.

The idea of writing this book and the accompanying book on hydrosystems engineering uncertainty/reliability analysis was born in 1989 to 1991 when the authors cochaired the ASCE Subcommittee which eventually resulted in an edited book, *Uncertainty and Reliability Analysis in Design of Hydraulic Structures*, published in 1993. Since then, there have been great advancements in the development of uncertainty analysis techniques and widespread applications of the techniques in hydrosystems engineering. For years, the authors strongly felt the need for a book that could systematically introduce readers to the water resource engineering profession, where we are now in this area of development. We are hoping, as an old Chinese saying goes, to use our work of "bricks" to attract more scholars and professionals to contribute their works of "jades."

Acknowledgments

The materials in this book are based on many years of teaching and research by the authors. We, first and foremost, would like to thank God for giving us His strength needed to complete the book. He surrounds us with people who do not hold back their support for us and that makes the task tolerable. We are grateful to our families, especially our wives Ruth and Be-ling, for giving us the love and peace of mind needed for writing the book and for their patience and understanding.

Over the years of preparing this book, our students have offered enormous help in various forms including manuscript reading from a student's perspectives, preparing figures, gathering the literature, and developing solution manuals. For that, we are truly thankful to Chen Xingyuang, Lu Zhihua, Wang Ying, Eddy YF Lau, and Wu Shiang-Jen. We also are thankful to Shue-Fen Tung and Shue-Wen Tung (daughters of the first author) for proofreading the earlier version of the book. Especially, the authors are deeply indebted to Dr. Steve C. Mechling of the University of Marquette for meticulously going through the entire manuscript, which immensely improved the book in many ways. Of course, any errors that remain are entirely our fault and negligence. We would also like to acknowledge our institutions, which provided support for the preparation of the book.

As the first author, I would take this opportunity to thank many of my colleagues for their encouragement and friendship all through the years. In particular, my deepest appreciation goes to two individuals, Professor Larry W. Mays of Arizona State University and the late Professor Ben C. Yen, the coauthor. During my entire career, Professors Mays and Yen (before his untimely demise) have been unflagging supporters and mentors to me. I would like to thank my dear friend, Ms. Joanne Lam, for her constant encouragement and prayer during the course of writing this book. Last, but not the least, gratitude is extended to McGraw-Hill for supporting the publication of the book, to Mr. Larry Hager, the editor, for his advice for preparing the book, and, in particular, to Ms. Mona Tiwary and her team, for their professional assistance in producing the book.

Yeou-Koung Tung
Hong Kong

Uncertainties in Hydrosystems Engineering and Management

One cannot avoid uncertainties in life. Neither can uncertainties be avoided in most engineering projects. In hydrosystem engineering infrastructural designs, uncertainties arise in various aspects. For example, uncertainties in designing a hydrosystem may include, but are not limited to, hydraulic, hydrologic, structural, environmental, and socio-economic aspects.

1.1 Definition of Uncertainty

Uncertainty is attributed to the lack of perfect information concerning the phenomena, processes, and data involved in problem definition and resolution. Uncertainty could simply be defined as the occurrence of events that are beyond one's control (Mays and Tung 1992). In practically all engineering designs and operations, decisions are frequently made under uncertainty. As such, the reliability and safety of engineering projects are closely related to the level of uncertainty involved.

1.2 Types and Sources of Uncertainty

Yen and Ang (1971) classified uncertainties into two types—*objective uncertainties* associated with any random process or deducible from statistical samples, and *subjective uncertainties* for which no quantitative factual information is available. These uncertainties are sometimes referred to as *aleatory* and *epistemic uncertainties* (NRC 2000). Halder and Mahadevan (2000) referred to them as *noncognitive* and *cognitive* uncertainties, respectively. Yevjevich (1972) distinguished between the *basic risk* due to the inherent randomness of the process and uncertainty due to various other sources. The *overall risk* in an engineering system is the result of the combined effect of basic risk and uncertainties. As shown in Fig. 1.1, uncertainties come from two groups of sources—natural

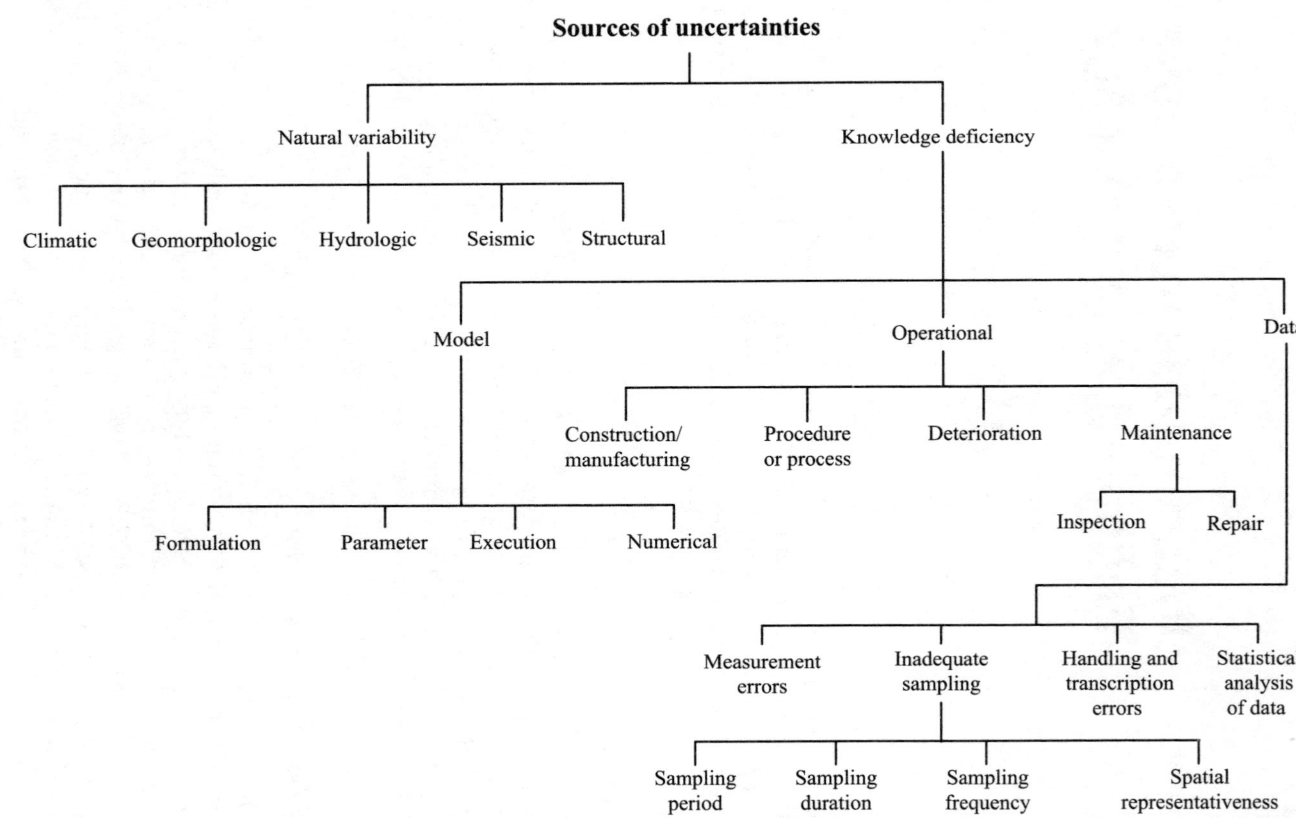

Figure 1.1 Sources of uncertainty.

variability and knowledge deficiency. In general, the uncertainty due to the inherent randomness of geophysical processes cannot be eliminated and one has to live with it although the improvement of the physical system may reduce the uncertainty. As noted by Plate (1986), "The design engineer therefore must live with a large random residue even if he had a perfect model of the hydrologic cycle, and of the fluid mechanical transformation." On the other hand, uncertainties associated with the knowledge deficiency about processes, models, parameters, and the like, can be reduced through research, data collection, and careful manufacturing.

In general, uncertainties in hydrosystem infrastructures can be divided into five basic categories: geophysical, transmissional, structural, operational, and economic. *Structural uncertainty* refers to failure from structural weaknesses. Physical failure of structures in an infrastructural system can be caused by many things such as water saturation and loss of soil stability, erosion or hydraulic soil failure, wave action, overloading, and structural collapse. *Economic uncertainty* can arise from uncertainties in construction costs, damage costs, projected revenue, operation and maintenance cost, inflation, project life, and other intangible benefit and cost items. Yen, Cheng, and Melching (1986) classified various sources of uncertainty in the analyses and designs of hydraulic engineering systems including natural variability, model uncertainties, parameter uncertainties, data uncertainties, and operational uncertainties that are equally relevant for other civil engineering infrastructural systems.

Natural variability is associated with the inherent randomness of natural geophysical processes such as the occurrence of precipitation, floods, high winds, and earthquakes. The occurrence of geophysical events often displays variations in time and space. Figures 1.2 to 1.5 are examples that illustrate the inherent randomness of the different geophysical variables encountered in civil engineering infrastructural systems design. For example, the number of storm events and their magnitudes vary from location to location and from time to time. Their occurrences and magnitudes cannot be predicted precisely in advance. Table 1.1 shows the range of flow characteristics affecting river-water quality.

Among the uncertainties due to knowledge deficiency, the most important are those of model, operation, and data. Engineering designs often require the use of models. Beck (1987) noted that uncertainties affect primarily four problem areas that must be addressed to improve the accuracy and usefulness of models:

1. Uncertainty about model structure or formulation, i.e., what are the basic processes involved, how do they interact, and how can these processes and interactions be mathematically characterized in an efficient and parsimonious manner

2. Uncertainty in the model parameters, i.e., parameter identification and calibration problems

3. Uncertainty associated with estimates of the future behavior of the system, i.e., aggregation of uncertainties in model structure or formulation, model

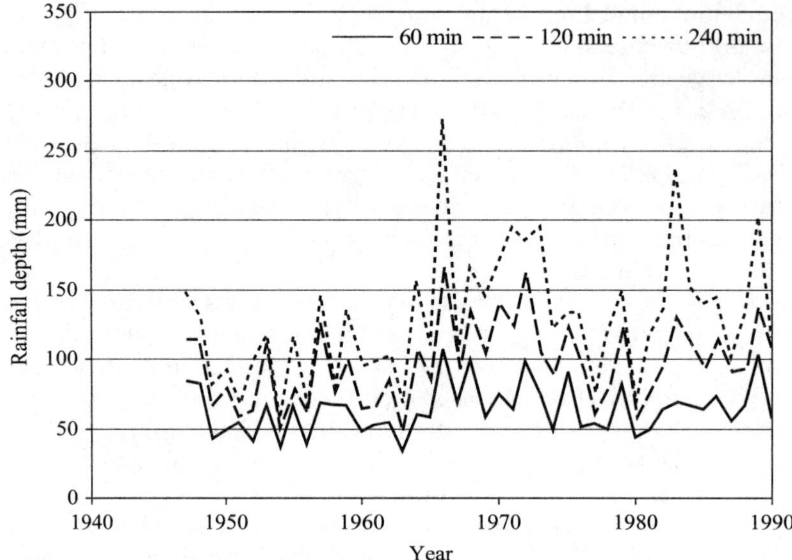

Figure 1.2 Annual maximum rainfall series of different durations (1947–1990) at Hong Kong Observatory, Hong Kong.

parameters, and in the definition of design or decision scenario into overall estimation uncertainty

4. Reduction of critical modeling uncertainties through carefully designed experiments and monitoring programs

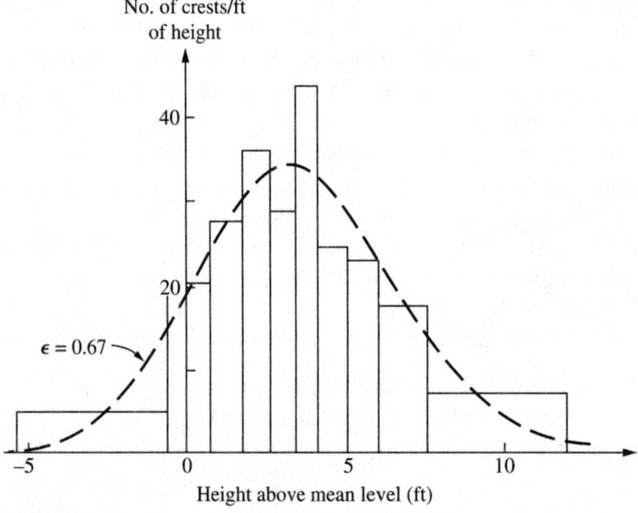

Figure 1.3 Histogram of wave height above mean sea level (Cartwright and Longuet-Higgins 1956).

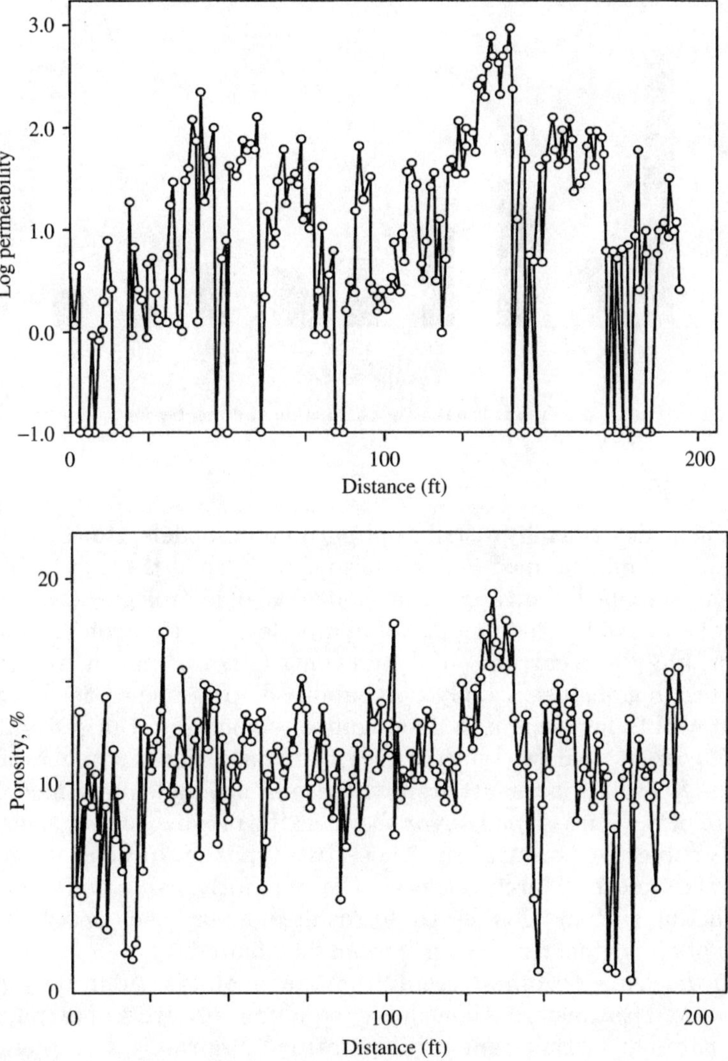

Figure 1.4 Spatial variation of log-transformed permeability (milli-darcy) and porosity measured based on core samples from a borehole in the Mt. Simon sandstone aquifer in Illinois (Bakr 1976).

Model formulation varies over a wide spectrum, ranging from simple empirical equations to sophisticated partial differential equations with computer simulations. It should be recognized that a model is only an abstraction of reality, which generally involves certain degrees of simplifications and idealizations. *Model formulation uncertainty* reflects the inability of the model or design procedure to represent precisely the system's true physical behavior. For example, in hydrologic modeling, runoff from rainfall varies temporally and spatially and

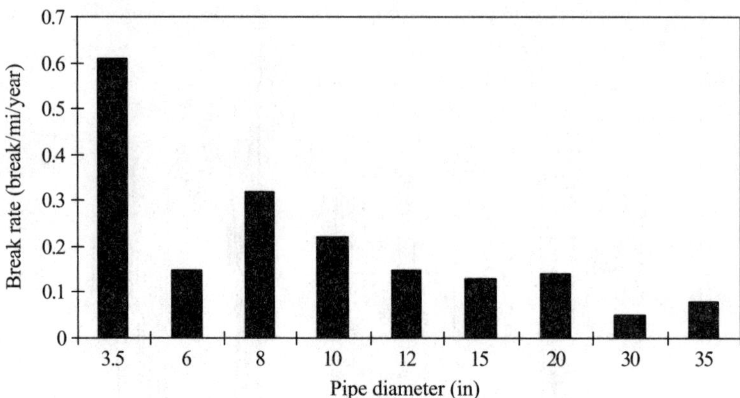

Figure 1.5 Histogram of pipe break rate as function of diameter for St. Louis (Goodrich et al. 1989).

should be described by distributed-parameter models. However, in engineering practice simple, lumped-parameter models, such as the unit hydrograph model, are often applied as an approximation. Also, in hydrologic flood frequency analysis, due to the limited amount of available data, the true random mechanism describing the occurrence of flood events is never known. Yet, it is a common practice that engineers apply a probability distribution model for estimating rare flood events for designing flood control structures. Burges and Lettenmaier (1975) categorized two types of uncertainty associated with mathematical modeling. *Type I error* results from the use of an inadequate model with correct parameter values. *Type II error* assumes the use of a perfect model with parameters subject to uncertainty. These two types of uncertainty simultaneously exist in almost all circumstances. Consequently, using an imperfect model for predicting system behavior could result in error based on which the performance of an engineering design cannot be ensured.

Parameter uncertainties result from the inability to accurately quantify model inputs and parameters. All models or equations involve several physical or empirical parameters that cannot be quantified accurately. Parameter uncertainty

TABLE 1.1   Range of Water Quality Parameters for Different Types of Stream Environment (Chadderton, Miller, and McDonnell 1982)

Stream class description	Reareation coefficient (per day)	Deoxygenation coefficient (per day)	Average flow velocity (m/s)	Water depth (m)
Sluggish	0.05–0.10	0.033–0.08	0.03–0.15	3.05–6.10
Low velocity	0.1–1.0	0.05–0.67	0.03–0.15	0.92–3.05
Moderate velocity	1.0–5.0	0.5–2.5	0.15–0.61	0.61–1.52
Swift	1.0–10.0	0.2–3.33	0.61–1.83	0.61–3.05

could be caused by changes in the operational conditions of infrastructure, the inherent variability of inputs and parameters in time and space, and a lack of sufficient data. Consequently, quantities obtained from a model, such as the average flow velocity in the channel and the peak discharge of urban runoff, cannot be assessed with certainty. Examples of parameter uncertainty are: (1) statistical parameters, such as mean and standard deviation, in a probability distribution model that cannot be estimated accurately due to limited amounts of sample data; (2) physical parameters, such as channel slope, roughness coefficient, and bed material properties that can vary both in space and time; and (3) coefficients in empirical equations that are developed on the basis of a limited amount of sample data through calibration or fitting of the model to the data.

Data uncertainties include (1) measurement errors, (2) inconsistency and nonhomogeneity of data, (3) data handling and transcription errors, and (4) inadequate representation of data samples due to time and space limitations.

Operational uncertainties include those associated with construction, manufacture, procedure, deterioration, maintenance, and human activities. Construction and manufacturing tolerances may result in a difference between the "nominal" and actual values. For example, a so-called 1000-mm diameter circular pipe obtained from a manufacturer may not be exactly 1000 mm in diameter; there could be an error of ±5 mm associated with it. The magnitude of this type of uncertainty is largely dependent on the workmanship and quality control during construction and manufacturing. As another example of construction quality uncertainty, Melching and Yen (1986) found that for 80 storm sewers in Tempe, Arizona, ranging in slope from 0.001 to 0.0055 ft/ft, comparison of "as built" slope data with design slopes yielded a standard construction error of 0.0008 ft/ft. For shallow-sloped sewers (0.001 to 0.002 ft/ft), this construction error had a large effect on the reliability of the sewer. The roughness coefficient of a storm sewer pipe could change over its service life due to the settlement of sediment materials and blockage by debris. Progressive deterioration due to the lack of proper maintenance could lead to changes in resistance coefficients and sewer flow capacity. This results in additional uncertainty in the design and evaluation of infrastructure performance.

Ang and Tang (1984) classify model *prediction error* into systematic errors and random errors. *Measurement errors* can also be categorized into systematic and random errors (British Standard Institution 1998; Rabinovich 2000). *Systematic errors* may arise from factors not accounted for in the model. Hence, model prediction tends to produce biased results that consistently overpredict or underpredict the outcomes of the process. *Random errors* are associated with the range of possible errors primarily due to sampling error. In general, systematic errors associated with model prediction can be removed by multiplying a bias-correction factor to or by subtracting the bias from the model output. Figures 1.6(*a*) to (*c*) show different cases of error involved in model prediction. Figure 1.6(*a*) demonstrates predictions involving only random errors. For models that produce unbiased predictions, the one associated with a smaller random error is more accurate. In Fig. 1.6(*b*), model prediction includes only systematic errors. Two

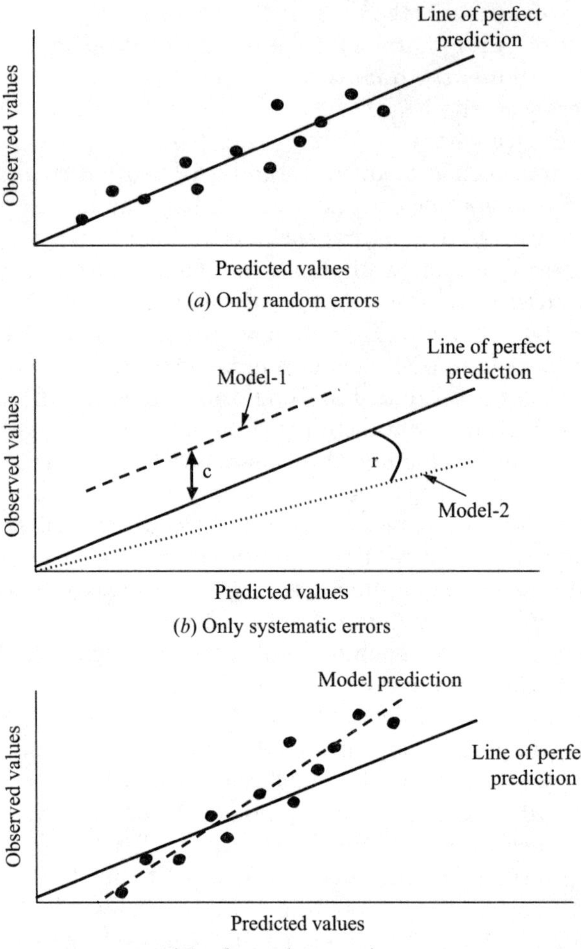

Figure 1.6   Types of prediction errors.

types of systematic errors can be observed in the figure. Model-1 consistently overpredicts the true values by a constant "c," whereas model-2 consistently underpredicts the true values by a ratio "r." In such cases, correction to systematic errors can be made to model-1 by subtracting the constant "c" from the prediction, and multiplying the ratio "r" can correct the prediction of model-2. Figure 1.6(c) shows the case in which both random and systematic errors coexist.

1.3 Purposes of Uncertainty Analysis

The main purpose of uncertainty analysis is to quantify the statistical features of system outputs or responses as affected by the stochastic basic parameters in the system. In engineering system design, analysis, and modeling, design

TABLE 1.2   Generation of Stochastic Output (Yen 1988)

Input	System	Output
Stochastic	Deterministic	Stochastic
Deterministic	Stochastic	Stochastic
Stochastic	Stochastic	Stochastic

quantities and system outputs are functions of several system parameters not all of which can be quantified with absolute accuracy. As shown in Table 1.2, for any system involving an input-output process, stochastic output with uncertainty is obtained unless the input and system process are both deterministic.

The task of uncertainty analysis is to determine the uncertainty features of the system responses as a function of uncertainties associated with the system model itself and the stochastic basic parameters involved. Uncertainty analysis provides a formal and systematic framework to quantify the uncertainty associated with system outputs. Furthermore, it offers the designer useful insights into the contribution of each stochastic basic parameter to the overall uncertainty of system outputs. Such knowledge is essential in identifying the "important" parameters to which more attention should be given so as to have a better assessment of their values and, accordingly, to reduce the overall uncertainty of the system outputs.

When a system involves basic parameters whose values cannot be certain, the conventional approach is to conduct a *sensitivity analysis* by which the rate of change in model output due to a unit change in a basic parameter is quantified. As shown in Chap. 4, sensitivity analysis provides partial information needed for conducting an uncertainty analysis. Therefore, performing an uncertainty analysis would generally encompass sensitivity analysis.

1.4   Measures of Uncertainty

Several expressions have been used to describe the degree of uncertainty associated with a parameter, a function, a model, or a system. In general, the uncertainty associated with the latter three results from the combined effects of the uncertainties of the contributing parameters.

A practical way to quantify the level of uncertainty for a parameter is to use the statistical moments (see Sec. 2.3) associated with a quantity subject to uncertainty. In particular, the second-order moment called *variance* is a measure of the dispersion of a random variable. Either the variance or standard deviation (the positive square root of variance) can be used. Sometimes, the coefficient of variation, which is the ratio of standard deviation to the mean, offers a normalized measure useful and convenient for comparison and for combining uncertainties of different variables.

The most complete and ideal description of the uncertainty features of a quantity can be given by the probability density function (PDF). Definition and operational properties of a PDF are described in Chap. 2. However, in most practical problems, such a probability function cannot be derived or found precisely.

A measure of the uncertainty over a range of a variable is to express uncertainty in terms of a reliability domain, such as the confidence interval. A *confidence interval* is a numerical interval that would capture the true value of a variable subject to uncertainty, with a specified probabilistic confidence. It is an expression frequently used for measurement errors (Rabinovich 2000). The methods used to evaluate the confidence interval of a parameter on the basis of data samples are well known and can be found in standard statistics and probability textbooks (e.g., Ang and Tang 1975). Nevertheless, the use of confidence intervals has two drawbacks: (1) the parameter population may not be normally distributed as required in the conventional procedures to determine the confidence interval—this problem is particularly important when the sample size is small; and (2) no means are available to directly combine the confidence intervals of individual contributing random components to give the overall confidence interval of the entire system.

1.5 Implications of Uncertainty

The existence of various uncertainties (including the inherent randomness of natural processes) is the main contributor to the potential failure of many infrastructural systems. Knowledge about the uncertainty features of an engineering system is essential for its reliability analysis. Therefore, uncertainty analysis is essential for reliability analysis.

Civil engineers are often involved in designing various infrastructural systems. In general, the determination of system capacity requires the specification of the loading condition under which the system is designed. Hence, making a forecast or prediction of future loading conditions is necessary. Due to random occurrences of geophysical events, any forecast of future loads, such as a 100-year flood peak discharge or a 50-year 1-h rainfall, is subject to uncertainty. Since "forecast" is to make a conjecture, an estimation of what is to come in the future without any conceived follow-up actions, it is generally reasonable to choose the middle or average value as the forecasted value. On the other hand, "design" implies formulating and planning a course of action to achieve some intended objectives. Therefore, determination of a design quantity would require forecasting a future event *and* taking into account the consequences of system performance corresponding to the uncertainty associated with the forecasted value. For example, to determine the flow-carrying capacity for accommodating a 100-year flow, the design capacity may be larger than the average 100-year discharge as estimated from a frequency analysis, if the consequence of the failure is significant.

In engineering system design and analysis, the decisions on the layout, capacity, and operation of the system largely depend on the infrastructural system response under some anticipated design conditions. When some of the elements in an infrastructural system are subject to uncertainty, the system's responses under the design conditions cannot be assessed with certainty. The presence of uncertainties makes the conventional deterministic design practice inappropriate because of its inability to account for the possible variation of system responses. In fact, the issues involved in the design and analysis of infrastructural systems under uncertainty are multidimensional. An engineer has to consider various criteria, including, but not limited to, the cost of the system, probability of failure, and consequence of failure, such that a proper design can be made for the system.

1.6 Overall View of Uncertainty Analysis Methods

Several statistical techniques can be applied to uncertainty analysis. Each technique has different levels of mathematical complexity and data requirements. Broadly speaking, these techniques can be classified into two categories: analytical approaches and approximate approaches. The appropriate technique to use depends on the nature of the problem at hand, including availability of information, model complexity, and type and accuracy of results desired.

Chapter 4 is devoted to several analytical approaches useful for uncertainty analysis; including derived distribution techniques and integral transform techniques. Some well-known integral transforms—the Fourier, Laplace, and exponential transforms, and a less known Mellin transform (Epstein 1948; Park 1987) are described. Although these analytical approaches are rather restrictive in practical applications due to the complexity of most practical problems, they are, however, powerful tools for deriving complete information about a stochastic process, including its distribution in some situations.

In Chap. 5, several approximation techniques are described. These techniques are particularly useful for problems involving complex functions whose uncertainty features cannot be analytically dealt with. They were primarily developed to estimate the statistical moments of the underlying random processes. One such approximation method is the first-order variance estimation (FOVE) method (Benjamin and Cornell 1970; Ang and Tang 1975). Yen, Chang, and Melching (1986) gave a very comprehensive evaluation and description of the FOVE method in uncertainty and reliability analyses. Four other techniques are the probabilistic point estimation methods developed by Rosenblueth (1975, 1981); Harr (1989) and its variation (Chang, Tung, and Yang 1995), and Li (1992). Each technique has its advantages and disadvantages, which are discussed in great detail in Chap. 5.

References

Ang, A. H. S., and W. H. Tang (1975). *Probability Concepts in Engineering Planning and Design Vol. I: Basic Principles*, John Wiley and Sons, New York.

Ang, A. H. S., and W. H. Tang (1984). *Probability Concepts in Engineering Planning and Design: Decision, Risk and Reliability, Vol. II: Decision, Risk, and Reliability*, John Wiley and Sons, New York.

Bakr, A. A. (1976). "Stochastic Analysis of the Effects of Spatial Variations of Hydraulic Conductivity on Groundwater Flow," Ph.D. Thesis, New Mexico Institute of Mining and Technology, Socorro, NM.

Beck, M. B. (1987). "Water Quality Modeling: A Review of the Analysis of Uncertainty," *Water Resources Research*, **23**(5):1393–1441.

Benjamin, J. R., and C. A. Cornell (1970). *Probability, Statistics, and Decisions for Civil Engineers*, McGraw-Hill, New York.

British Standard Institution (1998). *Measurement of Fluid Flow—Evaluation of Uncertainties*, BS ISO TR 5168.

Burges, S. J., and D. P. Lettenmaier (1975). "Probabilistic Methods in Stream Quality Management," *Water Resources Bulletin*, **11**:115–130.

Cartwright, D. E., and M. S. Longuet-Higgins (1956). "The Statistical Distribution of the Maxima of a Random Function," *Proceedings of Royal Society, Series-A*, **237**:212–232.

Chadderton, R. A., A. C. Miller, and A. J. McDonnell (1982). "Uncertainty Analysis of Dissolved Oxygen Model," *Journal of Environmental Engineering*, ASCE, **108**(5):1003–1012.

Chang, J. H., Y. K. Tung, and J. C. Yang, (1995). "Evaluating Performance of Probabilistic Point Estimates Methods," *Applied Mathematical Modelling*, **19**(2):95–105.

Epstein, B. (1948). "Some Applications of the Mellin Transform in Statistics," *Annals of Mathematical Statistics*, **19**:370–379.

Goodrich, J., L. W. Mays, Y. C. Su, and J. Woodburn (1989). "Chapter 4: Data Base Management Systems." In *Reliability Analysis of Water Distribution Systems*, L. W. Mays (ed.), ASCE, New York.

Haldar, A., and S. Mahadevan (2000). *Probability, Reliability, and Statistical Methods in Engineering Design*, John Wiley and Sons, New York.

Harr, M. E. (1989). "Probabilistic Estimates for Multivariate Analyses," *Applied Mathematical Modelling*, **13**:313–318.

Li, K. S. (1992). "Point Estimate Method for Calculating Statistical Moments," *Journal of Engineering Mechanics*, ASCE, **118**(7):1506–1511.

Mays, L. W., and Y. K. Tung (1992). *Hydrosystems Engineering and Management*, McGraw-Hill, New York.

Melching, C. S., and B. C. Yen (1986). "Slope Influence on Storm Sewer Risk," *Stochastic and Risk Analysis in Hydraulic Engineering*, B.C. Yen (ed.), 79–89, Water Resources Publications, Littleton, CO.

National Research Council (NRC) (2000). *Risk Analysis and Uncertainty in Flood Damage Reduction Studies*, National Academy Press, Washington, DC.

Park, C. S. (1987). "The Mellin Transform in Probabilistic Cash Flow Modeling," *The Engineering Economist*, **32**(2):115–134.

Plate, E. J. (1986). "Reliability Analysis in Hydraulic Design," *Stochastic and Risk Analysis in Hydraulic Engineering*, B. C. Yen (ed.), 37–47, Water Resources Publications, Littleton, CO.

Rabinovich, S. G. (2000). *Measurement Errors and Uncertainties—Theory and Practice*, 2d ed., Springer-Verlag, New York.

Rosenblueth, E. (1975). "Point Estimates for Probability Moments," *Proceedings, National Academy of Science*, **72**(10):3812–3814.

Rosenblueth, E. (1981). "Two-Point Estimates in Probabilities," *Applied Mathematical Modelling*, **5**:329–335.

Yen, B. C. (1988) "Risk Consideration in Storm Drainage," *Proceedings of U.S.-Italy Bilateral Seminar on Urban Storm Drainage*, Cagliari, Sardinia, Italy.

Yen, B. C., and A. H. S. Ang (1971). "Risk Analysis in Design of Hydraulic Projects." *Proceedings of First International Symposium on Stochastic Hydraulics*, University of Pittsburgh, C. L. Chiu (ed.), 694–701, Pittsburgh, PA.

Yen, B. C., S. T. Cheng, and C. S. Melching (1986). "First-Order Reliability Analysis," *Stochastic and Risk Analysis in Hydraulic Engineering*, B. C. Yen (ed.), 1–36, Water Resources Publications, Littleton, CO.

Yevjevich, V. (1972). *Probability and Statistics in Hydrology*, Water Resources Publications, Littleton, CO.

Fundamentals of Probability and Statistics for Uncertainty Analysis

Analysis of the uncertainty for a hydrosystem or its components requires the use of probability and statistics. The level of sophistication and the required mathematical skill vary. As most uncertainty analyses focus on the assessment of the statistical features of a system involving random variables, this chapter provides some reviews on the fundamentals of probability and statistics essential to uncertainty analysis.

2.1 Basic Concepts of Probability

In probability theory, an *experiment* represents the process of observation made on random phenomena. A phenomenon is considered *random* if the outcome of an observation cannot be predicted with absolute accuracy. The totality of all possible outcomes of an experiment constitutes the *sample space*. An *event* is any subset of outcomes contained in the sample space. Therefore, an event may be an empty (or null) set ($\varnothing$), subset of the sample space, or the sample space itself. Since events are sets, appropriate operators to be used are *union, intersection*, and *compliment*. The occurrence of events A or B is denoted as $A \cup B$, while the joint occurrence of events A and B is denoted as $A \cap B$ or simply (A, B). In this book, the complement of event A will be denoted as A'. If two events A and B contain no common elements in the sets, they are *mutually exclusive* or *disjoint events*, and are expressed as $(A, B) = \varnothing$. Venn diagrams illustrating the union and intersection of two events are shown in Fig. 2.1. If the occurrence of event A depends on that of event B, then the two events are called *conditional events* and are denoted by $A \mid B$. Some useful properties of set operations are:

a. Commutative rule: $A \cup B = B \cup A$; $A \cap B = B \cap A$

b. Associative rule: $(A \cup B) \cup C = A \cup (B \cup C)$; $(A \cap B) \cap C = A \cap (B \cap C)$

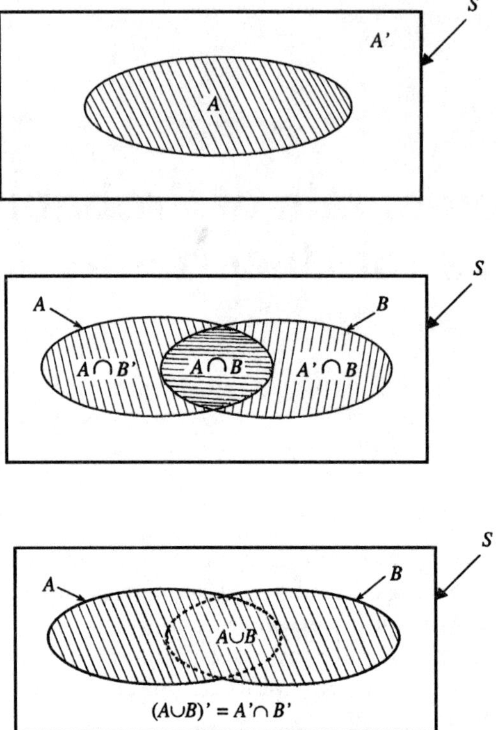

Figure 2.1   Venn diagrams for basic set operations.

c. Distributive rule: $A \cap (B \cup C) = (A \cap B) \cup (A \cap C)$; $A \cup (B \cap C) = (A \cup B) \cap (A \cap C)$

d. De Morgan's rule: $(A \cup B)' = A' \cap B'$; $(A \cap B)' = A' \cup B'$

Probability is a numeric measure of the likelihood of the occurrence of an event relative to a set of alternative events. Therefore, probability is a real-valued number that can be manipulated by using ordinary algebraic operators, such as addition, subtraction, multiplication, and division. Assignment of probability to an event may be based on (a) prior conditions (or deduced on the basis of prescribed assumptions), i.e., a priori determination; (b) the results of empirical observations, i.e., a posteriori determination; (c) a combination of the two.

There are three basic *axioms of probability* in probability computations: (a) *nonnegativity*: $P(A) \geq 0$; (b) *totality*: $P(S) = 1$ with S being the sample space; and (c) *additivity*: for two mutually exclusive events A and B, $P(A \cup B) = P(A) + P(B)$. Axioms (a) and (b) indicate that the value of probability must lie between 0 and 1, inclusive. Axiom (c) can be extended and generalized to consider

K mutually exclusive events as

$$P(A_1 \cup A_2 \cup \cdots \cup A_K) = P\left(\bigcup_{k=1}^{K} A_k\right) = \sum_{k=1}^{K} P(A_k) \tag{2.1}$$

An *empty set* is an impossible event and the corresponding probability is zero, that is, $P(\varnothing) = 0$. Therefore, for two mutually exclusive events A and B, $P(A, B) = P(\varnothing) = 0$.

Relaxing the requirement of mutual exclusiveness in axiom (c), the probability of the union of two events A and B can be evaluated as

$$P(A \cup B) = P(A) + P(B) - P(A, B) \tag{2.2}$$

which can further be generalized as

$$P\left(\bigcup_{k=1}^{K} A_k\right) = \sum_{k=1}^{K} P(A_k) - \sum \sum_{i < j} P(A_i, A_j)$$

$$+ \sum \sum \sum_{i < j < k} P(A_i, A_j, A_k) - \cdots + (-1)^K P(A_1, A_2, \ldots, A_K) \tag{2.3}$$

If all the events involved are mutually exclusive, all but the first summation term on the right-hand side of Eq. (2.3) vanish and it reduces to Eq. (2.1).

For two events that are said to be *statistically independent*, the occurrence of one event has no influence on the occurrence of the other event. Therefore, events A and B are independent if and only if $P(A, B) = P(A)P(B)$. The probability of the joint occurrence of K independent events can be generalized as

$$P\left(\bigcap_{k=1}^{K} A_k\right) = P(A_1) \times P(A_2) \times \cdots \times P(A_K) = \prod_{k=1}^{K} P(A_k) \tag{2.4}$$

Note that the mutual exclusiveness of two events does not, in general, imply independence, and vice versa, unless one of the events is an impossible event. If the two events A and B are independent, then A, A', B, and B' are all independent—but not mutually exclusive—events.

The probability that a conditional event occurs is called *conditional probability*. The conditional probability $P(A \mid B)$ can be computed as

$$P(A \mid B) = P(A, B) / P(B) \tag{2.5}$$

in which $P(A \mid B)$ is the occurrence probability of event A given that event B has occurred. In other words, $P(A \mid B)$ represents the reevaluation of the occurrence probability of event A in light of the information that event B has occurred. Intuitively, events A and B are independent of each other if and only if $P(A \mid B) = P(A)$. In many cases, it is convenient to compute the joint probability $P(A, B)$ by $P(A, B) = P(B)\,P(A \mid B)$ or $P(A, B) = P(A)\,P(B \mid A)$. The probability of the joint occurrence of K dependent events can be generalized as

$$P\left(\bigcap_{k=1}^{K} A_k\right) = P(A_1) \times P(A_2 \mid A_1) \times P(A_3 \mid A_2, A_1) \times \cdots \times P(A_K \mid A_{K-1}, \ldots, A_2, A_1) \quad (2.6)$$

Sometimes, the probability of the occurrence of an event E cannot be determined directly and easily. However, event E may occur along with other attribute events, A_k. Referring to Fig. 2.2, event E could occur jointly with K mutually exclusive and *collectively exhaustive* attribute events A_k, $k = 1, 2, \ldots, K$. That is, $A_i \cap A_j = \varnothing$ for $i \neq j$ and $(A_1 \cup A_2 \cup \cdots \cup A_K) = S$. Therefore, the probability of the occurrence of an event E, regardless of the attributes, can be computed as

$$P(E) = \sum_{k=1}^{K} P(E, A_k) = \sum_{k=1}^{K} P(E \mid A_k)\, P(A_k) \qquad (2.7)$$

Equation (2.7) is called the *total probability theorem.*

As stated in the total probability theorem, the occurrence of event E may be affected by a number of attribute events A_k, $k = 1, 2, \ldots, K$. In some situations one knows $P(E \mid A_k)$ and would like to determine the probability that a particular attribute event A_k is responsible for the occurrence of event E. In other words, one likes to find $P(A_k \mid E)$. Based on the definition of the conditional probability, Eq. (2.5), and the total probability theorem, Eq. (2.7), $P(A_k \mid E)$ can be computed as

$$P(A_k \mid E) = \frac{P(A_k, E)}{P(E)} = \frac{P(E \mid A_k)P(A_k)}{\sum_{k'=1}^{K} P(E \mid A_{k'})P(A_{k'})}, \quad k = 1, 2, \ldots, K \qquad (2.8)$$

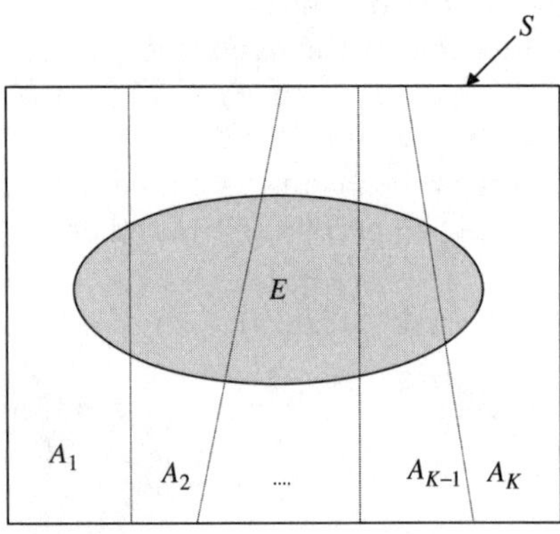

Figure 2.2 Schematic diagram of total probability theorem.

Equation (2.8) is called *Bayes' theorem* in which $P(A_k)$ is the *prior probability* representing the initial belief of the likelihood of the occurrence of attribute event A_k. $P(E \mid A_k)$ is the *likelihood function* and $P(A_k \mid E)$ is the *posterior probability* representing the new evaluation of A_k being responsible in light of the occurrence of event E. Hence, Bayes' theorem can be used to update and revise the calculated probability as more information becomes available.

2.2 Random Variables and Their Distributions

In analyzing statistical features of infrastructural system responses, many events of interest can be defined by the related random variables. A *random variable* is a real-valued function defined on the sample space. In other words, a random variable can be viewed as a mapping from the sample space to the real line, as shown in Fig. 2.3. The standard convention is to denote a random variable by an upper-case letter, while a lower-case letter is used to represent the realization of the corresponding random variable. For example, one may use Q to represent flow magnitude, a random variable, while q is used to represent the values that Q takes. A random variable can be discrete or continuous. Examples of discrete random variables encountered in hydrosystem infrastructural designs are the number of storm events occurring in a specified

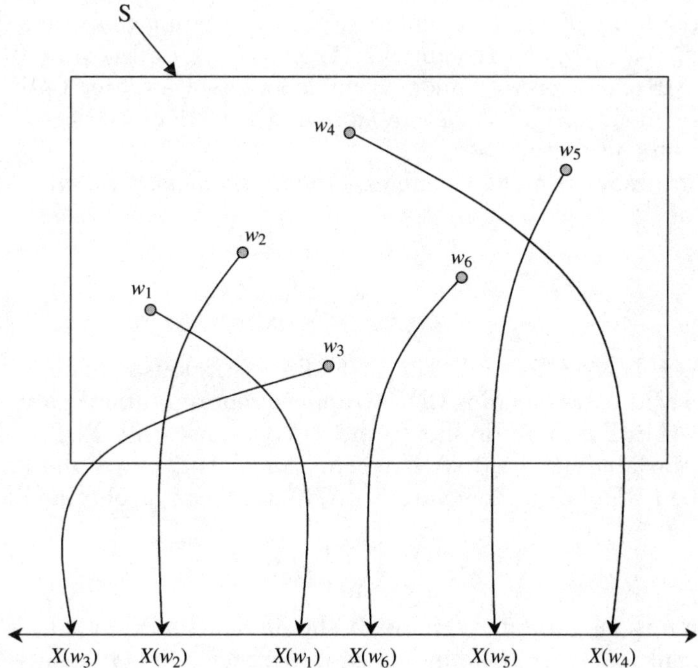

Figure 2.3 A random variable $X(w)$ as mapped from the sample space to the real line.

time period, the number of overtopping events per year for a levee system, and so on. On the other hand, examples of continuous random variables are flow rate, rainfall intensity, water surface elevation, roughness factor, pollution concentration, and the like.

2.2.1   Cumulative distribution function and probability density function

The *cumulative distribution function (CDF)* or simply *distribution function (DF)* of a random variable X is defined as

$$F_x(x) = P(X \leq x) \tag{2.9}$$

Therefore, CDF is the nonexceedance probability. CDF, $F_x(x)$, is a nondecreasing function of the argument x, that is, $F_x(a) \leq F_x(b)$ for $a < b$. As the argument x approaches the lower bound of the random variable X, the value of $F_x(x)$ approaches zero, that is, $\lim_{x \to -\infty} F_x(x) = 0$; on the other hand, the value of $F_x(x)$ approaches unity as its argument approaches the upper bound of X, $\lim_{x \to \infty} F_x(x) = 1$. With $a < b$, $P(a < X \leq b) = F_x(b) - F_x(a)$.

For a discrete random variable X, the *probability mass function (PMF)* is defined as

$$p_x(x) = P(X = x) \tag{2.10}$$

The PMF of any discrete random variable, according to axioms (a) and (b) in Sec. 2.1, must satisfy two conditions: (1) $p_x(x_k) \geq 0$ for all x_k's and (2) $\Sigma_{\text{all } k}\, p_x(x_k) = 1$. The PMF of a discrete random variable and its associated CDF are schematically sketched in Fig. 2.4. As can be seen, the CDF of a discrete random variable appears as a staircase.

For continuous random variables, *probability density function (PDF)* $f_x(x)$ is defined as

$$f_x(x) = \frac{dF_x(x)}{dx} \tag{2.11}$$

where $F_x(x)$ is the CDF of X. The PDF of a continuous random variable $f_x(x)$ is the slope of its corresponding CDF. Graphical representation of a PDF and a CDF are shown in Fig. 2.5. Similar to the discrete case, any PDF of a continuous random variable must satisfy two conditions: (1) $f_x(x) \geq 0$ and (2) $\int f_x(x)dx = 1$. Given the PDF of a random variable X, its CDF can be obtained as

$$F_x(x) = \int_{-\infty}^{x} f_x(u)\,du \tag{2.12}$$

in which u is the dummy variable. It should be noted that $f_x(u)$ is not a probability, it only takes on meaning when it is integrated between two points. The probability of a continuous random variable taking on a particular value is zero whereas it may not be the case for discrete random variables.

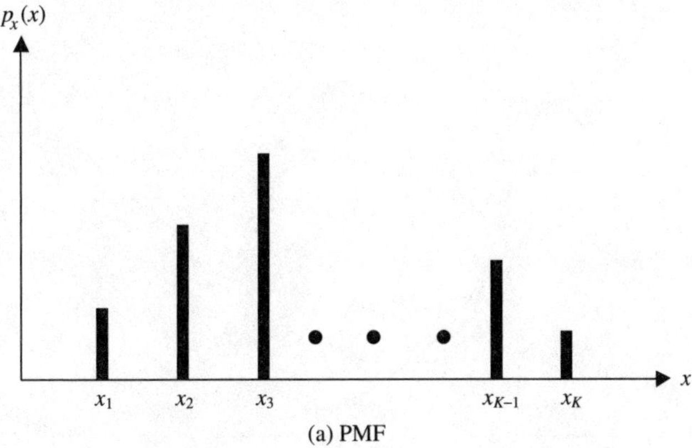

(a) PMF

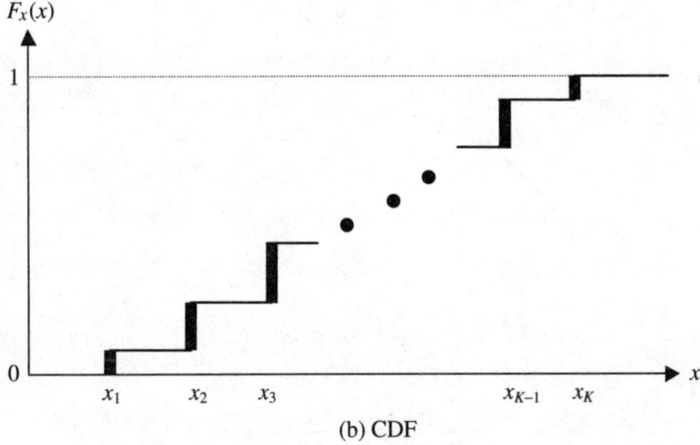

(b) CDF

Figure 2.4   PMF and CDF of a discrete random variable.

Example 2.1   The time-to-failure (T) of a pump in a water distribution system is a continuous random variable having the PDF of

$$f_t(t) = \frac{e^{-t/1250}}{\beta} \qquad \text{for } t \geq 0, \beta$$

in which t is the elapsed time (in hours) before the pump fails, and β is the parameter of the distribution function. Determine the constant β and the probability that the operating life of the pump is longer than 200 h.

Solution   The shape of the exponential PDF is shown in Fig. 2.6. If the function $f_t(t)$ is to serve as a PDF, it has to satisfy two conditions: (1) $f_t(t) \geq 0$ for all t and (2) the area under $f_t(t)$ must be equal to unity. The compliance of condition (1) can be easily proved. The value of constant β can be determined through condition (2) as

$$1 = \int_0^\infty f_t(t)\,dt = \int_0^\infty \frac{e^{-t/1250}}{\beta}\,dt = \left[\frac{-1250 e^{-t/1250}}{\beta} \right]_0^\infty = \frac{1250}{\beta}$$

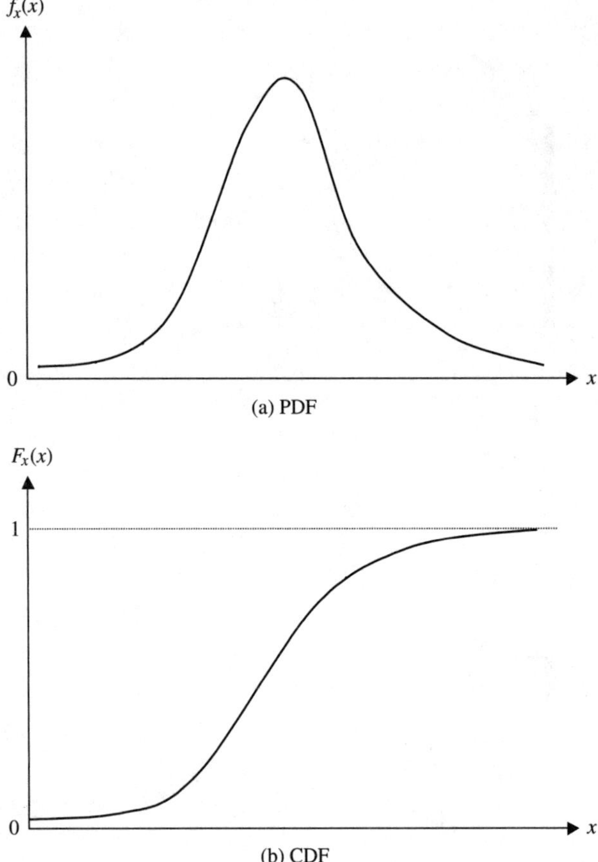

Figure 2.5   PDF and CDF of a continuous random variable.

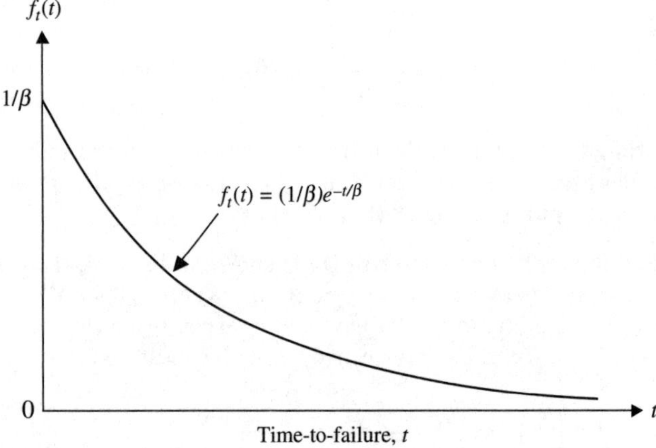

Figure 2.6   Exponential failure density curve.

Therefore, constant $\beta = 1250$ hours per failure. This particular PDF is called exponential distribution (see Sec. 2.4.3). To determine the probability that the operation life of the pump will exceed 200 h, one calculates $P(T \geq 200)$

$$P(T \geq 200) = \int_{200}^{\infty} \frac{e^{-t/1250}}{1250} \, dt = \left[-e^{-t/1250} \right]_{200}^{\infty} = e^{-200/1250} = 0.852$$

2.2.2 Joint, conditional, and marginal distributions

Analogous to the concepts of joint probability and conditional probability, *joint distribution* and *conditional distribution* are used when problems involve multiple random variables. For example, in the design and operation of a flood control reservoir, one often has to consider the flood peak and flood volume simultaneously. Other examples are the considerations of wind speed and earthquake in tall building design; and the severity and period of drought. In such cases, one would need to develop a *joint PDF* of flood peak and flood volume. For illustration purposes, the discussions are limited to the problem involving two random variables.

Joint PMF and *joint CDF* of two discrete random variables X and Y are defined, respectively, as

$$p_{x,y}(x, y) = P(X = x, \, Y = y) \tag{2.13a}$$

$$F_{x,y}(u, v) = P(X \leq u, Y \leq v) = \sum_{x \leq u} \sum_{y \leq v} p_{x,y}(x,y) \tag{2.13b}$$

Schematic diagrams of the joint PMF and joint CDF of two discrete random variables are shown in Fig. 2.7.

The joint PDF of two continuous random variables X and Y, denoted as $f_{x,y}(x, y)$, is related to its corresponding joint CDF as

$$f_{x,y}(x,y) = \frac{\partial^2 [F_{x,y}(x,y)]}{\partial x \, \partial y} \tag{2.14a}$$

$$F_{x,y}(x,y) = \int_{-\infty}^{x} \int_{-\infty}^{y} f_{x,y}(u,v) \, du \, dv \tag{2.14b}$$

Similar to the univariate case, $F_{x,y}(-\infty, -\infty) = 0$ and $F_{x,y}(\infty, \infty) = 1$. Two random variables X and Y are statistically independent if and only if $f_{x,y}(x, y) = f_x(x) \times f_y(y)$ and $F_{x,y}(x, y) = F_x(x) \times F_y(y)$. The same separability characteristics apply to the discrete random variable case. Consequently, a problem involving multiple independent random variables is, in effect, a univariate problem in which each individual random variable can be treated separately.

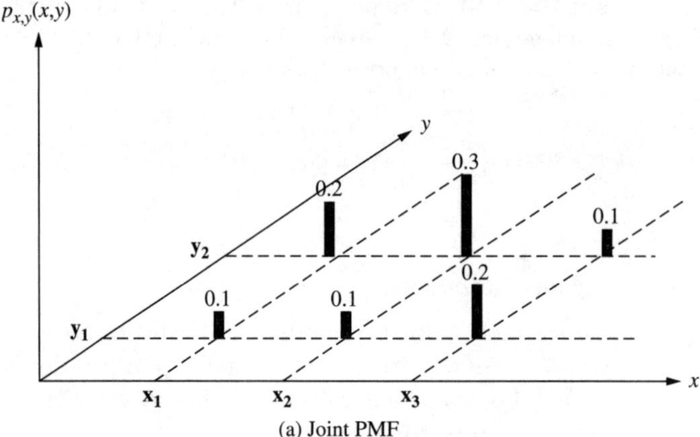

(a) Joint PMF

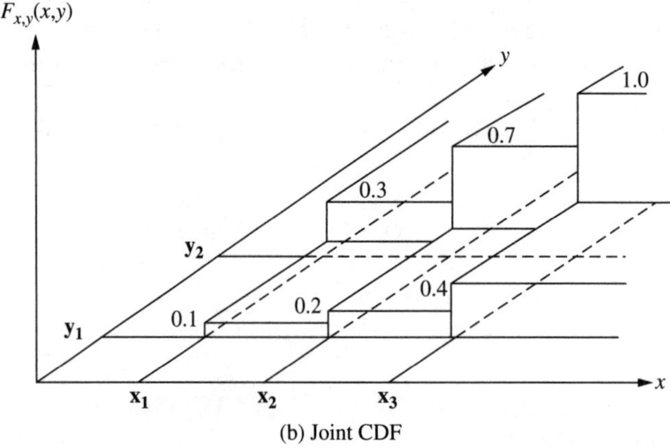

(b) Joint CDF

Figure 2.7   Joint PMF and CDF of two discrete random variables.

If one is interested in the behavior of one random variable regardless of all others, *marginal distribution* can be used. Given the joint PDF $f_{x,y}(x, y)$, the *marginal PDF* of a random variable X can be obtained as

$$f_x(x) = \int_{-\infty}^{\infty} f_{x,y}(x,y)\, dy \tag{2.15}$$

For continuous random variables, the *conditional PDF* for $X \mid Y$, similar to the conditional probability as shown in Eq. (2.5), can be defined as

$$f_x(x \mid y) = \frac{f_{x,y}(x,y)}{f_y(y)} \tag{2.16}$$

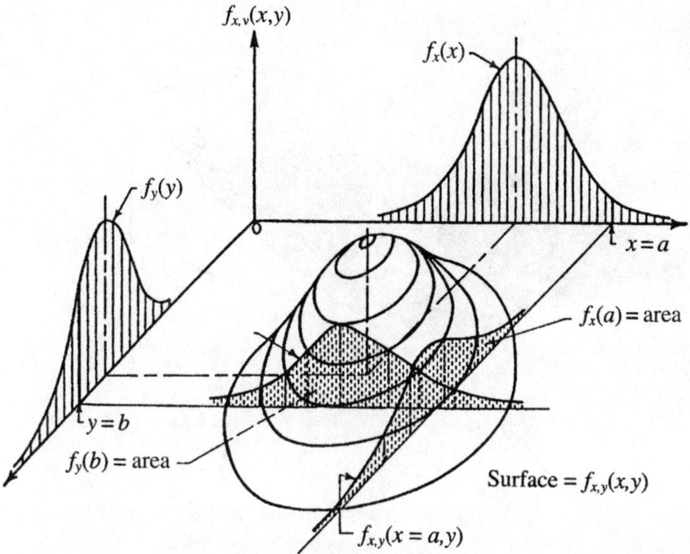

Figure 2.8   Joint and marginal PDFs of two continuous random variables (Ang and Tang 1975).

in which $f_y(y)$ is the marginal PDF of random variable Y. The *conditional PMF* for two discrete random variables can be similarly defined as

$$p_x(x \mid y) = \frac{p_{x,y}(x,y)}{p_y(y)} \tag{2.17}$$

Figure 2.8 shows the joint and marginal PDFs of two continuous random variables X and Y.

It is easily shown that when the two random variables are statistically independent, $f_x(x \mid y) = f_x(x)$. Equation (2.16) can alternatively be written as

$$f_{x,y}(x,y) = f_x(x \mid y) \times f_y(y) \tag{2.18}$$

which indicates that a joint PDF between two correlated random variables can be formulated by multiplying a conditional PDF and a suitable marginal PDF.

Example 2.2   Suppose that X and Y are two random variables that can take values only in the intervals $0 \le x \le 2$ and $0 \le y \le 2$. Suppose the joint CDF of X and Y for these intervals has the form of $F_{x,y}(x, y) = \alpha xy(x^2 + y^2)$. Find (a) the joint PDF of X and Y; (b) the marginal PDF of X; (c) the conditional PDF $f_y(y \mid x)$; and (d) $P(Y \le 1 \mid x = 1)$.

Solution   First, one has to find the constant α so that the function $F_{x,y}(x, y)$ is a legitimate CDF. It requires that the value of $F_{x,y}(x, y) = 1$ when both arguments are at their respective upper bounds, that is,

$$F_{x,y}(x = 2, y = 2) = 1 = \alpha (2)(2)(2^2 + 2^2)$$

Therefore, $\alpha = 1/32$. The resulting joint CDF is shown in Fig. 2.9(a).

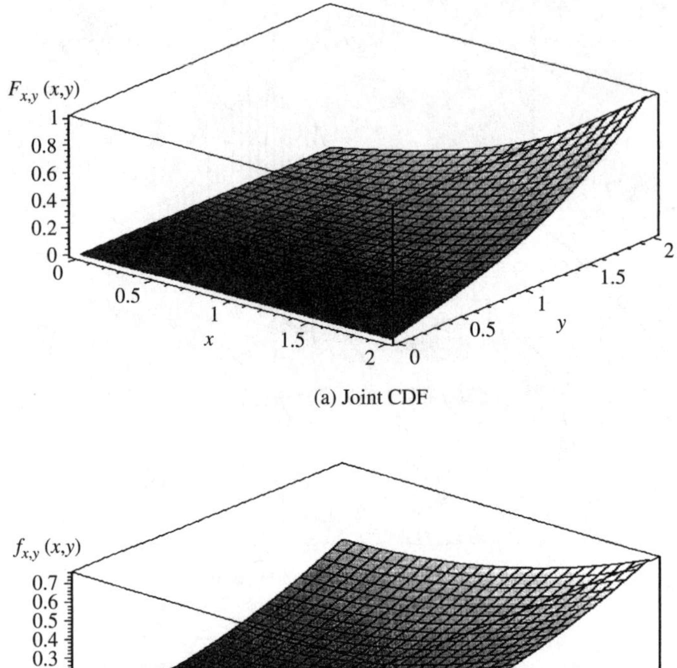

(a) Joint CDF

(b) Joint PDF

Figure 2.9   Joint CDF and PDF of Example 2.2.

(a) To derive the joint PDF, Eq. (2.14a) can be applied.

$$f_{x,y}(x,y) = \frac{\partial}{\partial x}\left[\frac{\partial}{\partial y}\left(\frac{xy(x^2+y^2)}{32}\right)\right] = \frac{\partial}{\partial x}\left[\frac{x^3+3xy^2}{32}\right] = \frac{3(x^2+y^2)}{32} \qquad \text{for } 0 \le x, y \le 2$$

A plot of joint PDF is shown in Fig. 2.9(*b*)

(b) To find the marginal distribution of *X*, Eq. (2.15) can be used.

$$f_x(x) = \int_0^2 \frac{3(x^2+y^2)}{32}\, dy = \frac{4+3x^2}{16} \qquad \text{for } 0 \le x \le 2$$

(c) The conditional distribution $f_y(y\,|\,x)$ can be obtained by following Eq. (2.16) as

$$f_y(y\,|\,x) = \frac{f_{x,y}(x,y)}{f_x(x)} = \frac{\frac{3(x^2+y^2)}{32}}{\frac{4+3x^2}{16}} = \frac{3(x^2+y^2)}{2(3x^2+4)}$$

(d) The conditional probability $P(Y \le 1\,|\,x = 1)$ can be computed as

$$P(Y \le 1\,|\,X = 1) = \int_0^1 f_y(y\,|\,x = 1)\,dy = \int_0^1 \frac{3(1+y^2)}{2(3+4)}\,dy = \frac{2}{7}$$

2.3   Statistical Properties of Random Variables

In statistics, the term *population* is synonymous with the sample space that describes the complete assemblage of all the values representative of a particular random process. A *sample* is any subset of population. Furthermore, *parameters* in a statistical model are quantities that are descriptive of the population. In this book, Greek letters are used to denote statistical parameters. *Sample statistics* or simply *statistics* are quantities calculated on the basis of sample observations.

2.3.1   Statistical moments of random variables

In practical statistical applications, descriptors that are commonly used to assess statistical properties of a random variable are categorized into three types that show: (1) the central tendency, (2) the dispersion, and (3) the asymmetry of a distribution. The frequently used descriptors in these three categories are related to the *statistical moments* of a random variable. Currently, two types of statistical moments are used in hydrosystem engineering applications—product-moments and L-moments. The former is a conventional one with a long history of practice whereas the latter has been receiving great attention recently by water resources engineers in analyzing hydrologic data (Stedinger, Vogel, and Foufoula-Georgoiu 1993). To be consistent with the current general practice and usage, the term "moments" or "statistical moments" in this book refers to the conventional product-moments, unless otherwise specified.

Product-moments.   The rth-order *product-moment* of a random variable X about any reference point $X = x_o$ is defined, for the continuous case, as

$$E[(X - x_o)^r] = \int_{-\infty}^{\infty} (x - x_o)^r f_x(x)\,dx = \int_{-\infty}^{\infty} (x - x_o)^r dF_x(x) \qquad (2.19a)$$

whereas for the discrete case

$$E[(X - x_o)^r] = \sum_{k=1}^{K} (X - x_o)^r p_x(x_k) \qquad (2.19b)$$

where $E[\cdot]$ is a *statistical expectation operator*. In practice, the first three moments ($r = 1, 2, 3$) are used to describe the central tendency, variability, and asymmetry of the distribution of a random variable. Without losing generality, the following discussions consider continuous random variables. For discrete random variables, the integral sign is replaced by the summation sign. Here, it is convenient to point out that, when the PDF in Eq. (2.19a) is replaced by a conditional PDF, as described in Sec. 2.2, the moments obtained are called *conditional moments*.

As the expectations operator $E[\cdot]$ is for determining the average value of the random terms in the bracket, the sample estimator for the product-moments, based on n available data ($x_1, x_2,..., x_n$), for $\mu'_r = E(X_r)$ can be written as

$$\widehat{\mu'_r} = \sum_{i=1}^{n} w_i(n)\, x_i^r \tag{2.20}$$

where $w_i(n)$ = weighting factor for sample data x_i, which depends on sample size n. Most commonly, $w_i(n) = 1/n$ for all $i = 1, 2,..., n$. The last column of Table 2.1 lists the formula in practice for computing some commonly used statistical moments.

Two types of product-moments are commonly used: *moments about the origin*, where $x_o = 0$, and *central moments*, where $x_o = \mu_x$ with $\mu_x = E[X]$. The rth-order central moment is denoted as $\mu_r = E[(X - \mu_x)^r]$, while the rth-order moment about the origin is denoted as $\mu'_r = E[X^r]$. It can be easily shown, through *binomial expansion*, that the central moments $\mu_r = E[(X - \mu_x)^r]$ can be obtained from the moments about the origin as

$$\mu_r = \sum_{i=0}^{r} (-1)^i\, C_{r,i}\, \mu_x^i\, \mu'_{r-i} \tag{2.21}$$

where $C_{r,i} = \binom{r}{i} = \frac{r!}{i!\,(r-i)!}$ is a *binomial coefficient* with ! representing factorial, that is, $r! = r \times (r - 1) \times (r - 2) \times \cdots \times 2 \times 1$. Conversely, the moments about the origin can be obtained from the central moments in a similar fashion as

$$\mu'_r = \sum_{i=0}^{r} C_{r,i} \mu_x^i\, \mu_{r-i} \tag{2.22}$$

Equation (2.21) enables one to compute central moments from moments about the origin while Eq. (2.22) does the opposite. Derivation for the expressions of the first four central moments in terms of the moments about the origin is left as exercises (Probs. 2.5 and 2.6).

The main disadvantages of the product-moments are: (1) estimation from sample observations is sensitive to the presence of extraordinary values (called *outliers*); and (2) the accuracy of sample product-moments deteriorate rapidly with the increase in the order of the moments. An alternative type of moments called *L-moments* can be considered to circumvent these disadvantages.

TABLE 2.1 Product-Moments of Random Variables

Moment	Measure of	Definition	Continuous variable	Discrete variable	Sample estimator
First	Central location	Mean, expected value $E(X) = \mu_x$	$\mu_x = \int_{-\infty}^{\infty} x f_x(x)\,dx$	$\mu_x = \sum_{allx's} x_k p(x_k)$	$\bar{x} = \sum x_i/n$
Second	Dispersion	Variance, $\mathrm{Var}(X) = \mu_2 = \sigma_x^2$	$\sigma_x^2 = \int_{-\infty}^{\infty} (x - \mu_x)^2 f_x(x)\,dx$	$\sigma_x^2 = \sum_{allx's} (x_k - \mu_x)^2 p_x(x_k)$	$s^2 = \dfrac{1}{n-1}\sum (x_i - \bar{x})^2$
		Standard deviation, σ_x	$\sigma_x = \sqrt{\mathrm{Var}(X)}$	$\sigma_x = \sqrt{\mathrm{Var}(X)}$	$s = \sqrt{\dfrac{1}{n-1}\sum (x_i - \bar{x})^2}$
		Coefficient of variation, Ω_x	$\Omega_x = \sigma_x/\mu_x$	$\Omega_x = \sigma_x/\mu_x$	$C_v = s/\bar{x}$
Third	Asymmetry	Skewness	$\mu_3 = \int_{-\infty}^{\infty} (x - \mu_x)^3 f_x(x)\,dx$	$\mu_3 = \sum_{allx's} (x_k - \mu_x)^3 p_x(x_k)$	$m_3 = \dfrac{n}{(n-1)(n-2)}\sum (x_i - \bar{x})^3$
		Skewness coefficient, γ_x	$\gamma_x = \mu_3/\sigma_x^3$	$\gamma_x = \mu_3/\sigma_x^3$	$g = m_3/s^3$
Fourth	Peakedness		$\mu_4 = \int_{-\infty}^{\infty} (x - \mu_x)^4 f_x(x)\,dx$	$\mu_4 = \sum_{allx's} (x_k - \mu_x)^4 p_x(x_k)$	$m_4 = \dfrac{n(n+1)}{(n-1)(n-2)(n-3)}\sum (x_i - \bar{x})^4$
		Kurtosis, κ_x	$\kappa_x = \mu_4/\sigma_x^4$	$\kappa_x = \mu_4/\sigma_x^4$	$k = m_4/s^4$
		Excess coefficient, ε_x	$\varepsilon_x = \kappa_x - 3$	$\varepsilon_x = \kappa_x - 3$	

Example 2.3   Referring to Example 2.1, determine the first two moments about the origin for the time-to-failure of the pump. Then, calculate the first two central moments.

Solution   From Example 2.1, the random variable T is the time-to-failure having an exponential PDF as

$$f_t(t) = \frac{e^{-t/\beta}}{\beta} \qquad \text{for } t \geq 0, \beta > 0$$

in which t is the elapsed time (in hours) before the pump fails and $\beta = 1250$ hours per failure. The moments about the origin, according to Eq. (2.19a), is

$$E[T^r] = \mu_r' = \int_0^\infty t^r \left[\frac{e^{-t/\beta}}{\beta}\right] dt$$

Using integration-by-part, the results of the above integration are

$$\text{for } r = 1 \qquad \mu_1' = E(T) = \mu_t = \beta = 1250 \text{ h, and}$$

$$\text{for } r = 2 \qquad \mu_2' = E(T^2) = 2\beta^2 = 3{,}125{,}000 \text{ h}^2$$

Based on the moments about the origin, the central moments can be determined, according to Eq. (2.21) or Prob. 2.5, as

$$\text{for } r = 1, \; \mu_1 = E(T - \mu_t) = 0$$

$$\text{for } r = 2, \; \mu_2 = E[(T - \mu_t)^2] = \mu_2' - \mu^2 = 2\beta^2 - \beta^2 = \beta^2 = 1{,}562{,}500 \text{ h}^2$$

L-moments.   The rth-order *L-moments* is defined as (Hosking 1986, 1990)

$$\lambda_r = \frac{1}{r} \sum_{k=0}^{r-1} (-1)^k \binom{r-1}{k} E[X_{r-k:r}] \qquad r = 1, 2, \ldots \tag{2.23}$$

in which $X_{k:n}$ is the kth-*order statistic* of a random sample of size n from the distribution $F_x(x)$, namely, $X_{(1)} \leq X_{(2)} \leq \cdots \leq X_{(k)} \leq \cdots \leq X_{(n)}$. The L in "L-moments" emphasizes that λ_r is a linear function of the expected order statistics. Therefore, sample L-moments can be made as a linear combination of the ordered data values. The definition of L-moments given in Eq. (2.23) may appear to be mathematically perplexing. Their computations, however, can be greatly simplified through their relations with *probability-weighted moments*, which is defined as (Greenwood et al. 1979)

$$M_{r,p,q} = E[X^r \{F_x(X)\}^p \{1 - F_x(X)\}^q] = \int_{-\infty}^\infty x^r \{F_x(x)\}^p \{1 - F_x(x)\}^q dF_x(x) \tag{2.24}$$

Compared with Eq. (2.19a), one observes that the conventional product-moments are the special cases of the probability-weighted moments with $p = q = 0$, that is, $M_{r,0,0} = \mu'_r$. The probability-weighted moments are particularly attractive when the closed-form expression for the CDF of the random variable is available.

To work with random variables linearly, $M_{1,p,q}$ can be used. In particular, two types of probability-weighted moments are commonly used in practice, that is,

$$\alpha_r = M_{1,0,r} = E[X\{1 - F_x(X)\}^r] \qquad r = 0, 1, 2,\ldots \tag{2.25a}$$

$$\beta_r = M_{1,r,0} = E[X\{F_x(X)\}^r] \qquad r = 0, 1, 2,\ldots \tag{2.25b}$$

In terms of α_r or β_r, the rth-order L-moment, λ_r, can be obtained as (Hosking 1986)

$$\lambda_{r+1} = (-1)^r \sum_{i=0}^{r} p^*_{r,i}\, \alpha_i = \sum_{i=0}^{r} p^*_{r,i}\, \beta_i \qquad r = 0, 1,\ldots \tag{2.26}$$

in which

$$p^*_{r,i} = (-1)^{r-i} \binom{r}{i}\binom{r+i}{i}$$

For example, the first four L-moments of random variable X are

$$\lambda_1 = \beta_0 = \mu'_1 = \mu_x \tag{2.27a}$$

$$\lambda_2 = \beta_1 - \beta_0 \tag{2.27b}$$

$$\lambda_3 = 6\beta_2 - 6\beta_1 + \beta_0 \tag{2.27c}$$

$$\lambda_4 = 20\beta_3 - 30\beta_2 + 12\beta_1 - \beta_0 \tag{2.27d}$$

To estimate sample α- and β-moments, random samples are arranged in ascending or descending order. For example, arranging n random observations in ascending order, that is, $X_{(1)} \leq X_{(2)} \leq \cdots \leq X_{(k)} \leq \cdots \leq X_{(n)}$, the rth-order β-moment, β_r, can be estimated as

$$\hat{\beta}_r = \frac{1}{n} \sum_{i=1}^{n} X_{(i)} \widehat{F(X_{(i)})} \tag{2.28}$$

where $\widehat{F(X_{(i)})}$ is an estimator for $F(X_{(i)}) = P[X \leq X_{(i)}]$ for which many so-called *plotting position formulas* have been used in practice (Stedinger, Vogel, and Foufoula-Georgoiu 1993). The one that is often used is the *Weibull plotting position formula*, that is, $\widehat{F(X_{(i)})} = i/(n + 1)$.

L-moments possess several advantages over the conventional product-moments. Estimators of L-moments are more robust against the outliers and are less biased. They approximate asymptotic normal distributions more rapidly and closely. Although, it has not been widely used in reliability applications as compared with the conventional production moments, the L-moments could have a great potential to improve reliability estimation. However, before more evidence becomes available, this book will limit its discussions to the uses of conventional product-moments.

Example 2.4   Referring to Example 2.3, determine the first two L-moments, that is, λ_1 and λ_2, of random time-to-failure, T.

Solution   To determine λ_1 and λ_2, one first calculates β_0 and β_1, according to Eq. (2.25b), as

$$\beta_0 = E[T\{F_t(T)\}^0] = E(T) = \mu_t = \beta$$

$$\beta_1 = E[T\{F_t(T)\}^1] = \int_0^\infty [t\,F_t(t)]\,f_t(t)\,dt = \int_0^\infty [t\,(1 - e^{-t/\beta})]\left(\frac{e^{-t/\beta}}{\beta}\right)dt = \frac{3}{4}\beta$$

From Eq. (2.27), the first two L-moments can be computed as

$$\lambda_1 = \beta_0 = \mu_t = \beta \qquad \lambda_2 = 2\beta_1 - \beta_0 = \frac{6\beta}{4} - \beta = \frac{\beta}{2}$$

2.3.2   Mean, mode, median, and quantiles

The central tendency of a continuous random variable X is commonly measured by its *expectation*, which is the 1st-order moment about the origin

$$E[X] = \mu_x = \int_{-\infty}^\infty x\,f_x(x)\,dx = \int_0^1 x\,dF_x(x) = \int_{-\infty}^\infty [1 - F_x(x)]\,dx \qquad (2.29)$$

This expectation is also known as the *mean* of a random variable. It can be easily seen that the mean of a random variable is the 1st-order L-moment, λ_1. Geometrically, the mean or expectation of a random variable is the location of the centroid of the PDF or PMF. The second and third integrations in Eq. (2.29) indicate that the mean of a random variable is the shaded area shown in Fig. 2.10.

The following two operational properties of the expectation are useful:

1. The expectation of the sum of several random variables equals the sum of the expectation of the individual random variables, that is,

$$E\left(\sum_{k=1}^K a_k X_k\right) = \sum_{k=1}^K a_k \mu_k \qquad (2.30)$$

in which $\mu_k = E[X_k]$ for $k = 1, 2,..., K$.

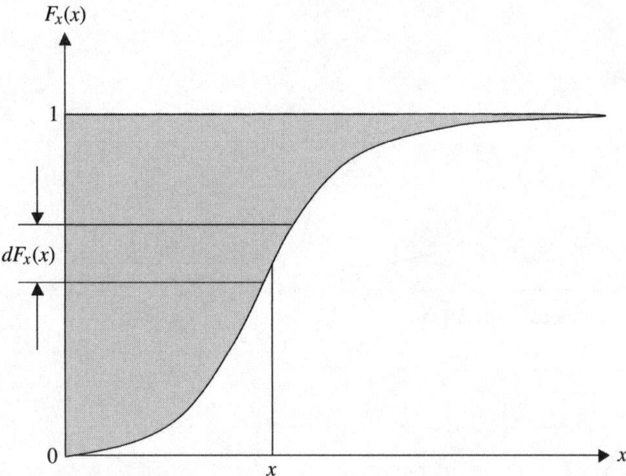

Figure 2.10   Geometric interpretation of the mean.

2. The expectation of the multiplication of several *independent* random variables equals the product of the expectation of the individual random variable, that is,

$$E\left(\prod_{k=1}^{K} X_k\right) = \prod_{k=1}^{K} \mu_k \qquad (2.31)$$

Two other types of measures of central tendency of a random variable are sometimes used in practice, namely, median and mode. The *median* of a random variable is the value that splits the distribution into two equal halves. Mathematically, the median x_{md} of a continuous random variable satisfies

$$F_x(x_{md}) = \int_{-\infty}^{x_{md}} f_x(x)\, dx = 0.5 \qquad (2.32)$$

The median, therefore, is the 50th *quantile* (or *percentile*) of a random variable X. In general, the $100p$th quantile of a random variable X is a quantity x_p that satisfies

$$P(X \le x_p) = F_x(x_p) = p \qquad (2.33)$$

Mode is the value of a random variable at which the value of a PDF has peaked. The mode of a random variable X, x_{mo}, can be obtained by solving the following equation

$$\left[\frac{\partial f_x(x)}{\partial x}\right]_{x=x_{mo}} = 0 \qquad (2.34)$$

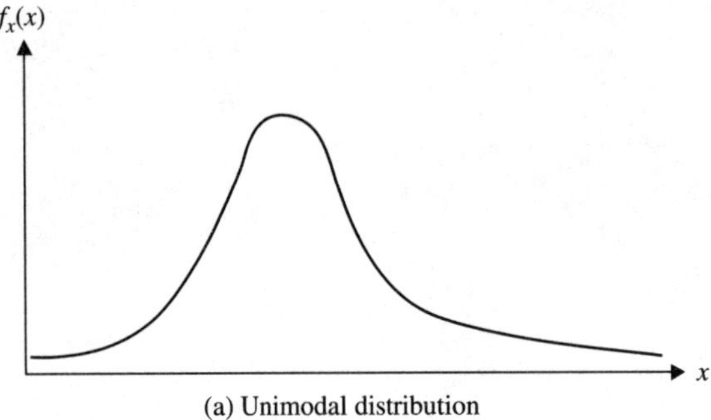

(a) Unimodal distribution

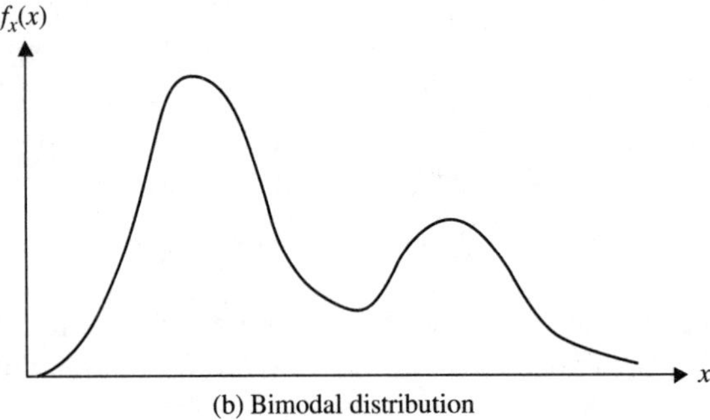

(b) Bimodal distribution

Figure 2.11   Unimodal and bimodal distributions.

Referring to Fig. 2.11, a PDF could be unimodal with a single peak, bimodal with double peaks, or multimodal with multiple peaks. Generally, the mean, median, and mode of a random variable are different, unless the PDF is symmetric and unimodal. Descriptors for the central tendency of a random variable are summarized in Table 2.1.

Example 2.5   Refer to Example 2.3 on the pump reliability problem. Find the mean, mode, median, and 10 percent quantile for the random time-to-failure (T).

Solution   The mean of time-to-failure, called *mean time-to-failure* ($MTTF$), is the 1st-order moment about the origin, which is $\mu_t = 1250$ h as calculated previously in Example 2.3. From the shape of the PDF for the exponential distribution as shown in Fig. 2.6, one can immediately identify that the mode, representing the most likely time of pump failure, is at the beginning of the pump operation, that is, $t_{mo} = 0$ h.

To determine the median time-to-failure of the pump, one can first derive the expression for the CDF from the given exponential PDF as

$$F_t(t) = P(T \leq t) = \int_0^t \frac{e^{-u/1250}}{1250}\, du = 1 - e^{-t/1250} \qquad \text{for } t \geq 0$$

in which u is a dummy variable. Then, the median time-to-failure (t_{md}) can be obtained, according to Eq. (2.32), by solving

$$F_t(t_{md}) = 1 - \exp(-t_{md}/1250) = 0.5$$

which yields $t_{md} = 866.43$ h.

Similarly, the 10 percent quantile ($t_{0.1}$), namely, the elapsed time the pump would fail with a probability of 0.1, can be found in the same way as the median except that the value of the CDF is 0.1, that is,

$$F_t(t_{0.1}) = 1 - \exp(-t_{0.1}/1250) = 0.1$$

which yields $t_{0.1} = 131.7$ h.

2.3.3 Variance, standard deviation, and coefficient of variation

The spreading of a continuous random variable over its range is measured by the *variance*, which is defined as

$$\text{Var}[X] = \mu_2 = \sigma_x^2 = E[(X - \mu_x)^2] = \int_{-\infty}^{\infty} (x - \mu_x)^2 f_x(x)\, dx \qquad (2.35)$$

The variance is the 2nd-order central moment. The positive squared root of the variance is called *standard deviation* (σ_x), which is often used as the measure of the degree of uncertainty associated with a random variable.

Standard deviation has the same units as random variable. To compare the degree of uncertainty of two random variables of different units, a dimensionless measure $\Omega_x = \sigma_x/\mu_x$, called *coefficient of variation*, is useful. By its definition, coefficient of variation indicates the variation of a random variable relative to its mean. Similar to the standard deviation, the 2nd-order L-moment λ_2 is a measure of dispersion of a random variable. The ratio of λ_2 to λ_1, that is, $\tau_2 = \lambda_2/\lambda_1$, is called *L-coefficient of variation*.

Three important properties of the variance are:

1. $\text{Var}[a] = 0$ when a is a constant. $\qquad\qquad\qquad\qquad\qquad\qquad$ (2.36)
2. $\text{Var}[X] = E[X^2] - E^2[X] = \mu'_2 - \mu_x^2$ $\qquad\qquad\qquad\qquad\qquad$ (2.37)
3. The variance of the sum of several *independent* random variables equals the sum of variance of the individual random variable, that is,

$$\text{Var}\left(\sum_{k=1}^{K} a_k X_k\right) = \sum_{k=1}^{K} a_k^2 \sigma_k^2 \qquad (2.38)$$

where a_k is a constant and σ_k is the standard deviation of random variable X_k, $k = 1, 2, ..., K$.

Example 2.6 (modified from Mays and Tung 1992)   Consider the mass balance of a surface reservoir over a 1-month period. The end-of-month storage S can be computed as

$$S_{m+1} = S_m + P_m + I_m - E_m - r_m$$

in which the subscript "m" is an indicator for month; S_m = initial storage volume in the reservoir; P_m = precipitation amount on the reservoir surface; I_m = surface runoff inflow; E_m = total monthly evaporation amount from the reservoir surface; and r_m = controlled monthly release volume from the reservoir.

It is assumed that, at the beginning of the month, the initial storage volume and the total release are known. The monthly total precipitation amount, surface runoff inflow, and evaporation are uncertain and are assumed to be independent random variables. The means and standard deviations of P_m, I_m, and E_m, from historical data for month m are estimated as

$$E(P_m) = 1000 \text{ m}^3 \qquad E(I_m) = 8000 \text{ m}^3 \qquad E(E_m) = 3000 \text{ m}^3$$

$$\sigma(P_m) = 500 \text{ m}^3 \qquad \sigma(I_m) = 2000 \text{ m}^3 \qquad \sigma(E_m) = 1000 \text{ m}^3$$

Determine the mean and standard deviation of the storage volume in the reservoir by the end of the month if the initial storage volume is 20,000 m^3 and the designated release for the month is 10,000 m^3.

Solution   From Eq.(2.30), the mean of the end-of-month storage volume in the reservoir can be determined as

$$E(S_{m+1}) = S_m + E(P_m) + E(I_m) - E(E_m) - r_m$$

$$= 20{,}000 + 1000 + 8000 - 3000 - 10{,}000 = 16{,}000 \text{ m}^3$$

Since the random hydrological variables are statistically independent, the variance of the end-of-month storage volume in the reservoir can be obtained from Eq. (2.38) as

$$\text{Var}(S_{m+1}) = \text{Var}(P_m) + \text{Var}(I_m) + \text{Var}(E_m) = (0.5)^2 + (2)^2 + (1)^2 = 5.25 \times (1000 \text{ m}^3)^2$$

The standard deviation and coefficient of variation of S_{m+1} then are

$$\sigma(S_{m+1}) = \sqrt{5.25} = 2290 \text{ m}^3 \qquad \text{and} \qquad \Omega(S_{m+1}) = 2290 / 16{,}000 = 0.143$$

2.3.4   Skewness coefficient and kurtosis

The asymmetry of the PDF of a random variable is measured by the *skewness coefficient*, γ_x, defined as

$$\gamma_x = \frac{\mu_3}{\mu_2^{1.5}} = \frac{E[(X - \mu_x)^3]}{\sigma_x^3} \tag{2.39}$$

The skewness coefficient is dimensionless and is related to the 3rd-order central moment. The sign of the skewness coefficient indicates the degree of symmetry of the probability distribution function. If $\gamma_x = 0$, the distribution is symmetric about its mean; if $\gamma_x > 0$, the distribution has a long tail to the right; if $\gamma_x < 0$, the distribution has a long tail to the left. Shapes of distribution functions with different skewness coefficients and the relative position of the mean, median, and mode are shown in Fig. 2.12.

Similarly, the degree of asymmetry can be measured by the *L-skewness coefficient* τ_3 defined as

$$\tau_3 = \frac{\lambda_3}{\lambda_2} \tag{2.40}$$

The value of the L-skewness coefficient for all feasible distribution functions must lie within the interval of $[-1, 1]$ (Hosking 1986).

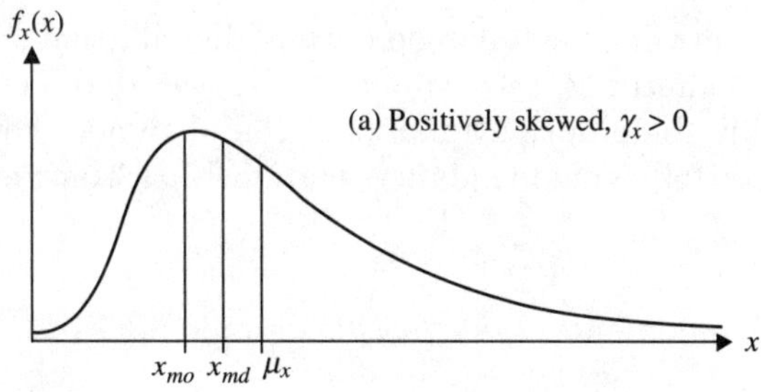

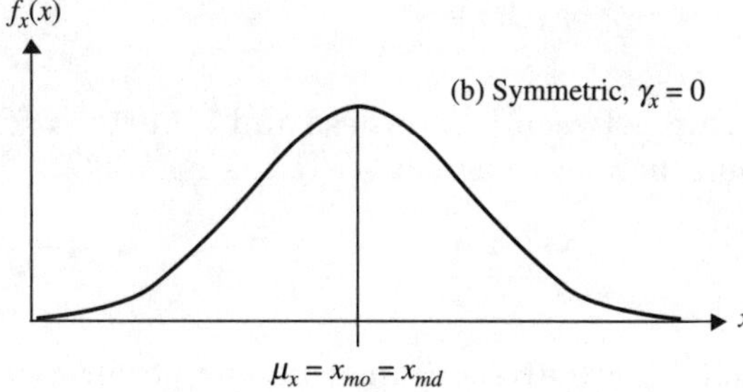

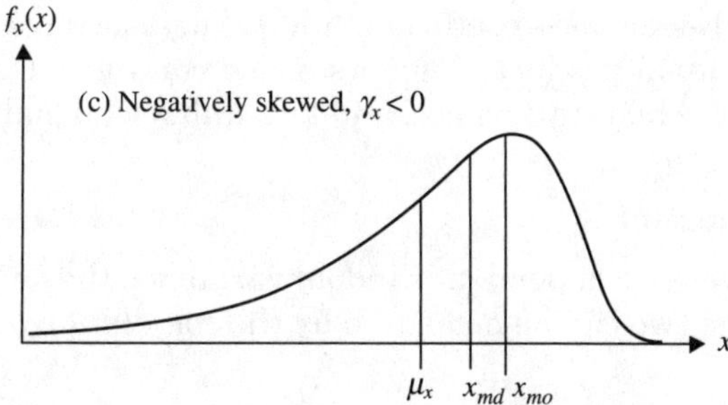

Figure 2.12   Relative locations of mean, median, and mode for positive-skewed, symmetric, and negative-skewed distributions.

Another indicator of the asymmetry is the *Pearson skewness coefficient* defined as

$$\gamma_1 = \frac{\mu_x - x_{mo}}{\sigma_x} \tag{2.41}$$

As can be seen, the Pearson skewness coefficient does not require computing the 3rd-order moment. In practice, product-moments higher than the 3rd-order are less used because they are unreliable and inaccurate when estimated from a small number of samples. Equations used to compute the sample estimates of the above product-moments are listed in the last column of Table 2.1.

Kurtosis (κ_x) is a measure of the peakedness of a distribution. It is related to the 4th-order central moment of a random variable as

$$\kappa_x = \frac{\mu_4}{\mu_2^2} = \frac{E[(X - \mu_x)^4]}{\sigma_x^4} \tag{2.42}$$

with $\kappa_x > 0$. For a random variable having a normal distribution (Sec. 2.4.1), its kurtosis is equal to 3. Sometimes, the *coefficient of excess*, defined as $\varepsilon_x = \kappa_x - 3$, is used. For all feasible distribution functions, the skewness coefficient and kurtosis must satisfy the following inequality relationship (Stuart and Ord 1987)

$$\gamma_x^2 + 1 \le \kappa_x \tag{2.43}$$

By the definition of L-moments, *L-kurtosis* is defined as

$$\tau_4 = \frac{\lambda_4}{\lambda_2} \tag{2.44}$$

Similarly, the relationship between L-skewness and L-kurtosis for all feasible probability distribution functions must satisfy (Hosking 1986)

$$\frac{5\tau_3^2 - 1}{4} \le \tau_4 < 1 \tag{2.45}$$

Royston (1992) conducted an analysis comparing the performance of sample skewness coefficient and kurtosis defined by product-moments and L-moments. Results indicated that L-skewness coefficient and L-kurtosis have clear advantages over the conventional product-moments-based counterparts in terms of easy to interpret, fairly robust to outliers, and less unbiased small samples.

2.3.5   Covariance and correlation coefficient

When a problem involves two dependent random variables, the degree of *linear dependence* between the two can be measured by the *correlation coefficient* $\rho_{x,y}$, which is defined as

$$\text{Corr}\,(X, Y) = \rho_{x,y} = \frac{\text{Cov}(X, Y)}{\sigma_x \sigma_y} \tag{2.46}$$

where $\text{Cov}(X, Y)$ is the *covariance* between random variables X and Y defined as

$$\text{Cov}(X, Y) = E[(X - \mu_x)(Y - \mu_y)] = E(XY) - \mu_x \mu_y \qquad (2.47)$$

There are various types of correlation coefficients developed in statistics for measuring the degree of association between random variables. The one defined by Eq. (2.46) is called the *Pearson product-moment correlation coefficient* or correlation coefficient for short in this book.

It can be easily shown that $\text{Cov}(X'_1, X'_2) = \text{Corr}(X_1, X_2)$ with X'_1 and X'_2 being the *standardized random variables*. In probability and statistics, a random variable can be standardized as

$$X' = \frac{(X - \mu_x)}{\sigma_x} \qquad (2.48)$$

Hence, a standardized random variable has zero mean and unit variance. Standardization will not affect its skewness coefficient and kurtosis of a random variable.

Figure 2.13 graphically illustrates several cases of the correlation coefficient. If the two random variables X and Y are statistically independent, then $\text{Corr}(X, Y) = \text{Cov}(X, Y) = 0$ (Fig. 2.13(c)). However, the reverse statement is not necessarily true as shown in Fig. 2.13(d). If the random variables involved are not statistically independent, Eq. (2.38)—for computing the variance of the sum of several random variables—can be generalized as

$$\text{Var}\left(\sum_{k=1}^{K} a_k X_k\right) = \sum_{k=1}^{K} a_k \, \sigma_k^2 + 2 \sum_{k=1}^{K-1} \sum_{j=k+1}^{K} a_k a_j \, \text{Cov}[X_k, X_j] \qquad (2.49)$$

Example 2.7   Perhaps the assumption of independence of P_m, I_m, and E_m in Example 2.6 may not be reasonable in reality. One examines the historical data closely and finds that there exist correlations among the three hydrological random variables. Analysis of data reveals that $\text{Corr}(P_m, I_m) = 0.8$, $\text{Corr}(P_m, E_m) = -0.4$, and $\text{Corr}(I_m, E_m) = -0.3$. Recalculate the standard deviation of the end-of-month storage volume.

Solution   By Eq. (2.49), the variance of the storage volume in the reservoir at the end of the month can be calculated as

$$\text{Var}(S_{m+1}) = \text{Var}(P_m) + \text{Var}(I_m) + \text{Var}(E_m) + 2\text{Cov}(P_m, I_m) - 2\text{Cov}(P_m, E_m) - 2\text{Cov}(I_m, E_m)$$

$$= \text{Var}(P_m) + \text{Var}(I_m) + \text{Var}(E_m) + 2\text{Corr}(P_m, I_m) \, \sigma(P_m) \, \sigma(I_m)$$

$$- 2\text{Corr}(P_m, E_m) \, \sigma(P_m) \, \sigma(E_m) - 2\text{Corr}(I_m, E_m) \, \sigma(I_m) \, \sigma(E_m)$$

$$= (500)^2 + (2000)^2 + (1000)^2 + 2(0.8)(500)(2000) - 2(-0.4)(500)(1000)$$

$$- 2(-0.3)(2000)(1000) = 8.45(1000 \text{ m}^3)^2$$

The corresponding standard deviation of the end-of-month storage volume is

$$\sigma(S_{m+1}) = \sqrt{8.45} \times 1000 = 2910 \text{ m}^3$$

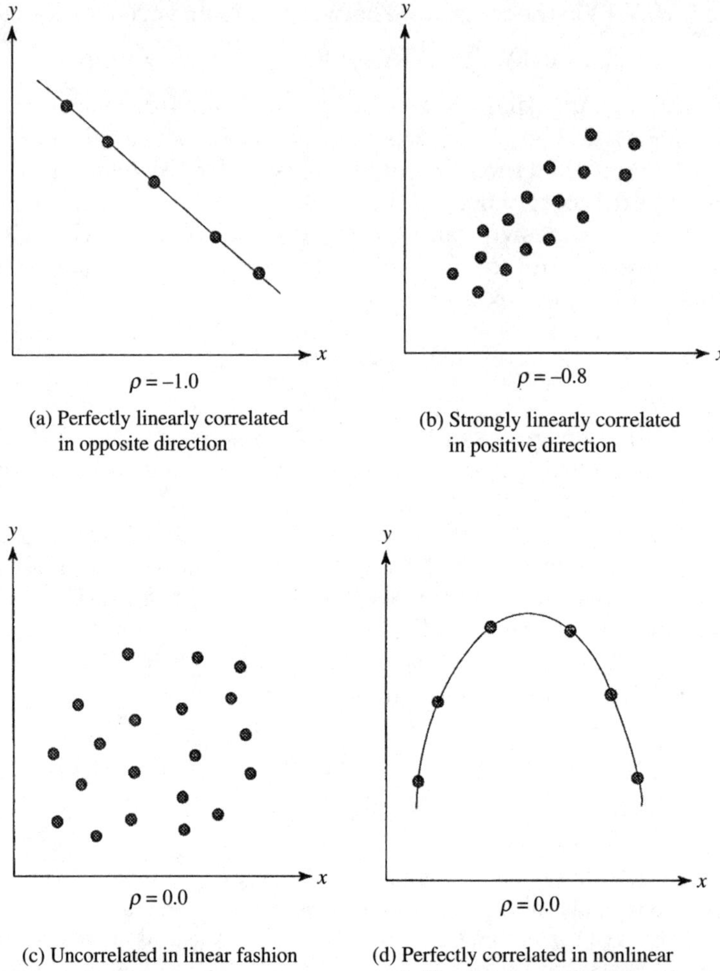

Figure 2.13   Different cases of correlation between two random variables (after Mays and Tung 1992).

In this case, the consideration of correlation increased the standard deviation 27 percent compared to the uncorrelated case in Example 2.6.

Example 2.8   Referring to Example 2.2, compute the correlation coefficient between X and Y.

Solution   Referring to Eqs. (2.46) and (2.47), the computation of correlation coefficient requires the determination of μ_x, μ_y, σ_x, and σ_y from the marginal PDFs of X and Y

$$f_x(x) = \frac{4 + 3x^2}{16} \qquad \text{for } 0 \le x \le 2; \qquad f_y(y) = \frac{4 + 3y^2}{16} \qquad \text{for } 0 \le y \le 2$$

as well as $E(XY)$ from their joint PDF obtained earlier

$$f_{x,y}(x,y) = \frac{3(x^2 + y^2)}{32} \qquad \text{for } 0 \le x, y \le 2$$

From the marginal PDFs, the first two moments of X and Y about the origin can be easily obtained as

$$\mu_x = E(X) = \int_0^2 x f_x(x)\,dx = \frac{5}{4} = E(Y) = \mu_y; \qquad E(X^2) = \int_0^2 x^2 f_x(x)\,dx = \frac{28}{15} = E(Y^2)$$

Hence, the variances of X and Y can be calculated as

$$\text{Var}(X) = E(X^2) - (\mu_x)^2 = 73/240 = \text{Var}(Y)$$

To calculate $\text{Cov}(X,Y)$, one could first compute $E(XY)$ from the joint PDF as

$$E(XY) = \int_0^2 \int_0^2 xy f_{x,y}(x,y)\,dx\,dy = \frac{3}{2}$$

Then, the covariance of X and Y, according to Eq. (2.47), as

$$\text{Cov}(X,Y) = E(XY) - \mu_x\,\mu_y = -1/16$$

The correlation between X and Y can be obtained as

$$\text{Corr}(X,Y) = \rho_{x,y} = \frac{-1/16}{73/240} = -0.205$$

2.4   Some Continuous Univariate Probability Distributions

Several continuous PDFs are frequently used in uncertainty analysis. They include normal, log-normal, gamma, Weibull, and exponential distributions. Other distributions, such as beta and extremal distributions, are also used sometimes. The relations among the various continuous distributions considered in this chapter and others are shown in Fig. 2.14.

2.4.1   Normal (gaussian) distribution

Normal distribution is a well-known probability distribution involving two parameters—the mean and variance. A normal random variable having mean μ_x and variance σ_x^2 is herein denoted as $X \sim \text{N}\,(\mu_x, \sigma_x)$ with the PDF

$$f_N(x \mid \mu_x, \sigma_x) = \frac{1}{\sqrt{2\pi}\,\sigma_x}\, \exp\left[-\frac{1}{2}\left(\frac{x - \mu_x}{\sigma_x}\right)^2\right] \qquad \text{for } -\infty < x < \infty \qquad (2.50)$$

The relationship between μ_x and σ_x and the L-moments are $\mu_x = \lambda_1$ and $\sigma_x = \sqrt{\pi}\,\lambda_2$.

Normal distribution is bell-shaped and symmetric with respect to the mean μ_x. Therefore, the skewness coefficient of a normal random variable is zero. Due to the symmetry of the PDF, all odd-order central moments are zero. The kurtosis of a normal random variable is $\kappa_x = 3.0$. Referring to Fig. 2.14, a linear

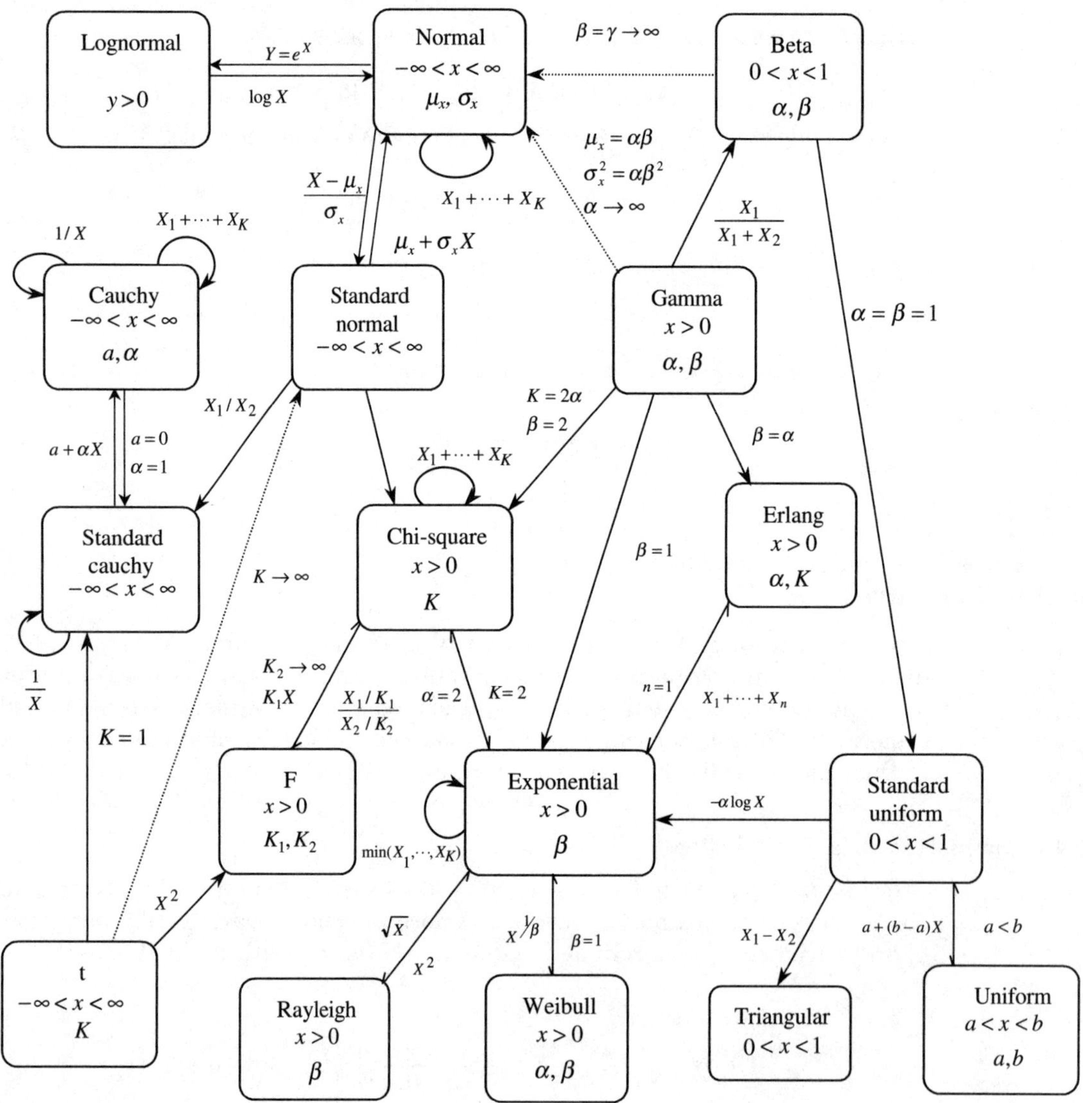

Figure 2.14 Relations among continuous univariate distributions (adapted from Leemis 1986).

function of several normal random variables is also normal. That is, the linear combination of K normal random variables, $W = a_1X_1 + a_2X_2 + \cdots + a_KX_K$, with $X_k \sim N(\mu_k, \sigma_k)$ for $k = 1, 2,\ldots, K$, is also a normal random variable with mean μ_w and variance σ_w^2, respectively, as

$$\mu_w = \sum_{k=1}^{K} a_k\mu_k; \qquad \sigma_w^2 = \sum_{k=1}^{K} a_k^2 \sigma_k^2 + 2 \sum_{k=1}^{K-1} \sum_{k'=k+1}^{K} a_k a_{k'} \text{Cov}(X_k, X_{k'})$$

The normal distribution sometimes provides a viable alternative to approximate the probability of a nonnormal random variable. Of course, the accuracy of such an approximation depends on how closely the distribution of the nonnormal random variable resembles the normal distribution. An important theorem relating to the sum of *independent* random variables is the *central limit theorem*, which loosely states that the distribution of the sum of a number of independent random variables, regardless of their individual distributions, can be approximated by a normal distribution, as long as none of the variables has a dominant effect on the sum. The larger the number of random variables involved in the summation, the better the approximation. Because many natural processes can be thought of as the summation of a large number of independent component processes, none dominating the others; the normal distribution is a reasonable approximation for these overall processes. Dowson and Wragg (1973) have shown that when only the mean and variance are specified, the maximum entropy distribution on the interval $(-\infty, +\infty)$ is the normal distribution (see Example 4.14 in Sec. 4.4.3), i.e., when only the first two moments and the interval are specified, the use of the normal distribution implies more information about the nature of the underlying process than any other distributions specified.

Probability computations for normal random variables are made by first transforming the original variable to a standardized normal variable Z by Eq. (2.48)

$$Z = \frac{X - \mu_x}{\sigma_x}$$

in which Z has a mean of zero and a variance of one. Since Z is a linear function of the normal random variable X, Z is, therefore, normally distributed, that is, $Z \sim N(\mu_z = 0, \sigma_z = 1)$. The PDF of Z, called *standard normal distribution*, can be easily obtained as

$$\phi(z) = \frac{1}{\sqrt{2\pi}} \exp\left(-\frac{z^2}{2}\right) \qquad \text{for } -\infty < z < \infty \tag{2.51}$$

For the standard normal random variable Z, the general expression for the product-moments are

$$E[Z^{2r}] = \frac{(2r)!}{2^r \times r!} \qquad \text{and} \qquad E[Z^{2r+1}] = 0 \qquad \text{for } r \geq 1 \tag{2.52}$$

Computations of probability for $X \sim N(\mu_x, \sigma_x)$ can be made as

$$P(X \le x) = P\left[\frac{X - \mu_x}{\sigma_x} \le \frac{x - \mu_x}{\sigma_x}\right] = P(Z \le z) = \Phi(z) \qquad (2.53)$$

where $\Phi(z)$ is the CDF of the standard normal random variable Z defined as

$$\Phi(z) = \int_{-\infty}^{z} \phi(z)\, dz \qquad (2.54)$$

Figure 2.15 shows the shape of the PDF of the standard normal random variable.

The integral result of Eq. (2.54) is not analytically available. A table of the CDF of Z, such as Table 2.2 or similar, can be found in many statistics textbooks (Abramowitz and Stegun 1972; Blank 1980; Devore 1987; Haan 1977). For numerical computation purposes, several highly accurate approximations are available for determining $\Phi(z)$. One such approximation is the polynomial approximation (Abramowitz and Stegun 1972).

$$\Phi(z) = 1 - \phi(z)\,(b_1 t + b_2 t^2 + b_3 t^3 + b_4 t^4 + b_5 t^5) \qquad \text{for } z \ge 0 \qquad (2.55)$$

in which $t = 1/(1 + 0.2316419z)$, $b_1 = 0.31938153$, $b_2 = -0.356563782$, $b_3 = 1.781477937$, $b_4 = -1.821255978$, and $b_5 = 1.33027443$. The maximum absolute error of the approximation is 7.5×10^{-8}, which is sufficiently accurate for most practical applications. Note that Eq. (2.55) is applicable to the nonnegative value z. For $z < 0$, the value of standard normal CDF can be computed as $\Phi(z) = 1 - \Phi(|z|)$ by the symmetry of $\phi(z)$. Approximation equations, such as Eq. (2.55), can be programmed easily for probability computations without needing the table of the standard normal CDF.

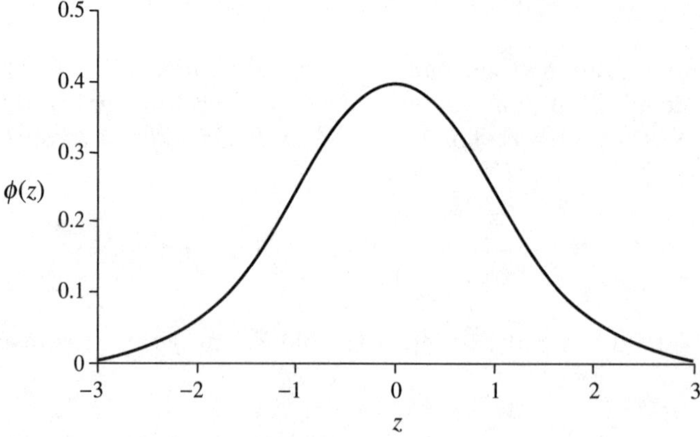

Figure 2.15   Probability density of the standard normal variable.

TABLE 2.2   Table of Standard Normal Probability, $\Phi(z) = P(Z \leq z)$

z	0.00	0.01	0.02	0.03	0.04	0.05	0.06	0.07	0.08	0.09
0.0	0.5000	0.5040	0.5080	0.5120	0.5160	0.5199	0.5239	0.5279	0.5319	0.5359
0.1	0.5398	0.5438	0.5478	0.5517	0.5557	0.5596	0.5636	0.5675	0.5714	0.5753
0.2	0.5793	0.5832	0.5871	0.5910	0.5948	0.5987	0.6026	0.6064	0.6103	0.6141
0.3	0.6179	0.6217	0.6255	0.6293	0.6331	0.6368	0.6406	0.6443	0.6480	0.6517
0.4	0.6554	0.6591	0.6628	0.6664	0.6700	0.6736	0.6772	0.6808	0.6844	0.6879
0.5	0.6915	0.6950	0.6985	0.7019	0.7054	0.7088	0.7123	0.7157	0.7190	0.7224
0.6	0.7257	0.7291	0.7324	0.7357	0.7389	0.7422	0.7454	0.7486	0.7517	0.7549
0.7	0.7580	0.7611	0.7642	0.7673	0.7704	0.7734	0.7764	0.7794	0.7823	0.7852
0.8	0.7881	0.7910	0.7939	0.7967	0.7995	0.8023	0.8051	0.8078	0.8106	0.8133
0.9	0.8159	0.8186	0.8212	0.8238	0.8264	0.8289	0.8315	0.8340	0.8365	0.8389
1.0	0.8413	0.8438	0.8461	0.8485	0.8508	0.8531	0.8554	0.8577	0.8599	0.8621
1.1	0.8643	0.8665	0.8686	0.8708	0.8729	0.8749	0.8770	0.8790	0.8810	0.8830
1.2	0.8849	0.8869	0.8888	0.8907	0.8925	0.8944	0.8962	0.8980	0.8997	0.9015
1.3	0.9032	0.9049	0.9066	0.9082	0.9099	0.9115	0.9131	0.9147	0.9162	0.9177
1.4	0.9192	0.9207	0.9222	0.9236	0.9251	0.9265	0.9279	0.9292	0.9306	0.9319
1.5	0.9332	0.9345	0.9357	0.9370	0.9382	0.9394	0.9406	0.9418	0.9429	0.9441
1.6	0.9452	0.9463	0.9474	0.9484	0.9495	0.9505	0.9515	0.9525	0.9535	0.9545
1.7	0.9554	0.9564	0.9573	0.9582	0.9591	0.9599	0.9608	0.9616	0.9625	0.9633
1.8	0.9641	0.9649	0.9656	0.9664	0.9671	0.9678	0.9686	0.9693	0.9699	0.9706
1.9	0.9713	0.9719	0.9726	0.9732	0.9738	0.9744	0.9750	0.9756	0.9761	0.9767
2.0	0.9772	0.9778	0.9783	0.9788	0.9793	0.9798	0.9803	0.9808	0.9812	0.9817
2.1	0.9821	0.9826	0.9830	0.9834	0.9838	0.9842	0.9846	0.9850	0.9854	0.9857
2.2	0.9861	0.9864	0.9868	0.9871	0.9875	0.9878	0.9881	0.9884	0.9887	0.9890
2.3	0.9893	0.9896	0.9898	0.9901	0.9904	0.9906	0.9909	0.9911	0.9913	0.9916
2.4	0.9918	0.9920	0.9922	0.9925	0.9927	0.9929	0.9931	0.9932	0.9934	0.9936
2.5	0.9938	0.9940	0.9941	0.9943	0.9945	0.9946	0.9948	0.9949	0.9951	0.9952
2.6	0.9953	0.9955	0.9956	0.9957	0.9959	0.9960	0.9961	0.9962	0.9963	0.9964
2.7	0.9965	0.9966	0.9967	0.9968	0.9969	0.9970	0.9971	0.9972	0.9973	0.9974
2.8	0.9974	0.9975	0.9976	0.9977	0.9977	0.9978	0.9979	0.9979	0.9980	0.9981
2.9	0.9981	0.9982	0.9982	0.9983	0.9984	0.9984	0.9985	0.9985	0.9986	0.9986
3.0	0.9987	0.9987	0.9987	0.9988	0.9988	0.9989	0.9989	0.9989	0.9990	0.9990
3.1	0.9990	0.9991	0.9991	0.9991	0.9992	0.9992	0.9992	0.9992	0.9993	0.9993
3.2	0.9993	0.9993	0.9994	0.9994	0.9994	0.9994	0.9994	0.9995	0.9995	0.9995
3.3	0.9995	0.9995	0.9995	0.9996	0.9996	0.9996	0.9996	0.9996	0.9996	0.9997
3.4	0.9997	0.9997	0.9997	0.9997	0.9997	0.9997	0.9997	0.9997	0.9997	0.9998

NOTE: $\Phi(-z) = 1 - \Phi(z)$, $z \geq 0$.

Equally practical is the inverse operation of finding the standard normal quantile z_p with the specified probability level p. The standard normal CDF table can be used, along with some mechanism of interpolation, to determine z_p. However, for practical algebraic computations with a computer, the following rational approximation can be used (Abramowitz and Stegun 1972),

$$z_p = t - \frac{c_0 + c_1 t + c_2 t^2}{1 + d_1 t + d_2 t^2 + d_3 t^3} \qquad \text{for } 0.5 < p \leq 1 \tag{2.56}$$

in which $p = \Phi(z_p)$, $t = \sqrt{-2\ln(1-p)}$, $c_0 = 2.515517$, $c_1 = 0.802853$, $c_2 = 0.010328$, $d_1 = 1.432788$, $d_2 = 0.189269$, and $d_3 = 0.001308$. The corresponding maximum absolute error by this rational approximation is 4.5×10^{-4}. Note that Eq. (2.56) is valid for the value of $\Phi(z)$ that lies between [0.5, 1]. When $p < 0.5$, one can still use Eq. (2.56) by letting $t = \sqrt{-2 \times \ln(p)}$ and attaching a negative sign to the computed quantile value. Vedder (1995) proposed a simple approximation for computing the standard normal cumulative probabilities and standard normal quantiles.

Example 2.9 (adapted from Mays and Tung 1992)   The annual maximum flood magnitude in a river has a normal distribution with the mean of 6000 ft^3/s and a standard deviation of 4000 ft^3/s. (a) What is the annual probability that the flood magnitude would exceed 10,000 ft^3/s? (b) Determine the flood magnitude with a return period of 100 years.

Solution

(a) Let Q be the random annual maximum flood magnitude. Since Q has a normal distribution with the mean $\mu_Q = 6000$ ft^3/s and standard deviation $\sigma_Q = 4000$ ft^3/s, the probability of the annual maximum flood magnitude exceeding 10,000 ft^3/s is

$$P(Q > 10,000) = 1 - P\,[Z \le (10,000 - 6000)/4000]$$

$$= 1 - \Phi(1.00) = 1 - 0.8413$$

$$= 0.1587$$

(b) A flood event with a 100-year return period represents the event whose magnitude has, on the average, an annual probability of 0.01 being exceeded. That is, $P(Q > q_{100}) = 0.01$ in which q_{100} is the magnitude of the 100-year flood. This part of the problem is to determine q_{100} from

$$P(Q \le q_{100}) = 1 - P(Q > q_{100}) = 0.99$$

$$\text{Since} \quad P(Q \le q_{100}) = P\{Z \le (q_{100} - \mu_Q)/\sigma_Q]\}$$

$$= P\{Z \le (q_{100} - 6000)/4000\}$$

$$= \Phi\{(q_{100} - 6000)/4000\} = 0.99$$

From Table 2.2 or Eq. (2.56), one can find that $\Phi(2.33) = 0.99$. Therefore,

$$[q_{100} - 6000]/4000 = 2.33$$

which gives that the magnitude of the 100-year flood event is $q_{100} = 15,320$ ft^3/s.

2.4.2 Lognormal distribution

The *lognormal distribution* is a commonly used continuous distribution for positively valued random variables. Lognormal random variables are closely related to normal random variables by which a random variable X has a lognormal distribution if its logarithmic transform $Y = \ln(X)$ has a normal distribution with mean $\mu_{\ln x}$ and variance $\sigma_{\ln x}^2$. The PDF of a lognormal random variable is

$$f_{LN}\!\left(x \mid \mu_{\ln x}, \sigma_{\ln x}^2\right) = \frac{1}{\sqrt{2\pi}\,\sigma_{\ln x}\,x}\,\exp\!\left[-\frac{1}{2}\left(\frac{\ln(x) - \mu_{\ln x}}{\sigma_{\ln x}}\right)^2\right] \qquad \text{for } x > 0 \qquad (2.57)$$

which can be derived from the normal PDF. Statistical properties of a lognormal random variable in the original scale can be computed from those of log-transformed variables as

$$\mu_x = \lambda_1 = \exp\left(\mu_{\ln x} + \frac{\sigma_{\ln x}^2}{2}\right) \tag{2.58a}$$

$$\sigma_x^2 = \mu_x^2 \left[\exp\left(\sigma_{\ln x}^2\right) - 1\right] \tag{2.58b}$$

$$\Omega_x^2 = \exp\left(\sigma_{\ln x}^2\right) - 1 \tag{2.58c}$$

$$\gamma_x = \Omega_x^3 + 3\Omega_x \tag{2.58d}$$

From Eq. (2.58d) one realizes that the shape of a lognormal PDF is always positively skewed (Fig. 2.16). Equations (2.58a) and (2.58b) can be easily derived by the moment-generating function described in Sec. 4.2. Conversely, the statistical moments of $\ln(X)$ can be computed from those of X by

$$\mu_{\ln x} = \frac{1}{2} \ln\left[\frac{\mu_x^2}{1 + \Omega_x^2}\right] = \ln(\mu_x) - \frac{1}{2}\sigma_{\ln x}^2 \tag{2.59a}$$

$$\sigma_{\ln x}^2 = \ln\left(1 + \Omega_x^2\right) \tag{2.59b}$$

It is interesting to note from Eq. (2.59a) that the variance of a log-transformed variable is dimensionless as is Ω_x.

In terms of L-moments, the 2nd-order L-moment for a two-parameter and three-parameter lognormal distribution is (Stedinger, Vogel, and Foufoula-Georgoiu 1993)

$$\lambda_2 = \exp\left(\mu_{\ln x} + \frac{\sigma_{\ln x}^2}{2}\right) \text{erf}\left(\frac{\sigma_{\ln x}}{2}\right) = \exp\left(\mu_{\ln x} + \frac{\sigma_{\ln x}^2}{2}\right)\left[2\Phi\left(\frac{\sigma_{\ln x}}{\sqrt{2}}\right) - 1\right] \tag{2.60}$$

in which erf($\cdot$) is an *error function* whose definitional relationship with $\Phi(z)$ is

$$\text{erf}(x) = \frac{2}{\sqrt{\pi}} \int_0^x e^{-u^2} du = \frac{2}{\sqrt{\pi}} \int_0^x e^{-z^2/2}\, dz = 2\Phi\left(\sqrt{2}x\right) - 1 \tag{2.61}$$

Hence, $L - Cv$ is $\tau_2 = 2\Phi(\sigma_{\ln x}/\sqrt{2})$. The relationship between 3rd- and 4th-order L-moment ratios can be approximated by the following polynomial function with the accuracy within 5×10^{-4} for $|\tau_2| < 0.9$ (Hosking 1991)

$$\tau_4 = 0.12282 + 0.77518\tau_3^2 + 0.12279\tau_3^4 - 0.13638\tau_3^6 + 0.11386\tau_3^8 \tag{2.62}$$

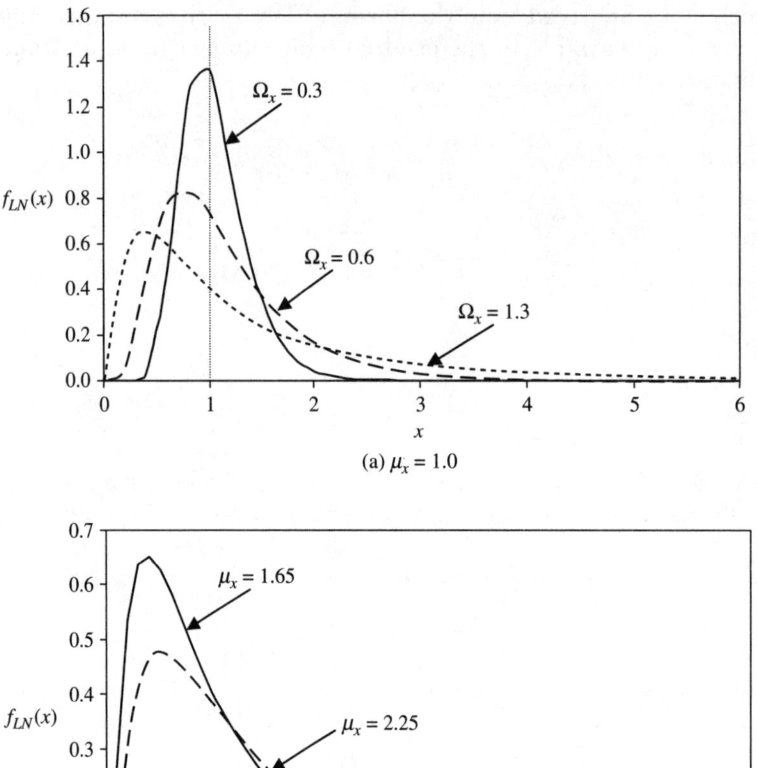

Figure 2.16   Shapes of lognormal probability density functions.

Since the sum of normal random variables is normally distributed, the product of lognormal random variables also is lognormally distributed (Fig. 2.14). The reproductive property of lognormal random variables that is useful can be stated as: if X_1, X_2,..., X_K are independent lognormal random variables $W = b_0 \, \Pi_{k=1}^{K} X_k^{b_k}$ has a lognormal distribution with mean and variance as

$$\mu_{\ln w} = \ln(b_0) + \sum_{k=1}^{K} b_k \, \mu_{\ln x_k}; \qquad \sigma_{\ln w}^2 = \sum_{k=1}^{K} b_k^2 \, \sigma_{\ln x_k}^2$$

If two lognormal random variables are correlated with a correlation coefficient $\rho_{x,y}$ in the original scale, the covariance terms in the log-transformed space must be included in calculating $\sigma^2_{\ln W}$. Given $\rho_{x,y}$, the correlation coefficient in the log-transformed space can be computed as

$$\text{Corr}(\ln X, \ln Y) = \rho_{\ln x, \ln y} = \frac{\ln(1 + \rho_{x,y}\,\Omega_x\,\Omega_y)}{\sqrt{\ln\left(1 + \Omega_x^2\right) \times \ln\left(1 + \Omega_y^2\right)}} \tag{2.63}$$

Equation (2.63) can be derived from Eq. (2.110) along with Eq. (2.59b).

Example 2.10   Re-solve Example 2.9 by assuming that the annual maximum flood magnitude in the river follows a lognormal distribution.

Solution

(a) Let Q be the annual maximum flood magnitude. Since Q has a lognormal distribution, $\ln(Q)$ is normally distributed with a mean and a variance that can be computed from Eqs. (2.59a) and (2.59b), respectively, as

$$\Omega_Q = 4000/6000 = 0.667$$

$$\sigma^2_{\ln Q} = \ln(1 + 0.667^2) = 0.368$$

$$\mu_{\ln Q} = \ln(6000) - 0.368/2 = 8.515$$

The probability of the annual maximum flood magnitude exceeding 10,000 ft^3/s is

$$P(Q > 10{,}000) = P[\ln Q > \ln(10{,}000)]$$

$$= 1 - P[\, Z \le (9.210 - 8.515)/\sqrt{0.368}\,]$$

$$= 1 - \Phi(1.146) = 1 - 0.8741 = 0.1259$$

(b) A 100-year flood q_{100} represents the event whose magnitude corresponds to $P(Q > q_{100}) = 0.01$ that can be determined from

$$P(Q \le q_{100}) = 1 - P(Q > q_{100}) = 0.99$$

$$\text{Since} \quad P(Q \le q_{100}) = P[\ln Q \le \ln(q_{100})]$$

$$= P\{Z \le [\ln(q_{100}) - \mu_{\ln Q}]/\sigma_{\ln Q}]\}$$

$$= P\{Z \le [\ln(q_{100}) - 8.515]/\sqrt{0.368}\}$$

$$= \Phi\{[\ln(q_{100}) - 8.515]/\sqrt{0.368}\} = 0.99$$

From Table 2.2 or Eq. (2.56), one can find that $\Phi(2.33) = 0.99$. Therefore

$$[\ln(q_{100}) - 8.515]/\sqrt{0.368} = 2.33$$

which yields $\ln(q_{100}) = 9.9284$. The magnitude of the 100-year flood event then is $q_{100} = 20{,}500$ ft^3/s.

2.4.3   Gamma distribution and variations

Gamma distribution is a versatile continuous distribution associated with a positive-valued random variable. *Two-parameter gamma distribution* has a

PDF defined as

$$f_G(x \mid \alpha, \beta) = \frac{1}{\beta \Gamma(\alpha)} (x/\beta)^{\alpha-1} e^{-x/\beta} \qquad \text{for } x > 0 \tag{2.64}$$

in which $\beta > 0$ and $\alpha > 0$ are the parameters and $\Gamma(\cdot)$ is a *gamma function* defined as

$$\Gamma(\alpha) = \int_0^\infty t^{\alpha-1} e^{-t} \, dt \tag{2.65}$$

The mean, variance, and skewness coefficient of a gamma random variable having a PDF as Eq. (2.64) are

$$\mu_x = \lambda_1 = \alpha\beta; \quad \sigma_x^2 = \alpha\beta^2; \quad \gamma_x = 2/\sqrt{\alpha} \tag{2.66}$$

In terms of L-moments, the 2nd-order L-moment is

$$\lambda_2 = \frac{\beta \, \Gamma(\alpha + 0.5)}{\sqrt{\pi} \, \Gamma(\alpha)} \tag{2.67}$$

and the relationship between the 3rd- and 4th-order L-moment ratios can be approximated as (Hosking 1991)

$$\tau_4 = 0.1224 + 0.30115\,\tau_3^2 + 0.95812\,\tau_3^4 - 0.57488\,\tau_3^6 + 0.19383\,\tau_3^8 \tag{2.68}$$

In case the lower bound of a gamma random variable is a positive quantity, the above two-parameter gamma PDF can be modified into a *three-parameter gamma PDF* as

$$f_G(x \mid \xi, \alpha, \beta) = \frac{1}{\beta\Gamma(\alpha)} \left(\frac{x - \xi}{\beta} \right)^{\alpha-1} e^{-(x-\xi)/\beta} \qquad \text{for } x > \xi \tag{2.69}$$

where ξ is the lower bound. The two-parameter gamma distribution can be reduced to a simpler form by letting $Y = X/\beta$ and the resulting *one-parameter gamma PDF* (called *standard gamma distribution*) is

$$f_G(y \mid \alpha) = \frac{1}{\Gamma(\alpha)} y^{\alpha-1} e^{-y} \qquad \text{for } y > 0 \tag{2.70}$$

Tables of the cumulative probability of the standard gamma distribution can be found in some statistics books (Dudewicz 1976). Shapes of some gamma distribution are shown in Fig. 2.17 to illustrate its versatility. If α is a positive integer in Eq. (2.70), it is called *Erlang distribution*.

When $\alpha = 1$, the 2-parameter gamma distribution reduces to the *exponential distribution* with the PDF

$$f_{\text{EXP}}(x \mid \beta) = e^{-x/\beta}/\beta \qquad \text{for } x > 0 \tag{2.71}$$

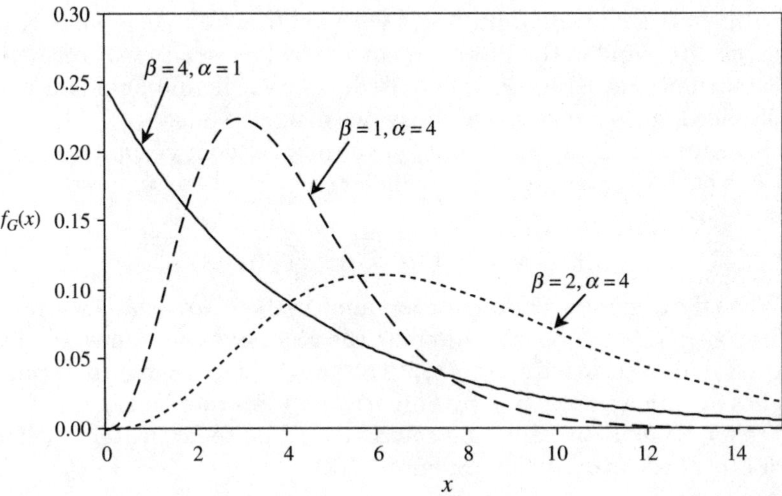

Figure 2.17   Shapes of the gamma probability density functions.

An exponential random variable with a PDF as Eq. (2.71) has the mean and standard deviation equal to β (Example 2.3). Therefore, the coefficient of variation of an exponential random variable is equal to unity. Exponential distribution is commonly used for describing the life span of various electronic and mechanical components. It plays an important role in reliability mathematics using time-to-failure analysis (Henley and Kumamoto 1981).

There are two variations of the gamma distribution that are frequently used in hydrologic frequency analysis; namely, the Pearson and log-Pearson type 3 distributions. In particular, the log-Pearson type 3 distribution is recommended for use by the U.S. Water Resources Council (1982) as the standard distribution for flood frequency analysis. A *Pearson type 3* random variables has the PDF

$$f_{P3}(x \mid \xi, \alpha, \beta) = \frac{1}{|\beta|\,\Gamma(\alpha)} \left(\frac{x - \xi}{\beta} \right)^{\alpha-1} e^{-(x-\xi)/\beta} \tag{2.72}$$

with $\alpha > 0$, $x \geq \xi$ when $\beta > 0$ and with $\alpha > 0$, $x \leq \xi$ when $\beta < 0$. When $\beta > 0$, the Pearson type 3 distribution is identical to the 3-parameter gamma distribution. However, the Pearson type 3 distribution has the flexibility to model negatively skewed random variables corresponding to $\beta < 0$. Therefore, the skewness coefficient of the Pearson type 3 distribution can be computed, from modifying Eq. (2.66), as $sign(\beta)\,2/\sqrt{\alpha}$.

Similar to the normal and lognormal relationship, the PDF of a *log-Pearson type 3* random variable is

$$f_{LP3}(x \mid \xi, \alpha, \beta) = \frac{1}{x\,|\beta|\,\Gamma(\alpha)} \left(\frac{\ln(x) - \xi}{\beta} \right)^{\alpha-1} e^{-(\ln(x)-\xi)/\beta} \tag{2.73}$$

with $\alpha > 0$, $x \geq e^{\xi}$ when $\beta > 0$ and with $\alpha > 0$, $x \leq e^{\xi}$ when $\beta < 0$. Numerous studies can be found in the literature about the Pearson type 3 and log-Pearson type 3 distributions. Kite (1977) and Stedinger, Vogel, and Foufoula-Georgoiu (1993) provide a good summary of these two distributions.

Evaluation of the probability of gamma random variables involves computations of the gamma function, which can be made by using the following recursive formula,

$$\Gamma(\alpha) = (\alpha - 1)\,\Gamma(\alpha - 1) \tag{2.74}$$

When the argument α is an integer number, then $\Gamma(\alpha) = (\alpha - 1)! = (\alpha - 1)(\alpha - 2)\cdots 1$. However, when α is a real number, the recursive relation would lead to $\Gamma(\alpha')$ as the smallest term with $1 < \alpha' < 2$. The value of $\Gamma(\alpha')$ can be determined by a table of the gamma function or by numerical integration on Eq. (2.65). Alternatively, the following approximation could be applied to accurately estimate the value of $\Gamma(\alpha')$ (Abramowitz and Stegun 1972)

$$\Gamma(\alpha') = \Gamma(x+1) = 1 + \sum_{i=1}^{5} a_i x^i \qquad \text{for } 0 < x < 1 \tag{2.75}$$

in which $a_1 = -0.577191652$, $a_2 = 0.988205891$, $a_3 = -0.897056937$, $a_4 = 0.4245549$, and $a_5 = -0.1010678$. The maximum absolute error associated with Eq. (2.75) is 5×10^{-5}.

2.4.4 Extreme value distributions

Hydrosystem engineering reliability analysis often focuses on the statistical characteristics of extreme events. For example, the design of flood control structures may be concerned with the distribution of the largest events over a recorded period. On the other hand, the establishment of a drought management plan or water-quality management scheme might be interested in the statistical properties of minimum flow over a specified period. In other words, statistics of extremes are concerned with the statistical characteristics of $X_{\max,n} = \max\{X_1, X_2,\ldots, X_n\}$ and $X_{\min,n} = \min\{X_1, X_2,\ldots, X_n\}$ in which $X_1, X_2,\ldots, X_n$ are observations of random processes. In fact, the exact distributions of extremes are functions of the underlying (or parent) distribution that generate the random observations $X_1, X_2,\ldots, X_n$ and the number of observations. Of practical interests are the asymptotic distributions of extremes. *Asymptotic distribution* means that the resulting distribution is the limiting form of $F_{\max,n}(y)$ or $F_{\min,n}(y)$ as the number of observations, n, approaches infinity. Asymptotic distributions of extremes turn out to be independent of the sample size n and the underlying distribution for random observations. That is,

$$\lim_{n \to \infty} F_{\max,n}(y) = F_{\max}(y) \qquad \lim_{n \to \infty} F_{\max,n}(y) = F_{\min}(y)$$

Furthermore, these asymptotic distributions of the extremes largely depend on the tail behavior of the parent distribution in either direction toward the extremes. The center portion of the parent distribution has little significance for defining the asymptotic distributions of extremes. The work on statistics of extremes was pioneered by Fisher and Tippett (1928) and was later extended by Gnedenko (1943). Gumbel (1958) dealt with various useful applications of $X_{max,n}$ and $X_{min,n}$ and other related issues.

Three types of asymptotic distributions of extremes are derived based on the different characteristics of the underlying distribution (Haan 1977)

Type I.   Parent distributions are unbounded in the direction of extremes and all statistical moments exist. Examples of this type of parent distribution are normal (for both the largest and smallest extremes), lognormal, and gamma distributions (for the largest extreme).

Type II.   Parent distributions are unbounded in the direction of extremes but all moments do not exist. One such distribution is the Cauchy distribution (Sec. 2.4.5). Thus, the type II extremal distribution has a few applications in practical engineering analysis.

Type III.   Parent distributions are bounded in the direction of the desired extreme. Examples of this type of underlying distributions are beta (for both the largest and smallest extremes), lognormal, and gamma distributions (for the smallest extreme).

Due to the fact that $X_{min,n} = -\max\{-X_1, -X_2,..., -X_n\}$, the asymptotic distribution functions of $X_{max,n}$ and $X_{min,n}$ satisfy the following relation (Leadbetter, Lindgren, and Rootzen 1983)

$$F_{min}(y) = 1 - F_{max}(-y) \tag{2.76}$$

Consequently, the asymptotic distribution of X_{min} can be obtained directly from that of X_{max}. Three types of asymptotic distributions of the extremes are listed in Table 2.3.

Extreme value type I distribution.   It is sometimes referred to as the *Gumbel distribution, Fisher-Tippett distribution*, or *double exponential distribution*. The CDF and PDF of the extreme value type I (EV1) distribution have, respectively,

TABLE 2.3   Three Types of Asymptotic CDFs of Extremes

Type	Maxima	Range	Minima	Range
I	$\exp(-e^{-y})$	$-\infty < y < \infty$	$1 - \exp(-e^{y})$	$-\infty < y < \infty$
II	$\exp(-y^{-\alpha})$	$\alpha < 0, y > 0$	$1 - \exp[-(-y)^{\alpha}]$	$\alpha < 0, y < 0$
III	$\exp[-(-y)^{\alpha}]$	$\alpha > 0, y < 0$	$1 - \exp(-y^{\alpha})$	$\alpha > 0, y > 0$

NOTE: $y = (x - \xi)/\beta$.

the following forms:

$$F_{EV1}(x \mid \xi, \beta) = \exp\left\{-\exp\left[-\left(\frac{x-\xi}{\beta}\right)\right]\right\} \qquad \text{for maxima}$$

$$= 1 - \exp\left\{-\exp\left[+\left(\frac{x-\xi}{\beta}\right)\right]\right\} \qquad \text{for minima} \qquad (2.77a)$$

$$f_{EV1}(x \mid \xi, \beta) = \frac{1}{\beta}\exp\left\{-\left(\frac{x-\xi}{\beta}\right) - \exp\left[-\left(\frac{x-\xi}{\beta}\right)\right]\right\} \qquad \text{for maxima}$$

$$= \frac{1}{\beta}\exp\left\{+\left(\frac{x-\xi}{\beta}\right) - \exp\left[+\left(\frac{x-\xi}{\beta}\right)\right]\right\} \qquad \text{for minima} \qquad (2.77b)$$

for $-\infty < x$, $\xi < \infty$, $\beta \geq 0$. The shapes of the EV1 distribution are shown in Fig. 2.18 in which transformed random variable $Y = (X - \xi)/\beta$ is used. As can be seen, the PDF associated with the largest extreme is a mirror image of the smallest extreme with respect to the vertical line passing through the common mode that happens to be the parameter ξ. The first three product-moments of an EV1 random variable are

$$\mu_x = \lambda_1 = \xi + 0.5772\,\beta \qquad \text{for the largest extreme}$$

$$= \xi - 0.5772\,\beta \qquad \text{for the smallest extreme} \qquad (2.78a)$$

$$\sigma_x^2 = 1.645\,\beta^2 \qquad \text{for both types} \qquad (2.78b)$$

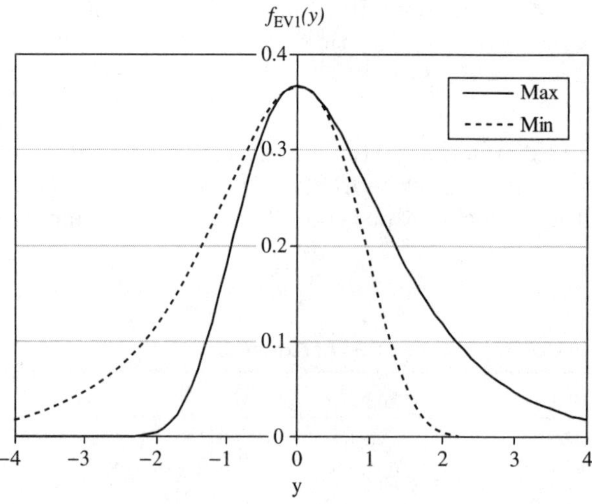

Figure 2.18   Probability density function of extreme-value type I random variables.

$$\gamma_x = 1.13955 \qquad \text{for the largest extreme}$$

$$= -1.13955 \qquad \text{for the smallest extreme} \tag{2.78c}$$

The 2nd- to 4th-order L-moments of the EV1 distribution are

$$\lambda_2 = \beta \ln(2); \qquad \tau_3 = 0.1699; \qquad \tau_4 = 0.1504 \tag{2.79}$$

Using the transformed variable $Y = (X - \xi)/\beta$, the CDFs of the EV1 for the maxima and minima are shown in Table 2.3. Shen and Bryson (1979) showed that if a random variable had an EV1 distribution, the following relationship is satisfied, when ξ is small

$$x_{T_1} \approx \left[\frac{\ln(T_1)}{\ln(T_2)} \right] x_{T_2} \tag{2.80}$$

where x_T is the quantile corresponding to the exceedance probability of $1/T$.

Example 2.11   Repeat Example 2.9 by assuming that the annual maximum flood follows the EV1 distribution.

Solution   Based on the values of a mean of 6000 ft^3/s and standard deviation of 4000 ft^3/s, the values of distributional parameters ξ and β can be determined as the following. From Eq. (2.77b), β is computed as

$$\beta = \frac{\sigma_Q}{\sqrt{1.645}} = \frac{4000}{1.2826} = 3118.72 \text{ ft}^3/\text{s}$$

and from Eq. (2.78a), one has

$$\xi = \mu_Q - 0.577\beta = 6000 - 0.577(3118.72) = 4200.50 \text{ ft}^3/\text{s}$$

(a) The probability of exceeding 10,000 ft^3/s, according to Eq. (2.77a), is

$$P(Q > 10,000) = 1 - F_{EV1}(10,000)$$

$$= 1 - \exp\left[-\exp\left(-\frac{10,000 - 4200.50}{3118.72} \right) \right]$$

$$= 1 - \exp\left[-\exp\left(-1.860 \right) \right]$$

$$= 1 - 0.8558 = 0.1442$$

(b) On the other hand, the magnitude of the 100-year flood event can be calculated as

$$y_{100} = \frac{q_{100} - \xi}{\beta} = -\ln[-\ln(1 - 0.01)] = 4.60$$

Hence, $q_{100} = 4200.50 + 4.60(3118.7) = 18,550 \text{ ft}^3/\text{s}$.

Extreme value type III distribution.   For the extreme value type III (EV3) distribution, the corresponding parent distributions are bounded in the direction of the desired extreme (Table 2.3). For many hydrologic and hydraulic random variables, the lower bound is zero and the upper bound is infinity. For this reason, the EV3 distribution for the maximum has limited applications. On the other hand, the EV3 distribution of the minima is widely used for modeling the smallest extremes, such as drought or low-flow condition. The EV3 distribution for the minima is also known as *Weibull distribution*, having a PDF defined as

$$f_W(x \mid \xi, \alpha, \beta) = \frac{\alpha}{\beta} \left(\frac{x - \xi}{\beta} \right)^{\alpha - 1} \exp\left[-\left(\frac{x - \xi}{\beta} \right)^{\alpha} \right] \quad \text{for } x \geq \xi \text{ and } \alpha, \beta > 0 \qquad (2.81)$$

When $\xi = 0$ and $\alpha = 1$, the Weibull distribution reduces to the exponential distribution. Figure 2.19 shows that the versatility of the Weibull distribution function. The CDF of Weibull random variables can be derived as

$$F_W(x \mid \xi, \alpha, \beta) = 1 - \exp\left[-\left(\frac{x - \xi}{\beta} \right)^{\alpha} \right] \qquad (2.82)$$

The mean and variance of a Weibull random variable can be derived as

$$\mu_x = \lambda_1 = \xi + \beta \Gamma\left(1 + \frac{1}{\beta} \right) \qquad (2.83a)$$

$$\sigma_x^2 = \beta^2 \left[\Gamma\left(1 + \frac{2}{\alpha} \right) - \Gamma^2\left(1 + \frac{1}{\alpha} \right) \right] \qquad (2.83b)$$

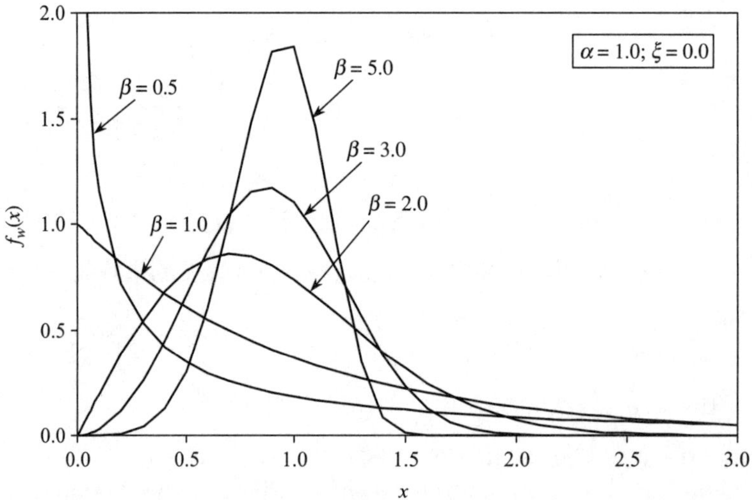

Figure 2.19   Probability density functions of Weibull random variable.

and the 2nd-order L-moment is

$$\lambda_2 = \beta(1 - 2^{-1/\alpha})\,\Gamma\!\left(1 + \frac{1}{\alpha}\right) \tag{2.84}$$

Generalized extreme value distribution. The *generalized extreme value (GEV) distribution* provides an expression that encompasses all three types of extreme value distributions. The CDF of a random variable corresponding to the maximum with a GEV distribution is

$$F_{\mathrm{GEV}}(x\,|\,\xi,\alpha,\beta) = \exp\left\{-\left[1 - \frac{\alpha\,(x - \xi)}{\beta}\right]^{1/\alpha}\right\} \qquad \text{for } \alpha \neq 0 \tag{2.85}$$

When $\alpha = 0$, Eq. (2.85) reduces to Eq. (2.77a) for the Gumbel distribution. For $\alpha < 0$, it corresponds to the EV2 distribution having a lower bound $x > \xi + \beta/\alpha$ whereas, on the other hand, for $\alpha > 0$, it corresponds to the EV3 distribution having an upper bound $x < \xi + \beta/\alpha$. For $|\alpha| < 0.3$, the shape of the generalized extreme value distribution is similar to the Gumbel distribution, except that the right-hand tail is thicker for $\alpha < 0$ and thinner for $\alpha > 0$ (Stedinger, Vogel, and Foufoula-Georgoiu 1993).

The first three moments of the GEV distribution, respectively, are

$$\mu_x = \lambda_1 = \xi + \left(\frac{\beta}{\alpha}\right)[1 - \Gamma(1 + \alpha)] \tag{2.86a}$$

$$\sigma_x^2 = \left(\frac{\beta}{\alpha}\right)^2 [\Gamma(1 + 2\alpha) - \Gamma^2(1 + \alpha)] \tag{2.86b}$$

$$\gamma_x = \mathrm{sign}(\alpha)\frac{-\Gamma(1 + 3\alpha) + 3\Gamma(1 + 2\alpha)\,\Gamma(1 + \alpha) - 2\Gamma^3(1 + \alpha)}{[\Gamma(1 + 2\alpha) - \Gamma^2(1 + \alpha)]^{1.5}} \tag{2.86c}$$

where $\mathrm{sign}(\alpha)$ is $+1$ or -1 depending on the sign of α. From Eqs. (2.86b) and (2.86c), one realizes that the variance of the GEV distribution exists when $\alpha > -0.5$ and the skewness coefficient exists when $\alpha > -0.33$. GEV distribution has recently been frequently used in modeling the random mechanism of hydrologic extremes, such as precipitation and floods.

The relationship between the L-moments and GEV model parameters are

$$\lambda_2 = \frac{\beta}{\alpha}(1 - 2^{-\alpha})\Gamma(1 + \alpha) \tag{2.87a}$$

$$\tau_3 = \frac{2(1 - 3^{-\alpha})}{(1 - 2^{-\alpha})} - 3 \tag{2.87b}$$

$$\tau_4 = \frac{1 - 5(4^{-\alpha}) + 10(3^{-\alpha}) - 6(2^{-\alpha})}{1 - 2^{-\alpha}} \tag{2.87c}$$

2.4.5   Beta distributions

Beta distribution is used for describing random variables having both lower and upper bounds. Random variables in hydrosystems that are bounded on both limits are reservoir storage and groundwater table for unconfined aquifers. The *nonstandard beta* PDF is

$$f_{NB}(x \mid a,b,\alpha,\beta) = \frac{1}{B(\alpha,\beta)(b-a)^{\alpha+\beta-1}}(x-a)^{\alpha-1}(b-x)^{\beta-1} \qquad \text{for } a \le x \le b \qquad (2.88)$$

in which a and b are the lower and upper bounds of the beta random variable, respectively; $\alpha > 0$, $\beta > 0$; and $B(\alpha, \beta)$ is a beta function defined as

$$B(\alpha,\beta) = \frac{\Gamma(\alpha)\Gamma(\beta)}{\Gamma(\alpha+\beta)} \qquad (2.89)$$

Using the new variable $Y = (X - a)/(b - a)$, the nonstandard beta PDF can be reduced to the *standard beta PDF* as

$$f_B(y \mid \alpha,\beta) = \frac{1}{B(\alpha,\beta)} y^{\alpha-1}(1-y)^{\beta-1} \qquad \text{for } 0 < y < 1 \qquad (2.90)$$

The beta distribution is also a very versatile distribution that can have many forms as shown in Fig. 2.20. The mean and variance of the standard beta random variable Y, respectively, are

$$\mu_Y = \frac{\alpha}{\alpha+\beta}; \qquad \sigma_Y^2 = \frac{\alpha\beta}{(\alpha+\beta+1)\,(\alpha+\beta)^2} \qquad (2.91)$$

When $\alpha = \beta = 1$, the beta distribution reduces to a uniform distribution as

$$f_U(x) = \frac{1}{b-a} \qquad \text{for } a \le x \le b \qquad (2.92)$$

2.4.6   Distributions used for hypothesis testing

Normal distribution has been playing an important role in the development of statistical theories. In particular, it is the backbone of many theories relating to the statistics of estimators from the sample. There are several distributions that are commonly used in describing the sampling distribution of an estimator. An *estimator*, or *sample statistic* is a function determined on the basis of random samples for estimating the unknown population parameter. The estimator of the population parameter θ is $\Theta(X_1, X_2,..., X_n)$ in which X's are random samples to be observed. Therefore, an estimator is also a random variable having a probability distribution associated with it. The probability distribution of an estimator is called a *sampling distribution*. After random samples are observed, the numerical value computed for the estimator of interest is called the *estimate*, which is no longer a random variable.

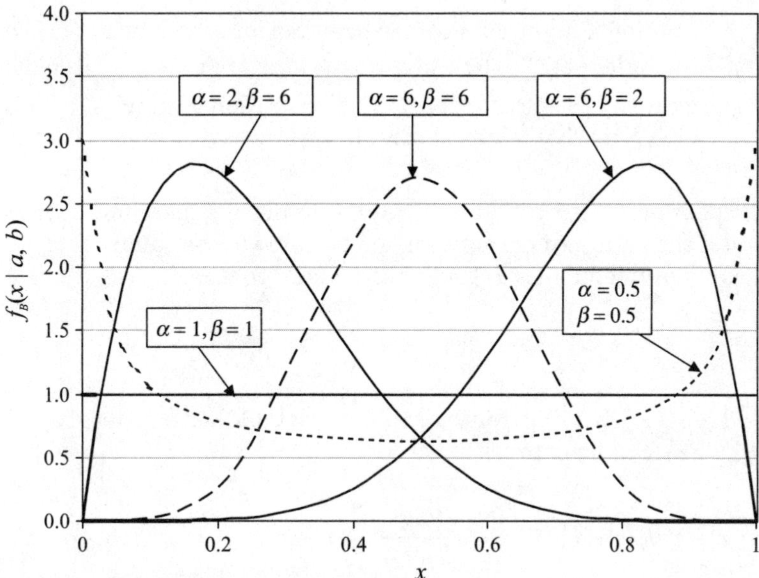

Figure 2.20 Shapes of standard beta PDFs (adapted from Johnson and Kotz 1972).

Chi-square (χ^2) distribution. The sum of the square of K independent, standard normal random variables results in a χ^2 (chi-square) random variable with K degrees of freedom, denoted as χ_K^2. In other words

$$\sum_{k=1}^{K} Z_k^2 \sim \chi_K^2 \tag{2.93}$$

in which Z_k's are standard normal random variables. The PDF of a χ^2 random variable with K degrees of freedom is

$$f_{\chi^2}(x \mid K) = \frac{1}{2^{K/2}\,\Gamma(K/2)}\, x^{(K/2-1)}\, e^{-x/2} \qquad \text{for } x > 0 \tag{2.94}$$

Referring to Eq. (2.94), one realizes that the χ^2 distribution is a special case of the gamma distribution with $\alpha = K/2$ and $\beta = 2$. The mean, variance, and skewness coefficient of a χ_K^2 random variable, respectively, are

$$\mu_x = K; \qquad \sigma_x^2 = 2K; \qquad \gamma_x = 2/\sqrt{K/2}$$

Thus, as the value of K increases, the χ^2 distribution approaches a symmetric distribution. If X_1, X_2,..., X_K are independent normal random variables with mean μ_x and variance σ_x^2, the χ^2 distribution is related to the sample of normal random variables as the following:

a. The sum of K squared standardized normal variables $Z_k = (X_k - \overline{X})/\sigma_x$, $k = 1, 2, \ldots, K$, has a χ^2 distribution with $(K-1)$ degrees of freedom.

b. The quantity $(K-1)S^2/\sigma_x^2$ has a χ^2 distribution with $(K-1)$ degrees of freedom in which S^2 is the unbiased sample variance computed according to Table 2.1.

t-distribution.   The random variable having a t-distribution results from the ratio of the standard normal random variable to the square root of the χ^2 random variable divided by its degrees of freedom, that is,

$$T_K = \frac{Z}{\sqrt{\chi_K^2/K}} \tag{2.95}$$

in which T_K is a t-distributed random variable with K degrees of freedom. The PDF of T_K can be expressed as

$$f_T(x \mid K) = \frac{\Gamma[(K+1)/2]}{\sqrt{-K}\ \Gamma(K/2)} \left(1 + \frac{x^2}{K}\right)^{-(K+1)/2} \qquad \text{for} -\infty < x < \infty \tag{2.96}$$

A t-distribution is symmetric with respect to the mean $\mu_x = 0$ when $K \geq 1$. Its shape is similar to the standard normal distribution, except that the tails of the PDF is thicker than $\phi(z)$. However, as $K \to \infty$, the PDF of a t-distributed random variable approaches the standard normal distribution. It should be noted that when $K = 1$, the t-distribution reduces to the *Cauchy distribution*, all product-moments do not exist. The mean and variance of the t-distributed random variable with K degrees of freedom are

$$\mu_x = 0 \qquad \sigma_x^2 = K/(K-2) \qquad \text{for } K \geq 3$$

When the population variance of normal random variables is known, the sample mean $\overline{X}$ of K normal random samples from $N(\mu_x, \sigma_x^2)$ is a normal distribution with mean μ_x and variance σ_x^2/K. However, when the population variance is unknown, but is estimated by S^2 according to Table 2.1, then the quantity, $\sqrt{K}(\overline{X} - \mu_x)/S$, which is the standardized sample mean using the sample variance, has a t-distribution with $(K-1)$ degrees of freedom.

F-distribution.   The random variable F having an F-distribution results from the ratio of two independent χ^2 random variables as

$$F = \frac{\chi_1^2/K_1}{\chi_2^2/K_2} > 0 \tag{2.97}$$

in which K_1 and K_2 are the two parameters of the F-distribution representing the degrees of freedom for the two χ^2 random variables, respectively. An important relation between the F-distribution and normal random populations is that when S_1^2 and S_2^2 are sample variances of two independent normal samples of sizes K_1 and K_2, the variable $(S_1^2/\sigma_1^2)/(S_2^2/\sigma_2^2)$ has an F-distribution with (K_1-1) and (K_2-1) degrees of freedom.

2.5   Commonly Used Multivariate Probability Distributions

Multivariate probability distributions are extensions of univariate probability distributions that jointly account for more than one random variable. *Bivariate* and *trivariate* distributions are special cases when two and three random variables, respectively, are involved. The general fundamental basis of multivariate probability distributions is described in Sec. 2.2.2. In probability and statistics, the available multivariate distribution models are significantly less than those available for univariate cases. Due to their frequent use in multivariate modeling and uncertainty and reliability analysis, two multivariate distributions, namely, multivariate normal and multivariate lognormal, are presented in this section. Treatments of some multivariate nonnormal random variables are described in Sec. 6.5.3. For other specific types of multivariate distributions, readers are referred to Johnson and Kotz (1976) and Johnson (1987).

There are several ways to construct a multivariate distribution and detailed descriptions can be found in Johnson and Kotz (1976) and Hutchinson and Lai (1990). Based on the joint distribution discussed in Sec. 2.2.2, the straightforward way of deriving a joint PDF involving K multivariate random variables is to extend Eq. (2.18) as

$$f_{\boldsymbol{x}}(\boldsymbol{x}) = f_1(x_1) \times f_2(x_2 \mid x_1) \times \cdots \times f_K(x_K \mid x_1, x_2, \ldots, x_{K-1}) \qquad (2.98)$$

in which $\boldsymbol{x} = (x_1, x_2, \ldots, x_K)^t$ is a vector containing variates K random variables with the subscript t indicating the transpose of a matrix and vector. Applying Eq. (2.98) requires knowing the conditional PDFs of the random variables, which may not be easily obtainable.

One simple way for constructing a joint PDF of two random variables is by mixing. Morgenstern (1956) suggested that the joint CDF of two random variables could be formulated, according to their respective marginal CDFs, as

$$F_{1,2}(x_1, x_2) = F_1(x_1)\, F(x_2)\, \{1 + \theta\,[1 - F_1(x_1)]\,[1 - F_2(x_2)]\} \qquad \text{for} -1 \le \theta \le 1 \quad (2.99)$$

in which $F_k(x_k)$ is the marginal CDF of the random variable X_k and θ is the weighting constant. When the two random variables are independent, the weighting constant $\theta = 0$. Furthermore, the sign of θ indicates the positiveness or negativeness of the correlation between the two random variables. The above equation was later extended by Farlie (1960) to the following form

$$F_{1,2}(x_1, x_2) = F_1(x_1)\, F(x_2)\{1 + \theta f_1(x_1) f_2(x_2)\} \qquad \text{for} -1 \le \theta \le 1 \quad (2.100)$$

in which $f_k(x_k)$ is the marginal PDF of the random variable X_k. Once the joint CDF is obtained, the joint PDF can be derived according to Eq. (2.14a).

TABLE 2.4   Valid Range of Correlation Coefficients for Bivariate Distribution Using Morgenstern Formula

Marginal distribution	N	U	SE	SR	T1L	T1S	LN	GM	T2L	T3S
N	0.318									
U	0.326	0.333								
SE	0.282	0.289	0.25							
SR	0.316	0.324	0.28	0.314						
T1L	0.305	0.312	0.27	0.303	0.292					
T1S	0.305	0.312	0.27	0.303	0.292	0.292				
LN	<0.318	<0.326	<0.282	<0.316	<0.305	<0.305	<0.318			
GM	<0.318	<0.326	<0.282	<0.316	<0.305	<0.305	<0.318	<0.381		
T2L	<0.305	<0.312	<0.270	<0.303	<0.292	<0.292	<0.305	<0.305	<0.292	
T3S	<0.305	<0.312	<0.270	<0.303	<0.292	<0.292	<0.305	<0.305	<0.292	<0.292

NOTE: N = Normal; U = Uniform; SE = Shifted exponential; SR = Shifted Rayleigh; T1L = Type I largest value; T1S = Type I smallest value; LN = Lognormal; GM = Gamma; T2L = Type II largest value; T3S = Type III smallest value.
SOURCE: Adapted from Liu and Der Kiureghian 1986.

Constructing a bivariate PDF by the mixing technique is simple because it only requires knowledge about the marginal distributions of the random variables involved. However, it should be pointed out that the joint distribution obtained from Eq. (2.99) or (2.100) does not necessarily cover the entire range of the correlation coefficient, $[-1, 1]$, for the two random variables under consideration. This is illustrated in Example 2.12. Liu and Der Kiureghian (1986) derived the range of the valid correlation coefficient value for the bivariate distribution, according to Eq. (2.99), from various combinations of marginal PDFs and the results are shown in Table 2.4.

Nataf (1962), Mardia (1970a,b), and Vale and Maurelli (1983) proposed other ways to construct a bivariate distribution for any pair of random variables. This was done by finding the transforms $Z_k = t(X_k)$ for $k = 1, 2$, such that Z_1 and Z_2 are standard normal random variables. Then, a bivariate normal distribution is ascribed to Z_1 and Z_2. One such transformation is $z_k = \Phi^{-1}[F_k(x_k)]$ for $k = 1, 2$. A detailed description of such a normal transform is given in Sec. 6.5.3.

Example 2.12   Consider two correlated random variables X and Y each of which has a marginal PDF of an exponential distribution type as

$$f_x(x) = e^{-x} \quad \text{for } x \geq 0; \qquad f_y(y) = e^{-y} \quad \text{for } y \geq 0$$

To derive a joint distribution for X and Y, one could apply the Morgenstern formula, i.e., Eq. (2.99). The marginal CDFs of X and Y can be easily obtained as

$$F_x(x) = 1 - e^{-x} \quad \text{for } x \geq 0; \qquad F_y(y) = 1 - e^{-y} \quad \text{for } y \geq 0$$

According to Eq. (2.99), the joint CDF of X and Y can be expressed as

$$F_{x,y}(x, y) = (1 - e^{-x})(1 - e^{-y})(1 + \theta e^{-x-y}) \quad \text{for } x, y \geq 0$$

Then, the joint PDF of X and Y can be obtained, according to Eq. (2.14a), as

$$f_{x,y}(x, y) = e^{-x-y} \left[1 + \theta(2e^{-x}-1)(2e^{-y}-1)\right] \qquad \text{for } x, y \geq 0$$

To compute the correlation coefficient between X and Y, one first computes the covariance of X and Y as $\text{Cov}(X,Y) = E(XY) - E(X)E(Y)$ in which $E(XY)$ is computed by

$$E(XY) = \int_0^\infty \int_0^\infty xy\, f_{x,y}(x, y)\, dx\, dy = 1 + \frac{\theta}{4} \qquad \text{for } x, y \geq 0$$

Referring to Eq. (2.71), since the exponential random variables X and Y currently considered are special cases of $\beta = 1$, therefore, $\mu_x = \mu_y = 1$ and $\sigma_x = \sigma_y = 1$. Consequently, the covariance of X and Y is $\theta/4$ and the corresponding correlation coefficient is $\theta/4$. Note that the weighing constant θ lies between $[-1, 1]$, the above bivariate exponential distribution obtained from the Morgenstern formula could only be valid for X and Y having a correlation coefficient in the range $[-1/4, 1/4]$.

2.5.1 Multivariate normal distributions

A *bivariate normal distribution* has a PDF defined as

$$f_{x_1, x_2}(x_1, x_2) = \frac{1}{2\pi\sigma_1\sigma_2\sqrt{1-\rho_{12}^2}} \exp\left[\frac{-Q}{2(1-\rho_{12}^2)}\right] \tag{2.101}$$

for $-\infty < x_1, x_2 < \infty$, in which

$$Q = \left(\frac{x_1 - \mu_1}{\sigma_1}\right)^2 + \left(\frac{x_2 - \mu_2}{\sigma_2}\right)^2 - 2\rho_{12}\left(\frac{x_1 - \mu_1}{\sigma_1}\right)\left(\frac{x_2 - \mu_2}{\sigma_2}\right)$$

where μ and σ are, respectively, the mean and standard deviation; the subscripts "1" and "2" indicate the random variables X_1 and X_2, respectively; and ρ_{12} is the correlation coefficient of the two random variables. Plots of the bivariate normal PDF in a three-dimensional form are shown in Fig. 2.21. The contour curves of the bivariate normal PDF of different correlation coefficients are shown in Fig. 2.22.

The marginal PDF of X_k can be derived, according to Eq. (2.15), as

$$f_k(x_k) = \frac{1}{\sigma_k\sqrt{2\pi}} \exp\left[-\frac{1}{2}\left(\frac{x_k - \mu_k}{\sigma_k}\right)^2\right] \qquad \text{for} -\infty < x_k < \infty$$

for $k = 1$ and 2. As can be seen, the two random variables having a bivariate normal PDF are, individually, normal random variables. It should be pointed

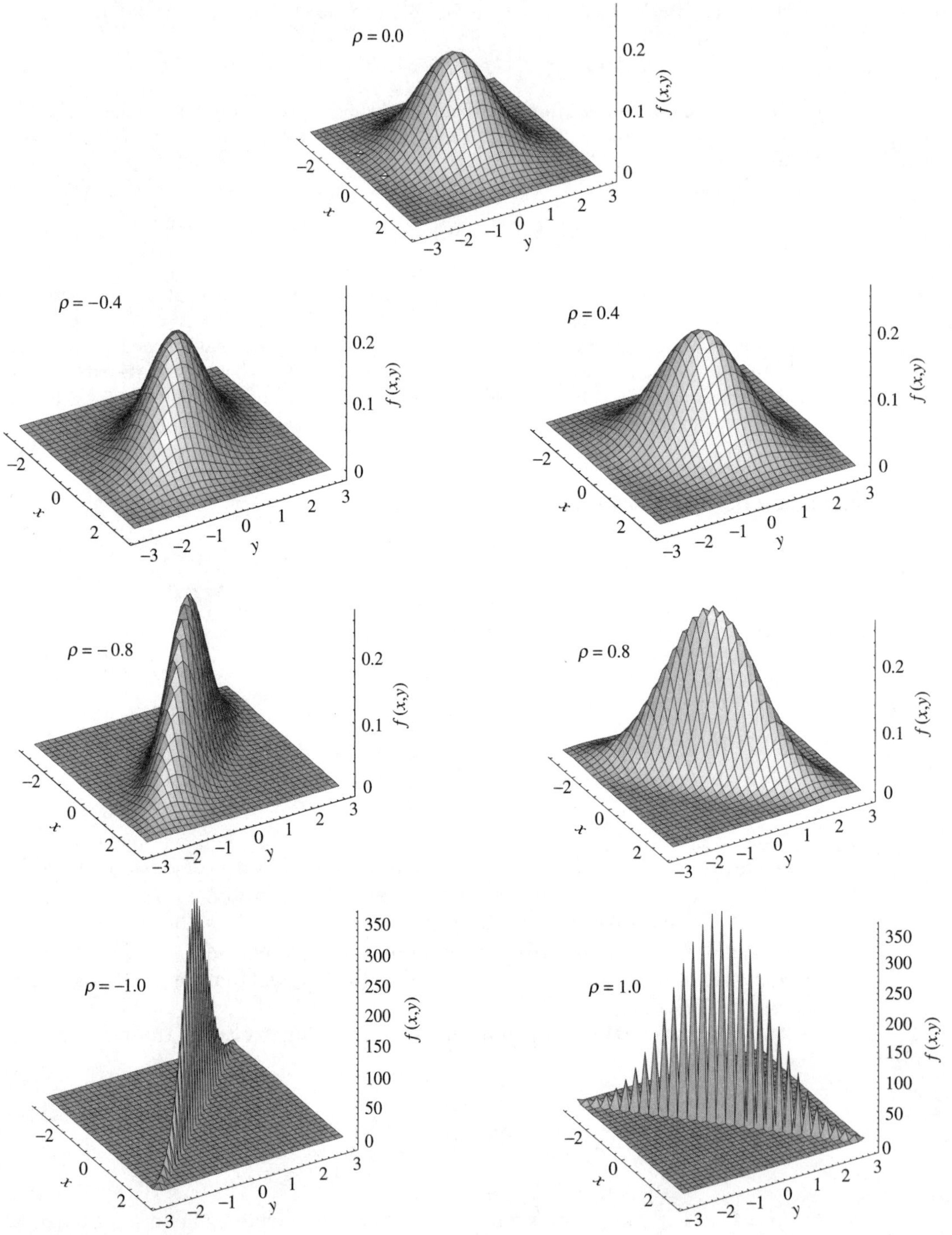

Figure 2.21 Three-dimensional plots of bivariate standard normal probability density functions (adapted from Johnson and Kotz 1976).

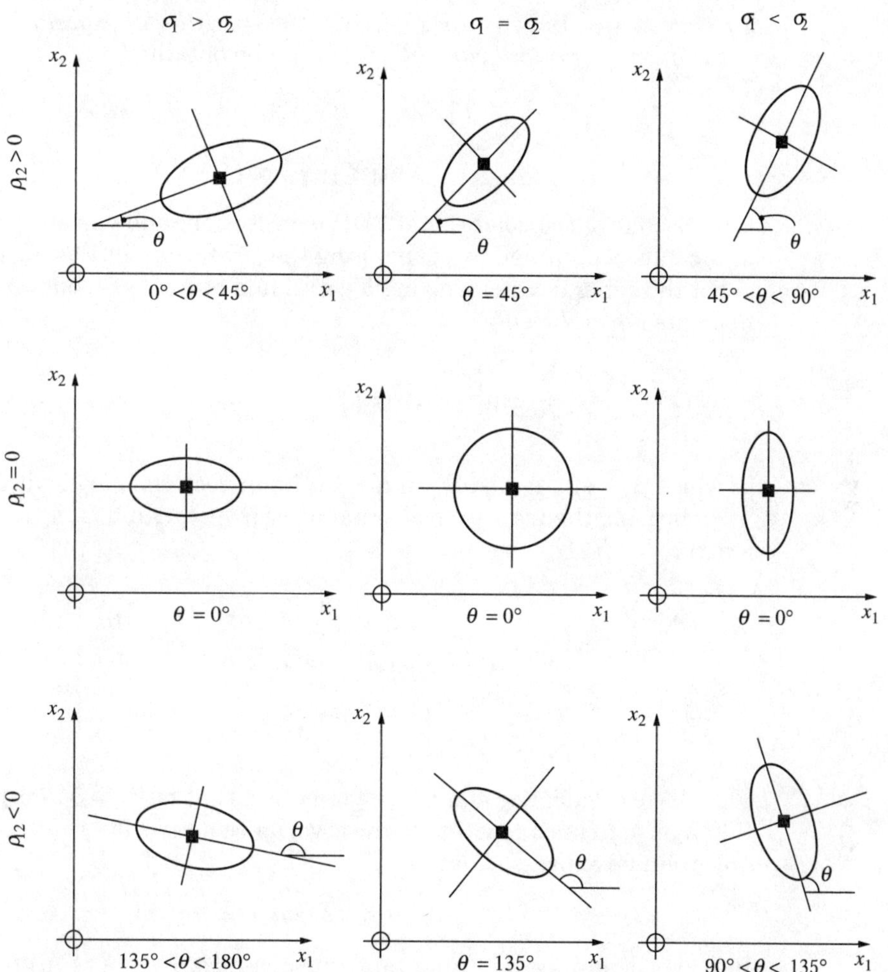

Figure 2.22   Contour of equal density of bivariate standard normal probability density functions (adapted from Johnson and Kotz 1976).

out that, given two normal marginal PDFs, one can construct a bivariate PDF that is not in the form of a bivariate normal as defined by Eq. (2.101).

According to Eq. (2.16), the *conditional normal PDF* of $X_1 \mid x_2$ can be obtained as

$$f_{x_1|x_2}(x_1 \mid x_2) = \frac{1}{\sigma_1\sqrt{2\pi(1-\rho_{12}^2)}} \exp\left[-\frac{1}{2}\left(\frac{(x_1-\mu_1) - \rho_{12}(\sigma_1/\sigma_2)(x_2-\mu_2)}{\sigma_1\sqrt{1-\rho_{12}^2}}\right)^2\right] \quad (2.102)$$

for $-\infty < x_1 < \infty$. Based on Eq. (2.102), the *conditional expectation and variance of the normal random variable* $X_1 \mid x_2$ can be obtained as

$$E[X_1|x_2] = \mu_1 + \rho_{12}(\sigma_1/\sigma_2)(x_2 - \mu_2) \tag{2.103}$$

$$\mathrm{Var}[X_1|x_2] = \sigma_1^2(1 - \rho_{12}^2) \tag{2.104}$$

Expressions of the conditional PDF, expectation, and variance for $X_2 \mid x_1$ can be immediately obtained by exchanging the subscripts in Eqs. (2.101) to (2.103).

For the general case involving K correlated normal random variables, the *multivariate normal PDF* is

$$f_x(x) = \frac{|\,C_x^{-1}\,|^{1/2}}{(2\pi)^{K/2}} \exp\left[-\frac{1}{2}(x - \mu_x)^t C_x^{-1}(x - \mu_x) \right] \qquad \text{for} -\infty < x < \infty \tag{2.105}$$

in which $\mu_x = (\mu_1, \mu_2, \dots, \mu_K)^t$ an $K \times 1$ column vector of mean with the superscript "t" indicating the transpose of a matrix or a vector, and C_x is a $K \times K$ covariance matrix

$$\mathrm{Cov}(X) = C_x = \begin{bmatrix} \sigma_{11} & \sigma_{12} & \cdots & \sigma_{1K} \\ \sigma_{21} & \sigma_{22} & \cdots & \sigma_{2K} \\ \vdots & \vdots & \vdots & \vdots \\ \sigma_{K1} & \sigma_{K2} & \cdots & \sigma_{KK} \end{bmatrix}$$

The above covariance matrix is symmetric, that is, $\sigma_{jk} = \sigma_{jk}$ for $j \neq k$ where $\sigma_{jk} = \mathrm{Cov}(X_j, X_k)$. In matrix notation, the covariance matrix for a vector of random variables can be expressed as

$$C_x = E\left[(X - \mu_x)(X - \mu_x)^t\right] \tag{2.106}$$

In terms of standard normal random variables, $Z_k = (X_k - \mu_k)/\sigma_k$, the *standardized multivariate normal PDF* can be expressed as

$$\phi_z(z) = \frac{|\,R_x^{-1}\,|^{1/2}}{(2\pi)^{K/2}} \exp\left[-\frac{1}{2} z^t R_x^{-1} z \right] \qquad \text{for} -\infty < z < \infty \tag{2.107}$$

in which $R_x = C_z = E(ZZ^t)$ is a $K \times K$ correlation matrix

$$\mathrm{Corr}(X) = \mathrm{Cov}(Z) = R_x = \begin{bmatrix} 1 & \rho_{12} & \cdots & \rho_{1K} \\ \rho_{21} & 1 & \cdots & \rho_{2K} \\ \vdots & \vdots & \vdots & \vdots \\ \rho_{K1} & \rho_{K2} & \cdots & 1 \end{bmatrix}$$

with $\rho_{jk} = \mathrm{Cov}(Z_j, Z_k)$ being the correlation coefficient between each pair of normal random variables, X_j and X_k. For bivariate standard normal variables, the following relationships of cross-product moments are useful (Hutchinson and Lai 1990)

$$E\left[Z_1^{2m} Z_2^{2n}\right] = \frac{(2m)!(2n)!}{2^{m+n}} \sum_{j=0}^{\min(m,n)} \frac{(2\rho_{12})^{2j}}{(m-j)!(n-j)!(2j)!}$$

$$E\left[Z_1^{2m+1} Z_2^{2n+1}\right] = \frac{(2m+1)!(2n+1)!}{2^{m+n}} \rho_{12} \sum_{j=0}^{\min(m,n)} \frac{(2\rho_{12})^{2j}}{(m-j)!(n-j)!(2j+1)!} \quad (2.108)$$

$$E\left[Z_1^{2m+1} Z_2^{2n}\right] = E\left[Z_1^{2m} Z_2^{2n+1}\right] = 0$$

for m and n being positive integer numbers.

2.5.2 Multivariate lognormal distributions

Similar to the univariate case, bivariate lognormal random variables have a PDF

$$f_{x_1,x_2}(x_1, x_2) = \frac{1}{2\pi\, x_1 x_2\, \sigma_{\ln x_1}\, \sigma_{\ln x_2} \sqrt{1-\rho_{12}'^2}} \exp\left[\frac{-Q'}{2(1-\rho_{12}'^2)}\right] \quad (2.109)$$

for $x_1, x_2 > 0$, in which

$$Q' = \frac{[\ln(x_1) - \mu_{\ln x_1}]^2}{\sigma_{\ln x_1}^2} + \frac{[\ln(x_2) - \mu_{\ln x_2}]^2}{\sigma_{\ln x_2}^2} - 2\rho_{12}' \frac{[\ln(x_1) - \mu_{\ln x_1}][\ln(x_2) - \mu_{\ln x_2}]}{\sigma_{\ln x_1}\, \sigma_{\ln x_2}}$$

where $\mu_{\ln x}$ and $\sigma_{\ln x}$ are the mean and standard deviation of log-transformed random variables; and subscripts "1" and "2" indicate the random variables X_1 and X_2, respectively; and $\rho_{12}' = \mathrm{Corr}(\ln X_1, \ln X_2)$ is the correlation coefficient of the two log-transformed random variables. After log transformation is made, properties of multivariate lognormal random variables follow exactly as for the multivariate normal case. Relationship between correlation coefficient in the original and log-transformed spaces can be derived using the moment generating function (see Sec. 4.2) as

$$\mathrm{Corr}(X_1, X_2) = \rho_{12} = \frac{\exp\left(\rho_{12}' \sigma_{\ln x_1}\, \sigma_{\ln x_2}\right) - 1}{\sqrt{\exp\left(\sigma_{\ln x_1}^2\right) - 1}\, \sqrt{\exp\left(\sigma_{\ln x_2}^2\right) - 1}} \quad (2.110)$$

Problems

2.1 Derive the PDF for a random variable having a triangular distribution with the lower bound "a," the mode "m," and the upper bound "b" as show in Fig. P2.1.

2.2 Show that $F_1(x_1) + F_2(x_2) - 1 \le F_{1,2}(x_1, x_2) \le \min[F_1(x_1), F_2(x_2)]$.

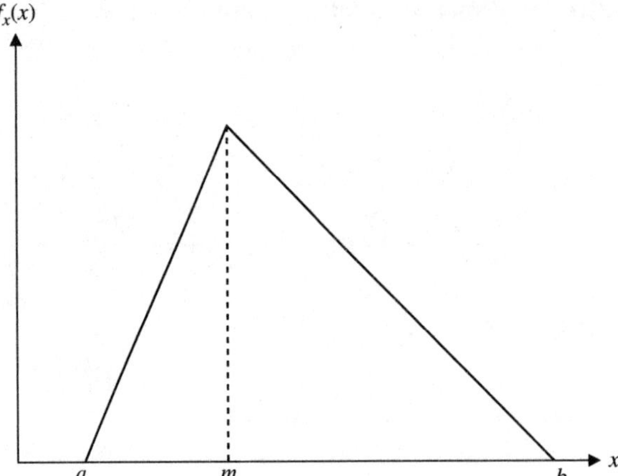

Figure P2.1   Triangular distribution.

2.3 The Farlie-Gumbel-Morgenstern bivariate uniform distribution has the joint CDF as (Hutchinson and Lai 1990)

$$F_{x,y}(x, y) = xy[1 + \theta\,(1 - x)(1 - y)] \qquad \text{for } 0 \le x, y \le 1$$

with $-1 \le \theta \le 1$. Do the following exercises: (a) derive the joint PDF; (b) obtain the marginal CDF and PDF of X and Y; and (c) derive the conditional PDFs $f_x(x\,|\,y)$ and $f_y(y\,|\,x)$.

2.4 Refer to Prob. 2.3. Compute (a) $P(X \le 0.5,\ Y \le 0.5)$; (b) $P(X \ge 0.5,\ Y \ge 0.5)$; and (c) $P(X \ge 0.5\,|\,Y = 0.5)$.

2.5 Apply Eq. (2.21) to show that the first four central moments in terms of moments about the origin are

$$\mu_1 = 0$$

$$\mu_2 = \mu_2' - \mu_x^2$$

$$\mu_3 = \mu_3' - 3\mu_x\mu_2' + 2\mu_x^3$$

$$\mu_4 = \mu_4' - 4\mu_x\mu_3' + 6\mu_x^2\mu_2' - 3\mu_x^4$$

2.6 Apply Eq. (2.22) to show that the first four moments about the origin could be expressed in terms of the first four central moments as

$$\mu_1' = \mu_x$$

$$\mu_2' = \mu_2 + \mu_x^2$$

$$\mu_3' = \mu_3 + 3\mu_x\,\mu_2 + \mu_x^3$$

$$\mu_4' = \mu_4 + 4\mu_x\,\mu_3 + 6\mu_x^2\,\mu_2 + \mu_x^4$$

2.7 Based on definitions of α- and β-moments, i.e., Eqs. (2.25a) and (2.25b), (a) derive the general expressions between the two moments; and (b) write out explicitly their relations for $r = 0, 1, 2$, and 3.

2.8 Refer to Example 2.4. Continue to derive the expressions for the 3rd and 4th L-moments of the exponential distribution.

2.9 A company plans to build a production factory by a river. You are hired by the company as a consultant to analyze the flood risk of the factory site. It is known that the magnitude of an annual flood has a lognormal distribution with a mean of 30,000 ft^3/s and standard deviation of 25,000 ft^3/s. It is also known that, from a field investigation, the stage-discharge relationship for the channel reach is $Q = 1500H^{1.4}$ where $Q =$ flow rate (in ft^3/s) and $H =$ water surface elevation (in feet) above a given datum. The elevation of a tentative location for the factory is 15 ft above the datum (Mays and Tung 1992).
(a) What is the annual risk that the factory site will be flooded?
(b) At this particular plant site, it is also known that the flood damage function can be approximated as

$$\text{Damage (in \$1000)} = \begin{cases} 0, & \text{if } H \le 15' \\ 40\,(\ln H + 8)(\ln H - 2.7), & \text{if } H > 15' \end{cases}$$

What is the annual expected flood damage? (Use the appropriate numerical approximation technique for calculations.)

2.10 Referring to Prob. 2.1, assume that Manning's roughness coefficient has a triangular distribution as shown in Fig. P2.1.
(a) Derive the expression for the mean and variance of Manning's roughness.
(b) Show that (i) for a symmetric triangular distribution, $\sigma = (b - m)/\sqrt{6}$; (ii) when the mode is at the lower or upper bound, $\sigma = (b - a)/3\sqrt{2}$.

2.11 Suppose that a random variable X has a uniform distribution (Fig. P2.2) with "a" and "b" being its lower and upper bounds, respectively. Show that
(a) $E(X) = \mu_x = (b + a)/2$;
(b) $\text{Var}(X) = (b - a)^2/12$; and
(c) $\Omega_x = (1 - a/\mu_x)/\sqrt{3}$

2.12 Referring to Fig. P2.2, (a) derive the expression for the first two probability-weighted moments; and (b) derive the expressions for the L-coefficient of variation.

2.13 Refer to Example 2.3. Based on the conditional PDF obtained in part (c), derive the conditional expectation, $E(Y|x)$, and the conditional variance, $\text{Var}(Y|x)$. Furthermore, plot the conditional expectation and conditional standard deviation of Y on x with respect to x.

2.14 Consider two random variables X and Y having the joint PDF of the form as

$$f_{x,y}(x,y) = c\left(5 - \frac{y}{2} + x^2\right) \qquad \text{for } 0 \le x, y \le 2$$

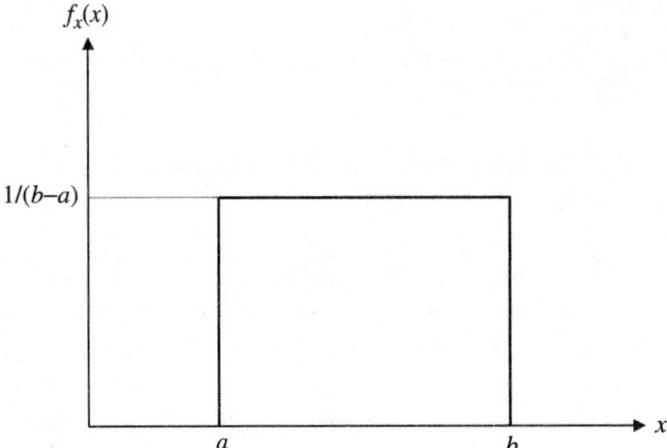

Figure P2.2   Uniform distribution.

(a) Determine the coefficient c

(b) Derive the joint CDF

(c) Find $f_x(x)$ and $f_y(y)$

(d) Determine the mean and variance of X and Y

(e) Compute the correlation coefficient between X and Y

2.15 Consider the following hydrologic model in which the runoff, Q, is related to the rainfall, R, by

$$Q = a + b\,R$$

in $a > 0$ and $b > 0$ are model coefficients. Ignoring uncertainties of model coefficients, show that $\mathrm{Corr}(Q, R) = 1.0$.

2.16 Suppose that the rainfall-runoff model in Prob. 2.15 has a model error and it can expressed as

$$Q = a + b\,R + \varepsilon$$

in which ε is the model error term that has a zero mean and standard deviation of σ_ε. Furthermore, the model error ε is independent of the random rainfall, R. Derive the expression for $\mathrm{Corr}(Q, R)$.

2.17 Let $X = X_1 + X_3$ and $Y = X_2 + X_3$. Find $\mathrm{Corr}(X,Y)$ assuming X_1, X_2, and X_3 are statistically independent.

2.18 The well-known *Thiem equation* can be used to compute the drawdown in a confined and homogeneous aquifer as

$$s_{ik} = \frac{\ln(r_{ok} / r_{ik})}{2\pi\,T}\,Q_k = \xi_{ik}\,Q_k$$

in which s_{ik} = drawdown at the ith observation location resulting from a pumpage of Q_k at the jth production well; r_{ok} = radius of influence of the kth production well; r_{ik} = distance between the ith observation point and the kth production well; and T = transmissivity of the aquifer. The overall effect of the aquifer drawdown at

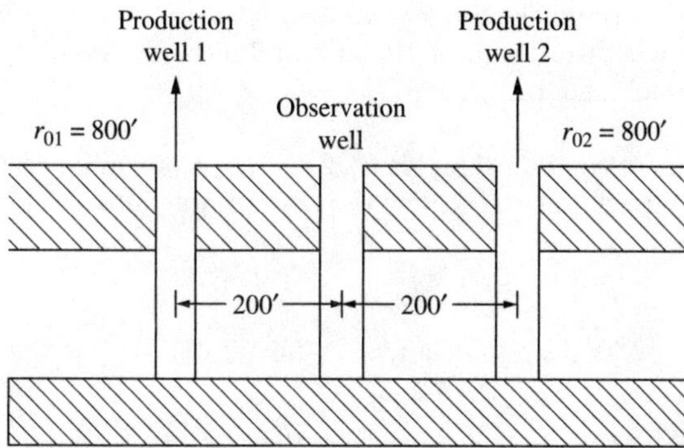

Figure P2.3 Locations of production and observation wells (after Mays and Tung 1992).

the ith observation point, when more than one production well is in operation, can be obtained, by the principle of linear superposition, as the sum of the responses caused by all production wells in the field, that is,

$$s_i = \sum_{k=1}^{K} s_{ik} = \sum_{k=1}^{K} \xi_{ik} Q_k$$

where K = total number of production wells in operation.

Consider a system consisting of two production wells and one observation well. The locations of the three wells, the pumping rates of the two production wells and their zones of influence are shown in Fig. P2.3. It is assumed that the transmissivity of the aquifer has a log-normal distribution with the mean $\mu_T = 4000$ gallons per day per foot (gpd/ft) and standard deviation $\sigma_T = 2000$ gpd/ft (Mays and Tung 1992).

(a) Prove that the total drawdown in the aquifer field is also lognormally distributed.

(b) Compute the exact values of the mean and variance of the total drawdown at the observation point when $Q_1 = 10,000$ gpd and $Q_2 = 15,000$ gpd.

(c) Compute the probability that the resulting drawdown at the observation point does not exceed 2 ft.

(d) If the maximum allowable probability of the total drawdown exceeding 2 ft is 0.10, find out the maximum allowable total pumpage from the two production wells.

2.19 A frequently used surface pollutant washoff model is based on a first-order decay function (Sartor and Boyd 1972)

$$M_t = M_0 e^{-cRt}$$

where M_o = initial pollutant mass at time $t = 0$; R = runoff intensity (mm/h); c = washoff coefficient (mm^{-1}); M_t = mass of the pollutant remaining on the street surface (kg); and t = time elapsed (in hours) since the beginning of the storm. The above model does not consider pollutant build-up and is generally appropriate for the within-storm event analysis.

Suppose that $M_o = 10,000$ kg and $c = 1.84$/cm. The runoff intensity R is a normal random variable with the mean of 10 cm/h and a coefficient of variation of 0.3. Determine the time, t, such that $P(M_t/M_o < 0.05) = 0.90$.

2.20 Consider n independent random samples $X_1, X_2, \ldots, X_n$ from an identical distribution with the mean μ_x and variance σ_x^2. Show that the sample mean $\overline{X}_n = \sum_{i=1}^{n} X_i / n$ has the following properties:

$$E[\overline{X}_n] = \mu_x \qquad \text{and} \qquad \text{Var } [\overline{X}_n] = \frac{\sigma_x^2}{n}$$

What would be the sampling distribution of $\overline{X}_n$ if random samples are normally distributed.

2.21 Consider that measured hydrologic quantity Y and its indicator for accuracy S are related to the unknown true quantity X as $Y = SX$. Assume that $X \sim LN(\mu_x, \sigma_x)$, $S \sim LN(\mu_s = 1, \sigma_s)$ and X is independent of S.

(a) What is the distribution function for Y? Derive the expressions for the mean and coefficient of variation of Y, that is, μ_y and Ω_y, in terms of those of X and S.

(b) Derive the expression for $r_p = y_p/x_p$ with $P(Y \le y_p) = P(X \le x_p) = p$ and plot r_p versus p.

(c) Define measurement error as $\varepsilon = Y - X$. Determine the minimum reliability of the measurement so that the corresponding relative absolute error, $|\varepsilon/X|$, does not exceed the require precision of 5 percent.

2.22 Consider that measure discharge Q' is subject to measurement error ε and both are related to the true, but unknown discharge Q as (Cong and Xu 1987)

$$Q' = Q + \varepsilon$$

It is common to assume that (1) $E(\varepsilon|q) = 0$; (2) $\text{Var}(\varepsilon|q) = [\alpha(q)q]^2$; and (3) random error ε is normally distributed, that is, $\varepsilon|q \sim N(\mu_{\varepsilon|q} = 0, \sigma_{\varepsilon|q})$.

(a) Show that $E(Q'|q) = q$; $E[(Q'/Q) - 1|q] = 0$; and $\text{Var}[(Q'/Q - 1)^2|q] = \alpha^2(q)$.

(b) Under $\alpha(q) = \alpha$, show that $E(Q') = E(Q)$; $\text{Var}(\varepsilon) = \alpha^2 E(Q^2)$; and $\text{Var}(Q') = (1 + \alpha^2)\text{Var}(Q) + \alpha^2 E^2(Q)$.

(c) Suppose it is required that 75 percent of measurements whose relative error lies in the range of ± 5 percent (precision level). Determine the corresponding value of $\alpha(q)$ assuming that the measurement error is normally distributed.

2.23 Analyzing the stream flow data from several flood events, it is found that the flood peak discharge (Q) and the corresponding volume (V) have the following relationship

$$\ln(V) = a + b \times \ln(Q) + \varepsilon$$

in which "a" and "b" are constants and ε is the model error terms. Suppose the model error terms ε has a normal distribution with mean 0 and standard deviation σ_ε. Then, show that the conditional PDF of $V|Q$, $h(v|q)$, is a lognormal distribution. Furthermore, suppose the peak discharge is a lognormal random variable. Show that the joint PDF of V and Q is bivariate lognormal.

2.24 Analyzing the stream flow data from 105 flood events at different locations in Wyoming, USA, Wahl and Rankl (1993) found that the flood peak discharge (Q, in ft^3/s) and the corresponding volume (V, in acre-feet) have the following relationship

$$\ln(V) = \ln(0.0655) + 1.011 \times \ln(Q) + \varepsilon$$

in which ε is the model error terms with the assumed $\sigma_\varepsilon = 0.3$. A flood frequency analysis of the North Platte River near Walden, Colorado. indicated that the annual maximum flood peak discharge has a lognormal distribution with mean $\mu_Q = 1380$ ft^3/s and $\sigma_Q = 440$ ft^3/s.
(a) Derive the joint PDF of V and Q for the annual maximum flood
(b) Determine the correlation coefficient between V and Q.

2.25 Let $X_1 = a_0 + a_1 Z_1 + a_2 Z_1^2$ and $X_2 = b_0 + b_1 Z_2 + b_2 Z_2^2$ in which Z_1 and Z_2 are bivariate standard normal random variables with correlation coefficient ρ, that is, $\mathrm{Corr}(Z_1, Z_2) = \rho$. Derive the expression for $\mathrm{Corr}(X_1, X_2)$ in terms of polynomial coefficients and ρ.

2.26 Let X and Y are bivariate lognormal random variables. Show that

$$\frac{\exp\!\left(-\sigma_{\ln x_1}\,\sigma_{\ln x_2}\right)-1}{\sqrt{\exp\!\left(\sigma_{\ln x_1}^2\right)-1}\,\sqrt{\exp\!\left(\sigma_{\ln x_2}^2\right)-1}} \le \mathrm{Corr}(X,Y) \le \frac{\exp\!\left(\sigma_{\ln x_1}\,\sigma_{\ln x_2}\right)-1}{\sqrt{\exp\!\left(\sigma_{\ln x_1}^2\right)-1}\,\sqrt{\exp\!\left(\sigma_{\ln x_2}^2\right)-1}}$$

What does this inequality indicate?

2.27 Derive Eq. (2.63) from Eq. (2.110)

$$\mathrm{Corr}(\ln X_1, \ln X_2) = \rho_{12}' = \frac{\ln(1 + \rho_{12}\,\Omega_1\,\Omega_2)}{\sqrt{\ln\!\left(1 + \Omega_1^2\right)}\,\sqrt{\ln\!\left(1 + \Omega_2^2\right)}}$$

where $\rho_{12} = \mathrm{Corr}(X_1, X_2)$; and $\Omega_k = $ coefficient of variation of X_k, $k = 1, 2$.

References

Abramowitz, M., and I. A. Stegun, eds. (1972). *Handbook of Mathematical Functions with Formulas, Graphs and Mathematical Tables*, 9th ed., Dover Publications, New York.

Ang, A. H. S., and W. H. Tang (1975). *Probability Concepts in Engineering Planning and Design, Volume I: Basic Principles*, John Wiley & Sons, New York.

Blank, L. (1980). *Statistical Procedures for Engineering, Management, and Science*, McGraw-Hill, New York.

Cong, S. Z., and Y. B. Xu (1987). "The effect of discharge measurement error in flood frequency analysis," *Journal of Hydrology*, **96**:237–254.

Devore, J. L. (1987). *Probability and Statistics for Engineering and Sciences*, 2d ed., Brooks/Cole, Monterey, CA.

Dowson, D. C., and A. Wragg (1973). "Maximum Entropy Distributions Having Prescribed First and Second Moments," *IEEE Transactions on Information Theory*, **19**(9):689–693.

Dudewicz, E. (1976). *Introduction to Statistics and Probability*, Holt, Rinehart, and Winston Publishing, New York.

Farlie, D. J. G. (1960). "The Performance of Some Correlation Coefficients for a General Bivariate Distribution," *Biometrika*, **47**:307–323.

Fisher, R. A., and L. H. C. Tippett (1928). "Limiting Forms of the Frequency Distribution of the Largest or Smallest Member of a Sample," *Proceedings of Cambridge Philosophical Society*, **24**:180–190.

Greenwood, J. A., J. M. Landwehr, N. C. Matalas, and J. R. Wallis (1979). "Probability Weighted Moments: Definitions and Relation to Parameters of Several Distribution Expressible in Inverse Form," *Water Resources Research*, AGU, **15**(6):1049–1054.

Gnedenko, B. V. (1943). "Sur la distribution limite du terme maximum d'une serie aleatoire," *Annals of Mathematics*, **44**:423–453.

Gumbel, E. J. (1958). *Statistics of Extremes*, Columbia University Press, New York.

Haan, C. T. (1977). *Statistical Methods in Hydrology*, Iowa State University Press, Ames, IA.

Henley, E. J., and H. Kumamoto (1981). *Reliability Engineering and Risk Assessment*, Prentice Hall, Englewood Cliffs, NJ.

Hosking, J. R. M. (1986). "The Theory of Probability Weighted Moments," *IBM Research Report*, No. 12210, October.

Hosking, J. R. M. (1990). "L-moments: Analysis and Estimation of Distributions Using Linear Combinations of Order Statistics," *Journal of Royal Statistical Society, Series B*, **52**(1):105–124.

Hosking, J. R. M. (1991). "Approximations for Use in Constructing L-Moment Ratio Diagram," *IBM Research Report*, No. 16635, T. J. Watson Research Center, Yorktown Heights.

Hutchinson, T. P., and C. D. Lai (1990). *Continuous Bivariate Distributions, Emphasizing Applications*, Rumsby Scientific Publishing, Adelaide, South Australia.

Johnson, M. E. (1987). *Multivariate Statistical Simulation*, John Wiley and Sons, New York.

Johnson, N. L., and S. Kotz (1972). *Distributions in Statistics: Continuous Univariate Distributions-2*, John Wiley and Sons, New York.

Johnson, N. L., and S. Kotz (1976). *Distributions in Statistics: Continuous Multivariate Distributions*, John Wiley and Sons, New York.

Kite, G. W. (1988). *Frequency and Risk Analysis in Hydrology*, Water Resources Publications, Littleton.

Leadbetter, M. R., G. Lindgren, and H. Rootzen (1983). *Extremes and Related Properties of Random Sequences and Processes*, Springer-Verlag, New York.

Leemis, L. M. (1986). "Relationships Among Common Univariate Distributions," *The American Statistician*, **40**(2):143–146.

Liu, P. L., and A. Der Kiureghian (1986). "Multivariate Distribution Models with Prescribed Marginals and Covariances," *Probabilistic Engineering Mechanics*, **1**(2):105–112.

Mardia, K. V. (1970a). *Families of Bivariate Distributions*, Charles Griffin Company, London.

Mardia, K. V. (1970b). "A Translation Family of Bivariate Distributions and Frechet's Bounds," *Sankhya, Series A*, **32**:119–121.

Mays, L. W. and Y. K. Tung (1992). *Hydrosystems Engineering and Management*, McGraw-Hill, New York.

Morgenstern, D. (1956). "Einfache Beispiele zweidimensionaler Verteilungen," *Mitteilingsblatt fur Mathematische Statistik*, **8**:234–235.

Nataf, A. (1962). "Determination des distributions dont les marges sont donnees," *Comptes Rendus de l'Academie des Sciences*, Paris, **225**:42–43.

Royston, P. (1992). "Which Measures of Skewness and Kurtosis are Best?" *Statistics in Medicine*, **11**:333–343.

Sartor, J. D., and G. B. Boyd (1972). "Water Pollution Aspects of Street Surface Contaminants," *Report* No. EPA-R2-72-081, U. S. Environmental Protection Agency, Washington, DC.

Shen, H.W., and M. Bryson (1979). "Impact of Extremal Distributions on Analysis of Maximum Loading," *Reliability in Water Resources Engineering*, E. A. McBean, K. W. Hipel, and T. E. Unny (eds.), Water Resources Publications, Littleton, CO.

Stedinger, J. R., R. M. Vogel, and E. Foufoula-Georgiou (1993). "Chapter 18: Frequency Analysis of Extreme Events," *Handbook of Hydrology*, D.R. Maidment (ed.), McGraw-Hill, New York.

Stuart, A., and J. K. Ord (1987). Kendall's Advanced Theory of Statistice, Vol. 1, Distribution Theory, 5th ed, Oxford University Press, New York.

U. S. Water Resources Council (now called the Interagency Advisory Committee on Water Data) (1982). *Guideline in Determining Flood Flow Frequency-Bulletin 17B*, Available from Office of Water Data Coordination, U. S. Geological Survey, Reston, VA.

Vale, C. D., and V. A. Maurelli (1983). "Simulating Multivariate Nonnormal Distributions," *Psychometrika*, **48**(3):65–471.

Vedder, J. D. (1995). "An Invertible Approximation to the Normal Distribution Function." *Computational Statistics and Data Analysis*, **16**(2):119–123.

Wahl, K. L., and J. G. Rankl (1993). "A Potential Problem with Mean Dimensionless Hydrographs at Ungaged Sites." *Proceedings of International Symposium on Engineering Hydrology*, San Francisco, CA, July 26–30.

Regression Analysis

3.1 Introduction

Most phenomena affecting design and analysis in hydrosystems engineering involve many interrelated factors. For example, hydrosystems engineers might be interested in relationships of rainfall and runoff (Fig. 3.1), stage and discharge (Fig. 3.2), or the operational cost of a small hydropower plant and its capacity (Fig. 3.3) for water resource planning and infrastructural designs. Due to the inherent complexity of system behavior and lack of full understanding of the involved processes, relationships among the various relevant factors or variables are often established empirically or semiempirically. Furthermore, due to the presence of uncertainties a deterministic functional relationship is generally not very appropriate or realistic. In hydrosystems engineering, many empirical and semiempirical relationships can be found and a few examples are given as follows:

- Time of concentration (travel time) of flow on an overland surface (Morgali and Linsley 1965)

$$t_o = 0.99 \frac{n^{0.605} \, L^{0.593}}{i^{0.388} \, S_o^{0.38}}$$

where n = Manning's roughness coefficient
L = overland flow length (ft)
i = rainfall intensity (in/h)
S_o = slope of the overland surface

- The Hazen-Williams formula in pipe flow computation

$$V = C_u \, C_{hw} \, R^{0.63} \, S^{0.54}$$

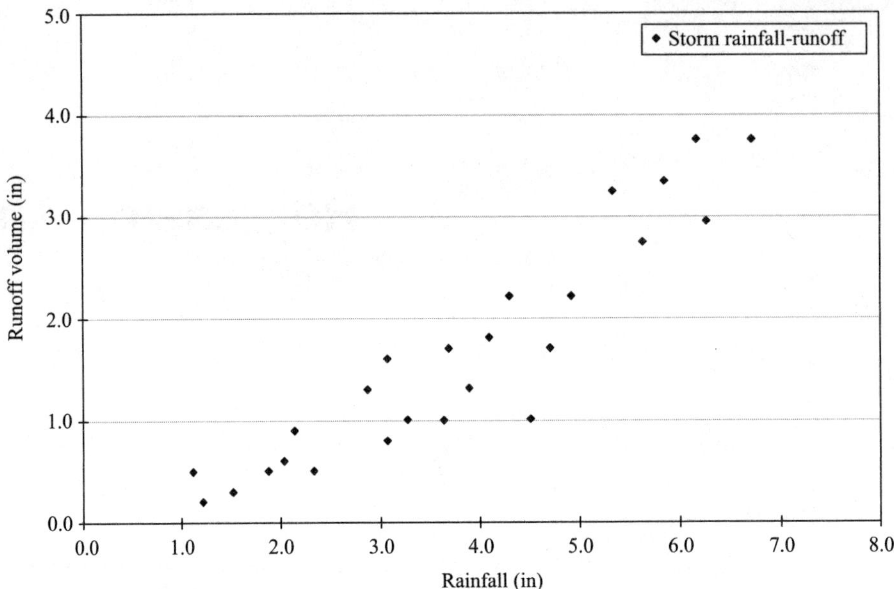

Figure 3.1   Total rainfall and effective runoff for a watershed (after Miller and Cronshey 1992).

where C_u = unit conversion factor (0.85 for SI, 1.32 for the U.S. customary unit)

V = average flow velocity in pipe (m/s; ft/s)

C_{hw} = Hazen-Williams roughness coefficient

R = hydraulic radius (m; ft), equaling $D/4$ with D being the pipe diameter (m; ft)

S = slope of the pipe

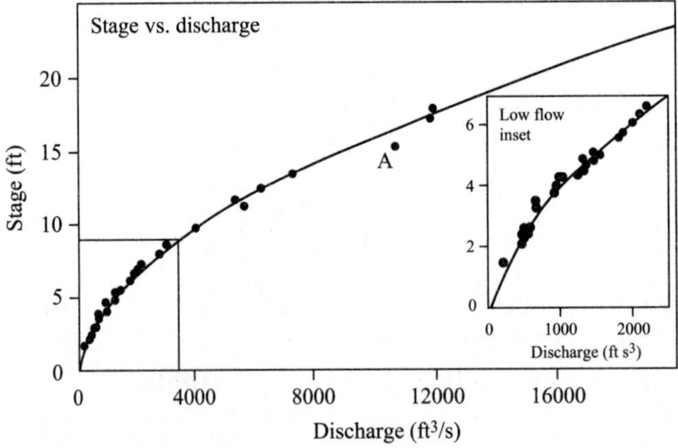

Figure 3.2   Rating curve for Grey River at Dobson, New Zealand (after Mosley and McKerchar 1992).

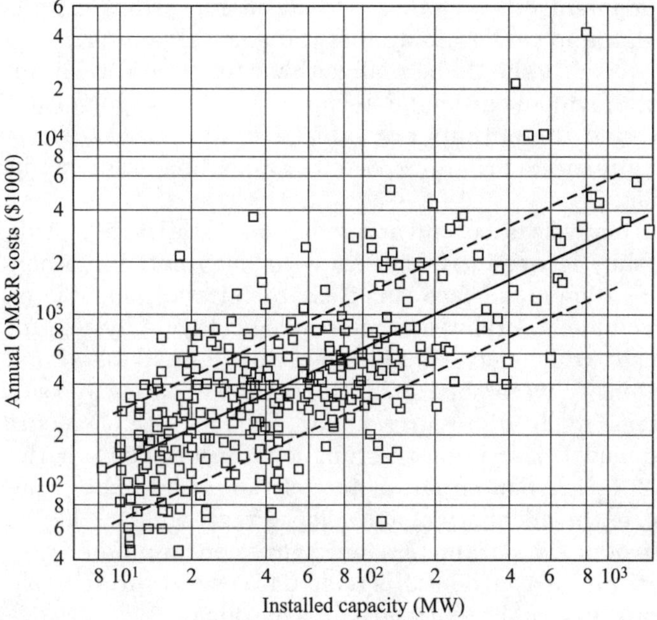

Figure 3.3 Annual operation, maintenance, and repair costs versus installed hydropower plant capacity (after Woods and Gulliver 1991).

- Sediment transport rate of a channel in arid regions (Zeller and Fullerton 1983)

$$q_s = \frac{0.0064\, n^{1.77}\, V^{4.32}\, G^{0.45}}{Y_b^{0.30}\, d_{50}^{0.61}}$$

    where q_s = unit sediment transport rate (ft³/s/ft)
          V = velocity (ft/s)
          G = gradation coefficient
          Y_b = hydraulic depth (ft)
         d_{50} = median diameter of bed material (mm)

- Time to peak of the 10-min unit hydrograph in urban areas (Espey and Altman 1978)

$$T_p = 3.1\, L^{0.23}\, S_c^{-0.25}\, I^{-0.18}\, \Psi^{1.57}$$

    where T_p = time to peak (min)
          L = total distance along the principal flow path from watershed
             boundary to the design point (ft)
         S_c = main channel slope (ft/ft)
          I = percentage of impervious area
         Ψ = dimensionless watershed conveyance factor

The development of an empirical relationship provides an engineer with useful insights and understanding of the general behavior of the concerned process. However, it should be realized that functional relationships such as those shown previously are not deterministic because the model itself, model parameters, and the outputs are subject to uncertainty. The model results can only be interpreted as an expected or a nominal value under some specified condition.

Regression analysis is a useful and widely used statistical tool for investigating the relationship between two or more variables related in a nondeterministic fashion. For example, surface runoff characteristics, such as peak discharge and runoff volume, are related to meteorological and physiographic characteristics of a basin, which may include, but are not limited to, rainfall volume, precipitation intensity, watershed size, extent of urbanization and so on. In general, the basic steps involved in regression analysis include (1) identification of an appropriate model describing the functional relationship between pertinent variables, (2) estimation of parameters of the model under consideration, (3) examination of the adequacy of a developed regression model, and (4) statistical inference of model parameters and regression equation.

Mathematically, if a variable Y is related to several variables $X_1, X_2,..., X_K$ and their relationships can be expressed, in general, as

$$Y = g(X_1, X_2,..., X_K) \tag{3.1}$$

in which the function $g(.)$ can be linear or nonlinear. Those variables X's whose values are used to estimate Y are called *independent (explanatory) variables* or *regressors* and the variable Y is called *dependent (response) variable*. In practical applications, the functional forms commonly used for establishing empirical relationships are

Additive: $$Y = \beta_0 + \beta_1 X_1 + \beta_2 X_2 + \cdots + \beta_K X_K \tag{3.2a}$$

Multiplicative: $$Y = \beta_0 \, X_1^{\beta_1} \, X_2^{\beta_2} \cdots X_K^{\beta_K} \tag{3.2b}$$

in which $\beta_0, \beta_1,..., \beta_K$ are model parameters, commonly called *regression coefficients*.

The functional equation forms of Eqs. (3.1) and (3.2) are deterministic because the value of the response variable Y can be uniquely determined for a given fixed value of $X_1 = x_1, X_2 = x_2,..., X_K = x_K$. When the reality says that Y is a random variable and its expected value is related to X's, then, for any fixed value of $X = x = (x_1, x_2,..., x_K)^t$, Eq. (3.1) can be modified as

$$Y = g(x_1, x_2,..., x_K) + \varepsilon \tag{3.3}$$

in which ε is a random variable with $E(\varepsilon) = 0$ and $\text{Var}(\varepsilon) = \sigma_\varepsilon^2$. Then, the conditional expected value and variance of Y for a given fixed independent variable x are, respectively,

$$E(Y \mid x) = g(x) \qquad \text{Var}(Y \mid x) = \sigma_\varepsilon^2$$

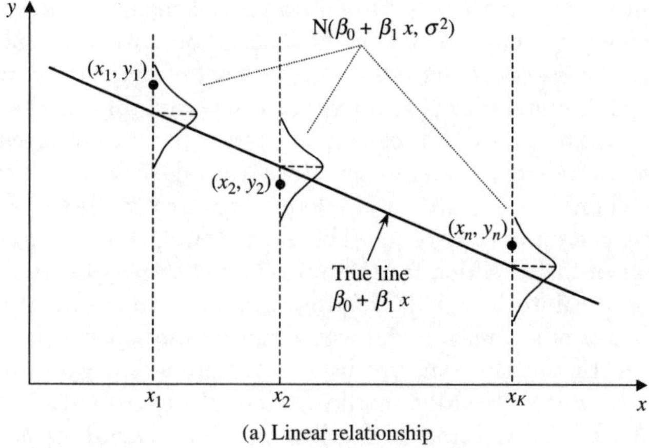

(a) Linear relationship

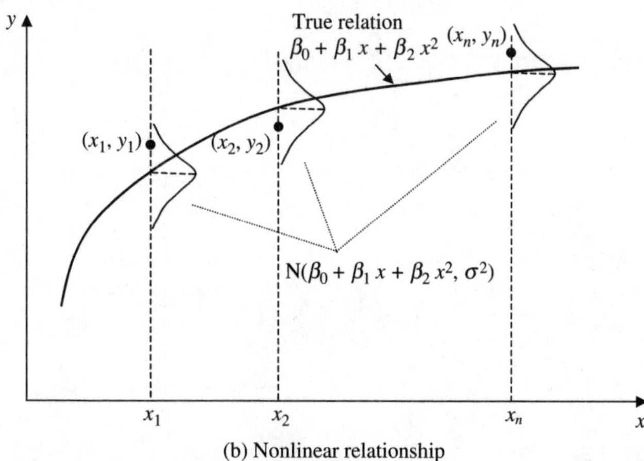

(b) Nonlinear relationship

Figure 3.4   Simple regression models.

When the regression model is linear as defined by Eq. (3.2a), the corresponding hyperplane would pass through the expected value of Y conditioned on various values of $\boldsymbol{x}$. As shown in Fig. 3.4 under a simple regression model with a single explanatory variable, at any given $X = x$, there is a distribution associated with Y. The same can be applied to any regression model of general form. Although Eq. (3.2b) is nonlinear, it can easily be transformed into a linear function by taking a log transformation on both sides of equality.

3.2   Identification of Appropriate Models

For physical processes, theories sometimes could help in selecting an appropriate functional relation for the regression model. In general, the true functional relationship is never known with regard to its equation form or model parameters.

What are generally available is a set of data for the response variable and some conceivably relevant explanatory variables based on which the model is postulated and its parameters estimated. Consider a set of data containing n pairs of measured response variable Y and explanatory variable X, that is, $\{(x_1, y_1),$ $(x_2, y_2),..., (x_n, y_n)\}$. The task in regression analysis is to determine a suitable functional relation, along with model parameters, that describes this set of data.

In the case when one has only a single explanatory variable, a "scatter diagram" of n observed pairs (x_i, y_i) should be plotted first. Such a diagram provides useful information as to which functional relation (linear or nonlinear) would be appropriate to relate Y to X. If the diagram shows a curvature as shown in Fig. 3.5(b), the use of a linear model would not be inappropriate.

For a given data set, there might be several models that could potentially be used to fit the data. Broadly speaking, model identification in regression analysis involves the selection of the independent variables and functional form by which these independent variables are related. As the true functional

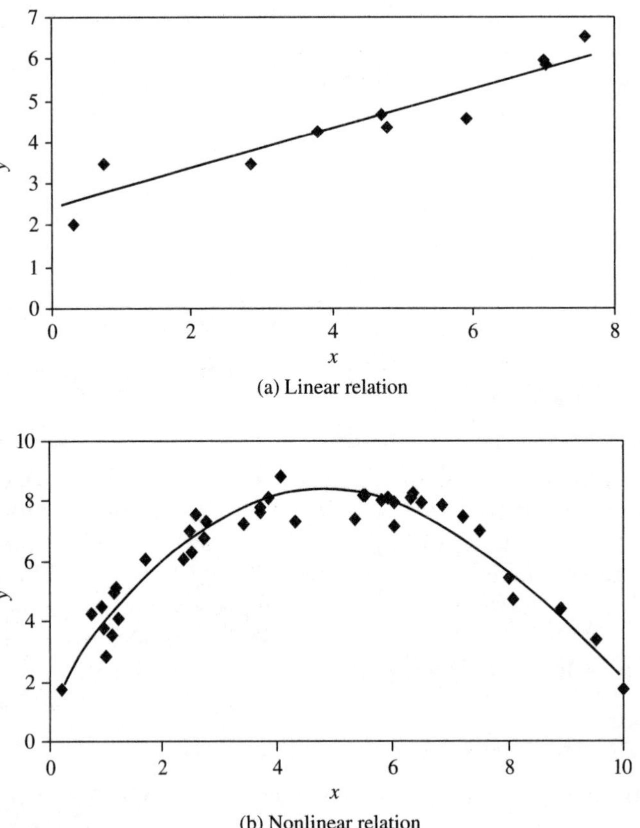

(a) Linear relation

(b) Nonlinear relation

Figure 3.5   Scatter diagrams.

relation between the dependent variable and the explanatory variables is unknown, identifying reasonable models is a subjective, trial-and-error process and selecting a plausible model should be based on the physical relevance, data availability, cost of data acquisition, model accuracy, and practical applicability. More will be discussed about the selection of "best" subset variables in Sec. 3.9.1.

3.3 Parameters Estimation by the Least Squares Method

Estimating parameters in a selected regression model based on a set of observations, in essence, is an exercise of curve fitting. To determine the model and its parameters that "best" fit the available data, one must adopt a goodness-of-fit criterion. The most natural indicator for goodness-of-fit is related to the residuals, defined as the difference between the observed values and those computed by the model, that is, $e_i = y_i - \hat{y}_i$ where y_i and $\hat{y}_i$ are, respectively, the observed and modeled values for the ith dependent variable.

Among various goodness-of-fit criteria used in model parameter estimation, the *least squares* (LS) criterion is the most widely used. For a selected model, the LS estimates of parameter values (i.e., β's) are the ones that minimize the sum of squared residuals, that is,

$$\underset{\beta's}{\text{Min}} \sum_{i=1}^{n} e_i^2 = \sum_{i=1}^{n} (y_i - \hat{y}_i)^2 = \sum_{i=1}^{n} [y_i - g(x_i \mid \beta_0, \beta_1, \ldots, \beta_K)]^2 \tag{3.4}$$

As can be seen, the determination of the LS parameters, in essence, is a problem of unconstrained minimization. Several optimum seeking procedures are available and the selection of suitable techniques for determining the optimal regression parameters depends on the form of the model, $g(.)$. In this chapter, the focus will be placed on regression models in which the response variable Y is linearly related to the regression coefficients, not necessarily to the explanatory variables.

Consider a *simple linear regression model* as Eq. (3.2a) with $K = 1$, the objective function by the LS criterion can be expressed as

$$\underset{\beta_0, \beta_1}{\text{Min}} D = \sum_{i=1}^{n} [y_i - (\beta_0 + \beta_1 x_i)]^2 \tag{3.5}$$

The optimal parameter values for the above unconstrained minimization problem must satisfy the following necessary conditions:

$$\frac{\partial D}{\partial \beta_0} = 0 \qquad \frac{\partial D}{\partial \beta_1} = 0$$

from which a system of linear equations (called *normal equations*) can be established

$$n\beta_0 + \left(\sum_{i=1}^{n} x_i\right)\beta_1 = \left(\sum_{i=1}^{n} y_i\right)$$

$$\left(\sum_{i=1}^{n} x_i\right)\beta_0 + \left(\sum_{i=1}^{n} x_i^2\right)\beta_1 = \left(\sum_{i=1}^{n} x_i y_i\right)$$

(3.6)

Hence, the LS model parameters β_0 and β_1 can be obtained by solving the above normal equations as

$$b_1 = \hat{\beta}_1 = \frac{n\sum x_i y_i - \left(\sum x_i\right)\left(\sum y_i\right)}{n\left(\sum x_i^2\right) - \left(\sum x_i\right)^2} = \frac{\sum x_i y_i - n\bar{x}\,\bar{y}}{\left(\sum x_i^2\right) - n\bar{x}^2} = \frac{S_{xy}}{S_{xx}}$$

$$b_0 = \hat{\beta}_0 = \frac{\sum y_i - \hat{\beta}_1\left(\sum x_i\right)}{n} = \bar{y} - \hat{\beta}_1\bar{x}$$

(3.7)

in which $\bar{x}$ and $\bar{y}$ are, respectively, the sample means of independent and dependent variables.

Example 3.1   Table 3.1 contains annual water use data for the City of Austin, Texas, from 1965 to 1985. Develop an appropriate regression model relating annual water use with the population in the city (adopted from Mays and Tung 1992).

TABLE 3.1   Water Use Data for the City of Austin, Texas (after Mays and Tung 1992)

Year	Population	Price ($/Kgal)	Income ($/person)	Annual rainfall (cm)	Water use (Mgal)
1965	216733	0.98	5919	103.05	12911.56
1966	223334	0.95	5970	63.98	13082.71
1967	230135	1.20	6521	85.19	14887.44
1968	237144	1.15	7348	102.69	13294.28
1969	244366	1.10	7965	85.32	14777.58
1970	251808	1.05	8456	77.83	16522.66
1971	259900	1.00	8713	63.37	18451.60
1972	268252	1.20	9286	66.22	18633.18
1973	276873	1.13	9694	102.77	18733.92
1974	285771	1.06	9542	91.97	20623.74
1975	294955	0.98	9684	93.50	18698.38
1976	304434	0.93	10152	99.49	16679.14
1977	314217	0.87	10441	56.24	22302.64
1978	324315	0.81	10496	78.66	22818.04
1979	334738	1.10	10679	95.25	21256.50
1980	345496	1.05	10833	69.55	25611.86
1981	354401	0.96	11060	116.15	24886.51
1982	368135	0.91	11338	67.64	28462.73
1983	383326	0.87	11752	86.31	26771.12
1984	399147	0.84	12763	66.80	31843.03
1985	424120	1.41	12748	82.52	31852.81

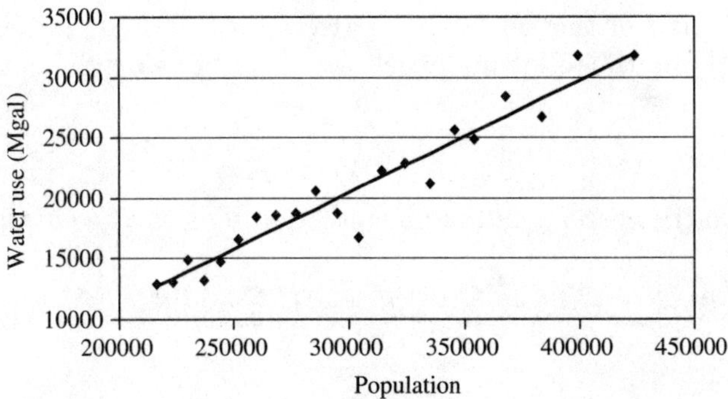

Figure 3.6 Scatter diagram of annual water use versus population for Austin, Texas, for 1965 to 1985.

Solution To establish the empirical relationship between the annual water use and population of the city, one can first plot the annual water use (a response variable, Q) against the population (an explanatory variable, POP) in Fig. 3.6. Sometimes, it is necessary to plot Q versus POP in different scales to identify the best functional relation between the two variables under consideration. For this example, the plot of annual water use and population in their original scales shows sufficient linearity between them and, therefore, the following model is plausible

$$Q = \beta_0 + \beta_1 \times POP + \varepsilon$$

Let $Y = Q$ and $X = POP$, based on Eq. (3.7) the LS estimate of β_1 can be computed as the following:

$$n = 21 \qquad \sum x_i = 6,341,600 \qquad \sum y_i = 433,101 \qquad \sum x_i^2 = 1,989,269,983,366$$

$$\sum x_i y_i = 137,689,768,550$$

$$\bar{x} = 301,981 \qquad \bar{y} = 20,623.88 \qquad S_{xx} = 74,227,525,747 \qquad S_{xy} = 6,901,385,637$$

Hence,

$$\hat{\beta}_1 = S_{xy}/S_{xx} = 0.093 \text{ Mgal/capita}$$
$$\hat{\beta}_0 = \bar{y} - \hat{\beta}_1 \bar{x} = -7453$$

The resulting regression model is

$$Q = -7453 + 0.093\ POP$$

and the corresponding line is shown in Fig. 3.6.

For a *multiple linear regression model* as Eq. (3.2a) in which more than one explanatory variable is involved, the objective function corresponding to the LS criterion can be expressed as

$$\operatorname*{Min}_{\beta_0, \beta_1, \dots, \beta_K} D = \sum_{i=1}^{n} [y_i - (\beta_0 + \beta_1 x_{i1} + \beta_2 x_{i2} + \cdots + \beta_K x_{iK})]^2 \qquad (3.8)$$

The LS estimation of the regression parameters, $\boldsymbol{\beta} = (\beta_0, \beta_1,..., \beta_K)^t$, can be obtained by solving the following necessary conditions for the optimum:

$$\frac{\partial D}{\partial \beta_j} = 0 \qquad j = 0, 1, 2,..., K \tag{3.9}$$

based on which the corresponding normal equations can be obtained as

$$n\beta_0 + \left(\sum_{i=1}^{n} x_{i1}\right)\beta_1 + \left(\sum_{i=1}^{n} x_{i2}\right)\beta_2 + \cdots + \left(\sum_{i=1}^{n} x_{iK}\right)\beta_K = \left(\sum_{i=1}^{n} y_i\right)$$

$$\left(\sum_{i=1}^{n} x_{i1}\right)\beta_0 + \left(\sum_{i=1}^{n} x_{i1}^2\right)\beta_1 + \left(\sum_{i=1}^{n} x_{i1}x_{i2}\right)\beta_2 + \cdots + \left(\sum_{i=1}^{n} x_{i1}x_{iK}\right)\beta_K = \left(\sum_{i=1}^{n} x_{i1}y_i\right) \tag{3.10}$$

$$\vdots$$

$$\left(\sum_{i=1}^{n} x_{iK}\right)\beta_0 + \left(\sum_{i=1}^{n} x_{iK}x_{i1}\right)\beta_1 + \left(\sum_{i=1}^{n} x_{iK}x_{i2}\right)\beta_2 + \cdots + \left(\sum_{i=1}^{n} x_{iK}^2\right)\beta_K = \left(\sum_{i=1}^{n} x_{iK}y_i\right)$$

or in matrix notation as

$$(X^t X)\,\boldsymbol{\beta} = X^t\,y \tag{3.11}$$

in which $X = (1, x_1, x_2,..., x_K)$, an $n \times (K+1)$ matrix, with $1_{n\times1} = (1,1,..., 1)^t$ a column vector of ones with the superscript t being the transpose of a vector or matrix; $x_j = (x_{1j}, x_{2j},..., x_{nj})^t$, an $n \times 1$ column vector of n observations of the jth explanatory variable; $\boldsymbol{\beta} = (\beta_0, \beta_1,..., \beta_K)^t$; and $y = (y_1, y_2,..., y_n)^t$, an $n \times 1$ column vector of n observations of the dependent variable. Then, the LS estimators of regression coefficients can be obtained as

$$\hat{\boldsymbol{\beta}} = (X^tX)^{-1}X^ty \tag{3.12}$$

provided that the matrix $X^t X$ is not singular.

3.4   Measures of Goodness-of-Fit

After the determination of the parameters for a selected model, several indicators can be used to quantify the accuracy and degree of goodness-of-fit. The variance associated with the regression model σ^2 (called *mean squared error, MSE*) can be estimated by

$$\hat{\sigma}^2 = s^2 = \frac{e^t e}{n-(K+1)} = \frac{\sum_{i=1}^{n} e_i^2}{n-(K+1)} \tag{3.13}$$

in which $e = (e_1, e_2,..., e_n)^t$ is a column vector of errors (or residuals). The denominator of Eq. (3.13) is called the *degree of freedom* obtained from subtracting the total number of unknown regression coefficients in the model from the total number of observations. The square root of the model variance (s) is called the *standard error of estimate*.

Another frequently used goodness-of-fit indicator is the *coefficient of determination* defined as

$$R^2 = 1 - \frac{\sum(y_i - \hat{y}_i)^2}{\sum(y_i - \bar{y})^2} = 1 - \frac{\sum e_i^2}{\sum(y_i - \bar{y})^2} = 1 - \frac{\text{SSE}}{\text{SST}} = \frac{\text{SSR}}{\text{SST}} \tag{3.14}$$

in which SSE is the *error sum of squares*; SST is the *total sum of squares* about the mean of the response variable Y; and SSR is the *regression sum of squares*. Hence, the relationship between the three sum-of-squares terms can be expressed as

$$\text{SST} = \sum(y_i - \bar{y})^2$$

$$= \sum(\hat{y}_i - \bar{y})^2 + \sum(y_i - \hat{y}_i)^2 = \text{SSR} + \text{SSE} \tag{3.15}$$

The term SSE or SST represents the portion of variability in the dependent variable Y that cannot be explained by the regression model under consideration. Hence, the value of the coefficient of determination is bounded between 0 and 1 representing the percentage of variability in the response variable Y explained by the regression model under consideration.

Example 3.2   Referring to Table 3.1, develop a regression model by relating the annual water use to population and water price in the following form:

$$Q = \beta_0 + \beta_1 \times POP + \beta_2 \times PRICE + \varepsilon$$

Solution   In the matrix form, the following matrices and vectors can be determined from the data:

$$\boldsymbol{\beta} = \begin{pmatrix} \beta_0 \\ \beta_1 \\ \beta_2 \end{pmatrix} \qquad \boldsymbol{Q} = \boldsymbol{y} = \begin{pmatrix} 12911.56 \\ 13082.71 \\ \vdots \\ 31852.81 \end{pmatrix} \qquad \boldsymbol{X} = \begin{bmatrix} 1 & 216733 & 0.98 \\ 1 & 223334 & 0.95 \\ \vdots & \vdots & \vdots \\ 1 & 424120 & 1.41 \end{bmatrix}$$

Based on Eq. (3.12), the following terms are calculated:

$$\boldsymbol{X}^t\boldsymbol{X} = \begin{bmatrix} 2.10 \times 10^1 & 6.34 \times 10^6 & 2.16 \times 10^1 \\ 6.34 \times 10^6 & 1.99 \times 10^{12} & 6.49 \times 10^6 \\ 2.16 \times 10^1 & 6.49 \times 10^6 & 2.25 \times 10^1 \end{bmatrix}$$

$$(\boldsymbol{X}^t\boldsymbol{X})^{-1} = \begin{bmatrix} 4.33 & -4.89 \times 10^{-6} & -2.73 \\ -4.89 \times 10^{-6} & 1.37 \times 10^{-11} & 7.33 \times 10^{-7} \\ -2.73 \times 10^1 & 7.33 \times 10^{-7} & 2.45 \end{bmatrix}$$

$$\boldsymbol{X}^t\boldsymbol{y} = \begin{pmatrix} 4.33 \times 10^5 \\ 1.38 \times 10^{11} \\ 4.42 \times 10^5 \end{pmatrix}$$

The LS-based model parameters can be calculated as

$$\hat{\boldsymbol{\beta}} = (\hat{\beta}_0, \hat{\beta}_1, \hat{\beta}_2)^t = (-7.57 \times 10^3,\ 9.30 \times 10^{-2},\ 1.05 \times 10^2)^t$$

with the following resulting equation

$$Q = -7570 + 0.093 \times POP + 105 \times PRICE$$

The standard error of estimate corresponding to the above regression equation can be calculated, according to Eq. (3.13), as

$$\hat{\sigma}_\varepsilon = s = \sqrt{\frac{\sum_{i=1}^n e_i^2}{n - (K+1)}} = \sqrt{\frac{46,109,623}{21 - (2+1)}} = 1601$$

Goodness-of-fit as indicated by R^2 can be calculated, according to Eq. (3.14), as

$$R^2 = 1 - \frac{\text{SSE}}{\text{SST}} = 1 - \frac{46,109,623}{687,777,359} = 0.933$$

which indicates that 93.3 percent of the variability in annual water use is explained by the above regression equation using population and the unit price of water.

3.5   Uncertainty Features of LS-Based Model Parameters

Note that the parameters β's in the regression models obtained from the LS method are estimated from the available sample data and, hence, may not be true values. As a result, LS-based regression model parameters are subject to uncertainty and should be treated as random variables. In conventional regression analysis, the statistical features of LS-based regression model parameters can be derived on the basis of the following assumptions:

a. Residuals ε_1, ε_2,..., ε_n are independently and identically distributed (*iid*) normal random variables, that is, $\varepsilon_i \overset{iid}{\sim} N(0, \sigma_\varepsilon^2)$ for $i = 1, 2,..., n$. If this is not true, it implies that additional information about the process could be determined from the data through a normal transformation as described in Sec. 3.9.2;

b. The values of explanatory variables are accurate without error;

c. Assumption (a) implies that the response variable Y under a given set of explanatory variables $\boldsymbol{X} = \boldsymbol{x}$ is a normal random variable with the mean of $E(Y \mid \boldsymbol{x}) = \beta_0 + \beta_1 x_1 + \beta_2 x_2 + \cdots + \beta_K x_K$, $\text{Var}(Y \mid \boldsymbol{x}) = \sigma_\varepsilon^2$, and $\text{Cov}(Y_i, Y_j) = 0$ for $i \neq j$.

On the basis of the above assumptions, the LS estimators of regression model parameters by Eq. (3.12), $\hat{\boldsymbol{\beta}}$ are the *best linear unbiased estimators* (often called *BLUE*) (Neter, Wasserman, and Kutner 1983). Referring to Eq. (3.12) where the LS estimators $\hat{\beta}_j$'s are linear functions of the response variables, it can be shown that the LS estimators of parameters in a linear regression model are multivariate normal random variables with the following statistical properties:

$$E(\hat{\boldsymbol{\beta}}) = \boldsymbol{\beta} \qquad Cov(\hat{\boldsymbol{\beta}}) = C_{\hat{\beta}} = \sigma_\varepsilon^2 (X^t X)^{-1} \tag{3.16}$$

Hence, the diagonal elements of the variance-covariance matrix $C_{\hat{\beta}}$ provide information about the variances of the LS-based estimators of regression model parameters, $\hat{\boldsymbol{\beta}}$, while the off-diagonal elements can be used to assess the correlation between different regression model parameters.

Example 3.3   The uncertainty features of the LS-based regression coefficient estimators in Example 3.2 can be expressed in terms of their sample estimated variance-covariance matrix, calculable by Eq. (3.16), as

$$\hat{C}_{\hat{\beta}} = \begin{bmatrix} \hat{\sigma}(\hat{\beta}_0, \hat{\beta}_0) & \hat{\sigma}(\hat{\beta}_0, \hat{\beta}_1) & \hat{\sigma}(\hat{\beta}_0, \hat{\beta}_2) \\ \hat{\sigma}(\hat{\beta}_1, \hat{\beta}_0) & \hat{\sigma}(\hat{\beta}_1, \hat{\beta}_1) & \hat{\sigma}(\hat{\beta}_1, \hat{\beta}_2) \\ \hat{\sigma}(\hat{\beta}_2, \hat{\beta}_0) & \hat{\sigma}(\hat{\beta}_2, \hat{\beta}_1) & \hat{\sigma}(\hat{\beta}_2, \hat{\beta}_2) \end{bmatrix}$$

$$= s^2 (X^t X)^{-1} = 1601^2 \begin{bmatrix} 4.33 & -4.89 \times 10^{-6} & -2.73 \\ -4.89 \times 10^{-6} & 1.37 \times 10^{-11} & 7.33 \times 10^{-7} \\ -2.73 & 7.33 \times 10^{-7} & 2.45 \end{bmatrix}$$

$$= \begin{bmatrix} 1.11 \times 10^7 & -1.25 \times 10^1 & -7.00 \times 10^6 \\ -1.25 \times 10^1 & 3.507 \times 10^{-5} & 1.879 \\ -7.00 \times 10^6 & 1.879 & 6.27 \times 10^6 \end{bmatrix}$$

in which $\hat{\sigma}(\hat{\beta}_i, \hat{\beta}_j)$ is the sample covariance between the two different regression coefficient estimators. Hence, the standard deviations corresponding to the regression coefficients are

$$\hat{\sigma}(\hat{\beta}_0) = \sqrt{11,088,900} = 3330$$

$$\hat{\sigma}(\hat{\beta}_1) = \sqrt{3.507 \times 10^{-5}} = 5.92 \times 10^{-3}$$

$$\hat{\sigma}(\hat{\beta}_2) = \sqrt{6,270,016} = 2504$$

The values of correlation coefficients between the regression coefficient estimators can be estimated by

$$\hat{\rho}(\hat{\beta}_0, \hat{\beta}_1) = \frac{\hat{\sigma}(\hat{\beta}_0, \hat{\beta}_1)}{\hat{\sigma}(\hat{\beta}_0) \times \hat{\sigma}(\hat{\beta}_1)} = 0.636 \qquad \hat{\rho}(\hat{\beta}_0, \hat{\beta}_2) = -0.839 \qquad \hat{\rho}(\hat{\beta}_1, \hat{\beta}_2) = 0.127$$

3.6 Statistical Inferences of Regression Coefficients

Under the normality assumption stated in Sec. 3.5, it has been shown that the LS estimators of regression model parameters $\hat{\boldsymbol{\beta}}$ are multivariate normal random variables with the mean and covariance matrix given in Eq. (3.16). Due to the fact that the true parameter values and the model variance are not known but have to be estimated, the sampling distributions for each of the LS estimators $\hat{\boldsymbol{\beta}}$ will follow a t-distribution according to its definition given in Eq. (2.95), that is,

$$T_j = \frac{\hat{\beta}_j - \beta_j}{s\sqrt{(\boldsymbol{X}^t\boldsymbol{X})_{jj}^{-1}}} \sim T_{n-(K+1)} \qquad \text{for } j = 0, 1, 2, ..., K \qquad (3.17)$$

in which s is the standard error of estimate; $(\boldsymbol{X}^t\boldsymbol{X})_{jj}^{-1}$ is the jth diagonal element of matrix $(\boldsymbol{X}^t\boldsymbol{X})^{-1}$ that is equal to $\mathrm{Var}(\hat{\beta}_j)$; and $T_{n-(K+1)}$ is the t-distributed random variable with $n-(K+1)$ degrees of freedom. Equation (3.17) can be used to conduct various statistical inferences about the regression model parameters, such as the test of hypothesis and construction of confidence intervals and prediction intervals (Sec. 3.7).

Due to the sampling errors, the LS-based estimates of regression model parameter corresponding to a particular term may not be zero, even if the true value could be zero. Therefore, in statistical inference of regression analysis, interests often is given to the issue about whether or not the true (but unknown) regression model parameter is zero. A zero-valued regression model parameter indicates that the corresponding explanatory variable term has either no influence on or contribution to the value of the response variable Y and, hence, could be deleted from the model. Therefore, the statistical hypothesis test can be cast in

Null hypothesis (H_0): $\beta_j = 0$ versus alternative hypothesis (H_a): $\beta_j \neq 0$ (3.18)

The task of the above hypothesis test is to decide whether to accept H_0 being true or to reject it. In this situation, the decision can be made on the basis of the value of the t-statistic with respect to H_0 as

$$t_j = \frac{\hat{\beta}_j}{s\sqrt{(\boldsymbol{X}^t\boldsymbol{X})_{jj}^{-1}}} \qquad \text{for } j = 0, 1, 2, ..., K \qquad (3.19)$$

At a preselected *significance level* of α (i.e., probability of rejecting H_0: $\beta_j = 0$, if H_0 is true), the null hypothesis H_0:$\beta_j = 0$ is rejected if $t_j > t_{n-(K+1),\alpha/2}$ or $t_j < -t_{n-(K+1),\alpha/2}$ in which $P[T_v \geq t_{v,\alpha}] = \alpha$ with T_v being the t-random variable having v degrees of freedom (Fig. 3.7(a)). A typical probability table for t-distribution is shown in Table 3.2.

To test the overall regression relation between the dependent variable Y and a set of independent variables X_1, X_2,..., X_K, one can conduct the following

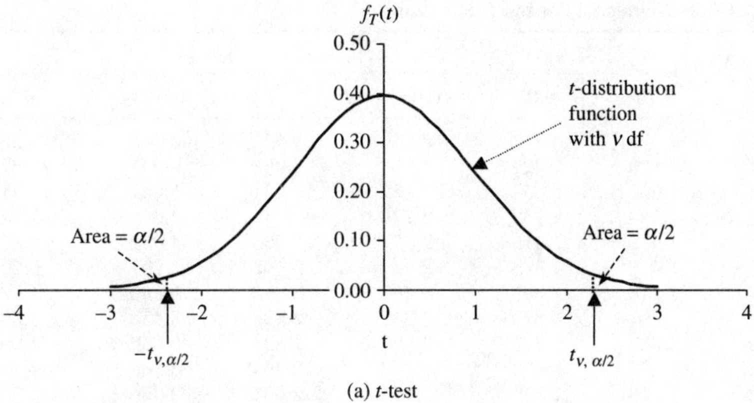

(a) t-test

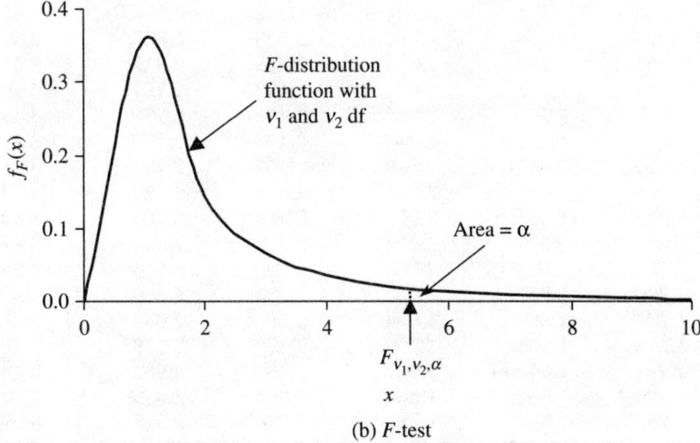

(b) F-test

Figure 3.7   Critical region for hypothesis tests.

hypothesis test:

$$H_0: \beta_1 = \beta_2 = \cdots = \beta_K = 0$$

$$\text{versus} \tag{3.20}$$

$$H_a: \text{Not all } \beta_k \ (k = 1, 2,\ldots, K) \text{ equal to } 0$$

Under the normality assumptions stipulated in Sec. 3.5, the following test statistic can be used

$$F = \frac{\text{MSR}}{\text{MSE}} = \frac{\text{SSR}/K}{\text{SSE}/(n-(K+1))} \tag{3.21}$$

where F has an F-distribution with K and $n - (K + 1)$ degrees of freedom (Eq. (2.97)); MSR and MSE are, respectively, regression and error mean squares. Referring to

TABLE 3.2   Values $t_{v,\alpha}$ for the t-distribution

df	Significance level α						
v	0.10	0.05	0.025	0.01	0.005	0.001	0.0005
1	3.078	6.314	12.706	31.821	63.656	318.289	636.578
2	1.886	2.920	4.303	6.965	9.925	22.328	31.600
3	1.638	2.353	3.182	4.541	5.841	10.214	12.924
4	1.533	2.132	2.776	3.747	4.604	7.173	8.610
5	1.476	2.015	2.571	3.365	4.032	5.894	6.869
6	1.440	1.943	2.447	3.143	3.707	5.208	5.959
7	1.415	1.895	2.365	2.998	3.499	4.785	5.408
8	1.397	1.860	2.306	2.896	3.355	4.501	5.041
9	1.383	1.833	2.262	2.821	3.250	4.297	4.781
10	1.372	1.812	2.228	2.764	3.169	4.144	4.587
12	1.356	1.782	2.179	2.681	3.055	3.930	4.318
14	1.345	1.761	2.145	2.624	2.977	3.787	4.140
16	1.337	1.746	2.120	2.583	2.921	3.686	4.015
18	1.330	1.734	2.101	2.552	2.878	3.610	3.922
20	1.325	1.725	2.086	2.528	2.845	3.552	3.850
22	1.321	1.717	2.074	2.508	2.819	3.505	3.792
24	1.318	1.711	2.064	2.492	2.797	3.467	3.745
26	1.315	1.706	2.056	2.479	2.779	3.435	3.707
28	1.313	1.701	2.048	2.467	2.763	3.408	3.674
30	1.310	1.697	2.042	2.457	2.750	3.385	3.646
35	1.306	1.690	2.030	2.438	2.724	3.340	3.591
40	1.303	1.684	2.021	2.423	2.704	3.307	3.551
45	1.301	1.679	2.014	2.412	2.690	3.281	3.520
50	1.299	1.676	2.009	2.403	2.678	3.261	3.496
60	1.296	1.671	2.000	2.390	2.660	3.232	3.460
70	1.294	1.667	1.994	2.381	2.648	3.211	3.435
80	1.292	1.664	1.990	2.374	2.639	3.195	3.416
90	1.291	1.662	1.987	2.368	2.632	3.183	3.402
100	1.290	1.660	1.984	2.364	2.626	3.174	3.390
∞	1.282	1.645	1.960	2.326	2.576	3.090	3.290

NOTE: $P(T_v \geq t_{v,\alpha}) = \alpha$.

Fig. 3.7(*b*), the decision leading to the rejection of H_0 in Eq. (3.20) for a pre-selected significance level α is when the value of the test statistic F is larger than $F_{K,n-(K+1),\ \alpha}$. A list of critical values for F-distribution with $\alpha = 5$ percent significance level is given in Table 3.3. Tables for other values of significance levels can be found in several statistics textbooks (e.g., Blank 1980; Devore 1987).

3.7   Confidence Interval and Prediction Interval

Once the model parameters are estimated by the LS method, it is feasible to use the model $\hat{\beta}_0 + \sum_{j=1}^{K} \hat{\beta}_j x_j$ to estimate $E(Y \mid x)$ and to predict the response variable Y as $\hat{y} = \hat{\beta}_0 + \sum_{j=1}^{K} \hat{\beta}_j x_j + e$ for a particular set of future values of explanatory variables.

TABLE 3.3 Critical Values $F_{v_1, v_2,\ \alpha = 0.05}$ for the *F*-distribution

	Degrees of freedom, v_1																	
v_2	1	2	3	4	5	6	7	8	9	10	12	15	20	25	30	40	60	120
1	161.5	199.5	215.7	224.6	230.2	234.0	236.8	238.9	240.5	241.9	243.9	246.0	248.0	249.3	250.1	251.1	252.2	253.3
2	18.51	19.00	19.16	19.25	19.30	19.33	19.35	19.37	19.38	19.40	19.41	19.43	19.45	19.46	19.46	19.47	19.48	19.49
3	10.13	9.55	9.28	9.12	9.01	8.94	8.89	8.85	8.81	8.79	8.74	8.70	8.66	8.63	8.62	8.59	8.57	8.55
4	7.71	6.94	6.59	6.39	6.26	6.16	6.09	6.04	6.00	5.96	5.91	5.86	5.80	5.77	5.75	5.72	5.69	5.66
5	6.61	5.79	5.41	5.19	5.05	4.95	4.88	4.82	4.77	4.74	4.68	4.62	4.56	4.52	4.50	4.46	4.43	4.40
6	5.99	5.14	4.76	4.53	4.39	4.28	4.21	4.15	4.10	4.06	4.00	3.94	3.87	3.83	3.81	3.77	3.74	3.70
7	5.59	4.74	4.35	4.12	3.97	3.87	3.79	3.73	3.68	3.64	3.57	3.51	3.44	3.40	3.38	3.34	3.30	3.27
8	5.32	4.46	4.07	3.84	3.69	3.58	3.50	3.44	3.39	3.35	3.28	3.22	3.15	3.11	3.08	3.04	3.01	2.97
9	5.12	4.26	3.86	3.63	3.48	3.37	3.29	3.23	3.18	3.14	3.07	3.01	2.94	2.89	2.86	2.83	2.79	2.75
10	4.96	4.10	3.71	3.48	3.33	3.22	3.14	3.07	3.02	2.98	2.91	2.85	2.77	2.73	2.70	2.66	2.62	2.58
11	4.84	3.98	3.59	3.36	3.20	3.09	3.01	2.95	2.90	2.85	2.79	2.72	2.65	2.60	2.57	2.53	2.49	2.45
12	4.75	3.89	3.49	3.26	3.11	3.00	2.91	2.85	2.80	2.75	2.69	2.62	2.54	2.50	2.47	2.43	2.38	2.34
13	4.67	3.81	3.41	3.18	3.03	2.92	2.83	2.77	2.71	2.67	2.60	2.53	2.46	2.41	2.38	2.34	2.30	2.25
14	4.60	3.74	3.34	3.11	2.96	2.85	2.76	2.70	2.65	2.60	2.53	2.46	2.39	2.34	2.31	2.27	2.22	2.18
15	4.54	3.68	3.29	3.06	2.90	2.79	2.71	2.64	2.59	2.54	2.48	2.40	2.33	2.28	2.25	2.20	2.16	2.11
16	4.49	3.63	3.24	3.01	2.85	2.74	2.66	2.59	2.54	2.49	2.42	2.35	2.28	2.23	2.19	2.15	2.11	2.06
18	4.41	3.55	3.16	2.93	2.77	2.66	2.58	2.51	2.46	2.41	2.34	2.27	2.19	2.14	2.11	2.06	2.02	1.97
20	4.35	3.49	3.10	2.87	2.71	2.60	2.51	2.45	2.39	2.35	2.28	2.20	2.12	2.07	2.04	1.99	1.95	1.90
30	4.17	3.32	2.92	2.69	2.53	2.42	2.33	2.27	2.21	2.16	2.09	2.01	1.93	1.88	1.84	1.79	1.74	1.68
40	4.08	3.23	2.84	2.61	2.45	2.34	2.25	2.18	2.12	2.08	2.00	1.92	1.84	1.78	1.74	1.69	1.64	1.58
50	4.03	3.18	2.79	2.56	2.40	2.29	2.20	2.13	2.07	2.03	1.95	1.87	1.78	1.73	1.69	1.63	1.58	1.51
70	3.98	3.13	2.74	2.50	2.35	2.23	2.14	2.07	2.02	1.97	1.89	1.81	1.72	1.66	1.62	1.57	1.50	1.44
100	3.94	3.09	2.70	2.46	2.31	2.19	2.10	2.03	1.97	1.93	1.85	1.77	1.68	1.62	1.57	1.52	1.45	1.38
120	3.92	3.07	2.68	2.45	2.29	2.18	2.09	2.02	1.96	1.91	1.83	1.75	1.66	1.60	1.55	1.50	1.43	1.35

NOTE: $P(F_{v_1, v_2} \geq F_{v_1, v_2, \alpha}) = \alpha$.

Consider a set of given values of explanatory variables $x_* = (1, x_{1*}, x_{2*},...,$ $x_{K*})^t$, the mean response variable $\bar{y}_* = \widehat{E(Y|x_*)}$ can be estimated as

$$\hat{\bar{y}}_* = x_*^t \hat{\beta} \tag{3.22}$$

The variance corresponding to $\hat{\bar{y}}_*$ can be estimated, by Eq. (2.49), as

$$\text{Var}(\hat{\bar{y}}_*) = x_*^t \hat{C}_{\hat{\beta}} x_* = s^2 x_*^t (X^t X)^{-1} x_* \tag{3.23}$$

Based on Eqs. (3.22) and (3.23), the $(1 - \alpha)$ percent *confidence interval* for the true mean response $\bar{y}_*$, conditioned on the set of explanatory variables x_*, can be obtained as

$$\left[\hat{\bar{y}}_* - t_{n-(K+1),\,\alpha/2} \times \sqrt{\text{Var}\left(\hat{\bar{y}}_*\right)} \quad \hat{\bar{y}}_* + t_{n-(K+1),\,\alpha/2} \times \sqrt{\text{Var}\left(\hat{\bar{y}}_*\right)} \right] \tag{3.24}$$

In the case of predicting the actual value, not the mean value, of the response variable $Y = y_*$ for a given set of explanatory variables x_*, the degree of uncertainty is larger than that of estimating the mean response. The additional uncertainty is contributed from the model error term, represented by ε. As a result, the variance of the predicted response y_* can then be estimated as

$$\text{Var}(y_*) = s^2 + x_*^t \hat{C}_{\hat{\beta}} x_* = s^2 (1 + x_*^t (X^t X)^{-1} x_*) \tag{3.25}$$

The corresponding $(1 - \alpha)$ percent *prediction interval* can be similarly derived as

$$\left[y_* - t_{n-(K+1),\,\alpha/2} \times \sqrt{\text{Var}(y_*)} \quad y_* + t_{n-(K+1),\,\alpha/2} \times \sqrt{\text{Var}(y_*)} \right] \tag{3.26}$$

Example 3.4   Determine the 95 percent confidence interval and prediction interval for the annual water use under population of 450,000 and unit water price of \$1.50.

Solution   Under the specified explanatory variables, that is, $x_* = (POP, PRICE)^t = (450,000, \$1.50)^t$, the estimated water use, according to the derived regression equation

$$Q = -7570 + 0.093 \times POP + 105 \times PRICE$$

is $\hat{\bar{Q}}_* = \hat{Q}_* = -7570 + 0.093 \times 450,000 + 105 \times 1.50 = 34,440$ Mgal. To calculate the variances of $\hat{\bar{Q}}_*$ and $\hat{Q}_*$, the term $x_*^t (X^t X)^{-1} x_*$ needs to be calculated as

$$x_*^t (X^t X)^{-1} x_* = (1 \;\; 450,000 \;\; 1.5) \begin{bmatrix} 4.33 & -4.89 \times 10^{-6} & -2.73 \\ -4.89 \times 10^{-6} & 1.37 \times 10^{-11} & 7.33 \times 10^{-7} \\ -2.73 & 7.33 \times 10^{-7} & 2.45 \end{bmatrix} \begin{pmatrix} 1 \\ 450,000 \\ 1.5 \end{pmatrix} = 1.0$$

With the standard error of $s = 1601$ (from Example 3.2), the variance of $\hat{\bar{Q}}_*$ and $\hat{Q}_*$ can be calculated, according to Eqs. (3.23) and (3.25), respectively, as

$$\text{Var}\left(\hat{\bar{Q}}_*\right) = s^2 \boldsymbol{x}_*^t (\boldsymbol{X}^t \boldsymbol{X})^{-1} \boldsymbol{x}_* = 1601^2 \times 1.0$$

$$\text{Var}(\hat{Q}_*) = s^2 \left(1 + \boldsymbol{x}_*^t (\boldsymbol{X}^t \boldsymbol{X})^{-1} \boldsymbol{x}_*\right) = 1601^2 \times 2.0$$

Hence, the 95 percent confidence interval and prediction interval can be obtained from Eqs. (3.24) and (3.26), respectively, with degrees of freedom, $v = n - (K + 1) = 21 - (2 + 1) = 18$ and $\alpha = 0.05$, that is, the 95 percent confidence interval for the mean water demand $\bar{Q}_*$ is

$$\left[\hat{\bar{Q}}_* - t_{18,0.025} \times \sqrt{\text{Var}\left(\hat{\bar{Q}}_*\right)} \quad \hat{\bar{Q}}_* + t_{18,0.025} \times \sqrt{\text{Var}\left(\hat{\bar{Q}}_*\right)}\right]$$

$$= \left[34,440 - 2.101 \times \sqrt{1.0 \times 1601^2} \quad 34,440 + 2.101 \times \sqrt{1.0 \times 1601^2}\right]$$

$$= [31,077 \text{ Mgal}, \ 37,804 \text{ Mgal}]$$

and the 95 percent prediction interval for $\hat{Q}_*$ is

$$\left[\hat{Q}_* - t_{18,0.025} \times \sqrt{\text{Var}\left(\hat{Q}_*\right)} \quad \hat{Q}_* + t_{18,0.025} \times \sqrt{\text{Var}\left(\hat{Q}_*\right)}\right]$$

$$= \left[34,440 - 2.101 \times \sqrt{2.0 \times 1601^2} \quad 34,440 + 2.101 \times \sqrt{2.0 \times 1601^2}\right]$$

$$= [29,683 \text{ Mgal}, \ 39,197 \text{ Mgal}]$$

3.8 Variance Contribution by Independent Variables

Consider a linear regression model consisting of multiple independent variables as Eq. (3.2a). Upon the completion of parameter estimation based on the LS criterion, it is useful to examine the contribution of each independent variable to the total variability of the dependent variable (or model output) Y. As stated in Sec. 3.4, the total variability in Y is represented by SST that consists of two components, namely, SSR and SSE and is represented as SST = SSR + SSE. The term SSR represents the overall contribution of all independent variables in the regression model in explaining the total variability in Y. For a linear regression model, as Eq. (3.2a), involving K independent variables, the overall SSR can further be decomposed (Neter, Wasserman, and Kutner 1983) as

$$\text{SSR} = \text{SSR}(X_1) + \text{SSR}(X_2 \mid X_1) + \text{SSR}(X_3 \mid X_1, X_2) + \cdots + \text{SSR}(X_K \mid X_1, X_2, \ldots, X_{K-1})$$

$$(3.27)$$

where

$$\text{SSR}(X_2 \mid X_1) = \text{SSR}(X_1, X_2) - \text{SSR}(X_1) = \text{SSE}(X_1) - \text{SSE}(X_1, X_2)$$

$$\text{SSR}(X_3 \mid X_1, X_2) = \text{SSR}(X_1, X_2, X_3) - \text{SSR}(X_1, X_2) = \text{SSE}(X_1, X_2) - \text{SSE}(X_1, X_2, X_3)$$

$$\vdots$$

with $\text{SSR}(X_1, X_2)$ being the value of SSR for a regression model involving both independent variables X_1 and X_2. Hence, $\text{SSR}(X_2 \mid X_1)$ represents the augmented contribution explaining the total variability by independent variable X_2, after X_1 has been included in the model. From Eq. (3.27) the percentage contribution by each individual independent variable X_k in a regression model and inherent model uncertainty can be assessed.

Note that the *sequential sum of squares* shown in the right-hand side of Eq. (3.27) for each independent variable may be affected by its order of arrangement in the model due to the existence of possible correlation among the explanatory variables. If the explanatory variables are truly statistically independent, then their order of arrangement does not affect the value of marginal component of SSR. Under such a condition, the percentage contribution of individual explanatory variable to the total variability can be accurately assessed. Otherwise, the assessment of variance contributions for correlated explanatory variables would become less clear.

3.9 Issues in Regression Analysis

The above descriptions on the theoretical background of regression analysis are based on the condition that the model form is specified. Other than the estimation of model parameters, examination of goodness-of-fit, and statistical inferences about the resulting regression equation, several practical and theoretical issues must be considered in the process of establishing a proper regression model. The main practical issues primarily focus on the choice of model and selection of explanatory variables. On the theoretical aspect, most issues are related to the compliance with the basic assumptions about the regression model that would dictate the statistical efficiency of the least squares estimators and the related statistical inferences. The following subsections briefly discuss the issues and some common practices applicable to situations where the basic assumptions in regression analysis are violated. More detailed treatments of the subjects can be found elsewhere (Seber 1977; Draper and Smith 1981; Montgomery and Peck 1982; Neter, Wasserman, and Kutner 1983).

3.9.1 Selection of explanatory variables

The exercise of regression analysis involves the tasks of choosing relevant explanatory variables, selecting proper model form, and determining model parameters. The first two tasks are largely dictated by the analyst's insights, prior experiences, or theoretical considerations about the processes involved. The coefficient of determination (R^2) and MSE can be used to select proper explanatory variables in the regression model. Indicators that are used, such as Mallow C_p statistic and others can be found in regression analysis textbooks (e.g., Montgomery and Peck 1982; Neter, Wasserman, and Kutner 1980).

The coefficient of determination, R^2, indicates the goodness-of-fit of a regression model to the data set. One would like to choose a model with high R^2 or use an explanatory variable that significantly enhance the value of R^2. The value of R^2 will continuously increase as more explanatory variable terms are added to the regression model. Increasing the number of explanatory variable terms, unless they are statistically relevant to the problem in hand, may result in increasing complexity of the model, increasing the burden of the data collection effort, and masking the essential features of the model. A main thrust of regression analysis is to identify minimum number of parameters that would best describe the data (the *principle of parsimony*). To avoid using an excessive number of explanatory variable terms, the *adjusted coefficient of determination* has been proposed

$$R^2_{\text{adj}} = \frac{(n-1)R^2 - K}{n-(K+1)} \tag{3.28}$$

The adjusted R^2 in Eq. (3.28) will have a lower value, with K terms, than that of $(K-1)$ terms if the added term does not significantly contribute to the explanation about the variation of the data. Other types of adjusted R^2 are also proposed, for instance,

$$R^2_{\text{adj}} = R^2 - \left(\frac{K-1}{n-1}\right)(1 - R^2) \tag{3.29}$$

The adjusted R^2 given in Eq. (3.29) will increase in value only when the t-statistic of the regression coefficient, Eq. (3.19), associated with the added explanatory variable term is larger than unity. Hence, by examining the change in value of an adjusted R^2 one is able to make decision on whether or not the introduction of a new term into the regression model is justifiable.

As for the MSE (s^2), one would choose a model and/or explanatory variables having low values of MSE. Referring to Eq. (3.13), adding a new independent variable would result in reductions in both the numerator (SSE) and the denominator (degrees of freedom). However, the net effect does not necessarily lead to a reduction in MSE if the added independent variable did not contribute to a significant reduction in the value of SSE to overcome its effect on reduction in degrees of freedom. Other selection criteria can be found elsewhere (Seber 1977)

To automatically search for the "best" subset of independent variables, *stepwise regression* procedures have been developed. The stepwise procedures can broadly be classified into forward-selection, backward-elimination, and the combination of the two. The *forward-selection procedure* adds one independent variable at a time according to the entering criterion set by the analyst. The procedure starts with an assessment of simple correlation among the dependent variable Y with individual explanatory variable X_k and considers the one having the highest correlation as the first variable to enter, provided that it passes the preset statistical significance criterion. With one independent variable chosen, the procedure continues to re-evaluate the correlation coefficient between the residuals of dependent variable and those of the remaining independent variables (called the *partial correlation coefficient*) for considering the next variable to enter the equation. The process is repeated until all independent variables are considered. The *backward-elimination procedure*, on the other hand, starts with a model considering all independent variables and then, one by one, removes those statistically insignificant independent variables from the equation. As the forward-selection procedure does not remove a variable once it is chosen, other algorithms have been developed to combine the two procedures in every step of the process. It should be noted that none of the algorithm guarantees the generation of the "best" subset. The order in which the independent variables enter or leave the model does not necessarily imply the order of their importance. In general, the backward-elimination procedure is less adversely affected by the multicollinearity (Sec. 3.9.3) of the independent variables.

3.9.2   Model linearization and adequacy check

To facilitate formal statistical inferences in regression analysis as shown previously, one would be concerned with the linear equation form, normality assumptions, statistical independence among explanatory variables, and constancy of variance.

Transformations to a linear equation.   By the LS criterion, linearity of the resulting normal equations requires that the response variable Y is linearly related to the unknown model parameters. Such linearity makes the determination of model parameters easier, but is not essential in model formulation. In the situation that nonlinearity is unavoidable, the use of LS criterion would result in a system of nonlinear normal equations, which requires the application of appropriate nonlinear optimization scheme for parameter estimation. The nonlinearity would make the validity of statistical inferences for model parameters described earlier questionable. In some cases, a nonlinear function can be expressed as a linear one through suitable transformation. Such linear models are called *intrinsically linear*. Figure 3.8 shows some of linearizable functions, their corresponding transformations, and the resulting linear forms. After a suitable transformation from nonlinear to a linear function, the LS method and statistical inferences described previously can be applied for parameter estimation on the basis of the transformed data. It should be remembered that the LS-based parameter estimates for the transformed model do not minimize the sum of squared residuals on the original data, but of the transformed data.

Constant variance and variance stabilizing transformation.   The LS procedure described in Sec. 3.3, specifically, is called the *ordinary LS (OLS)* procedure because equal weight is given to all data observations. As mentioned previously,

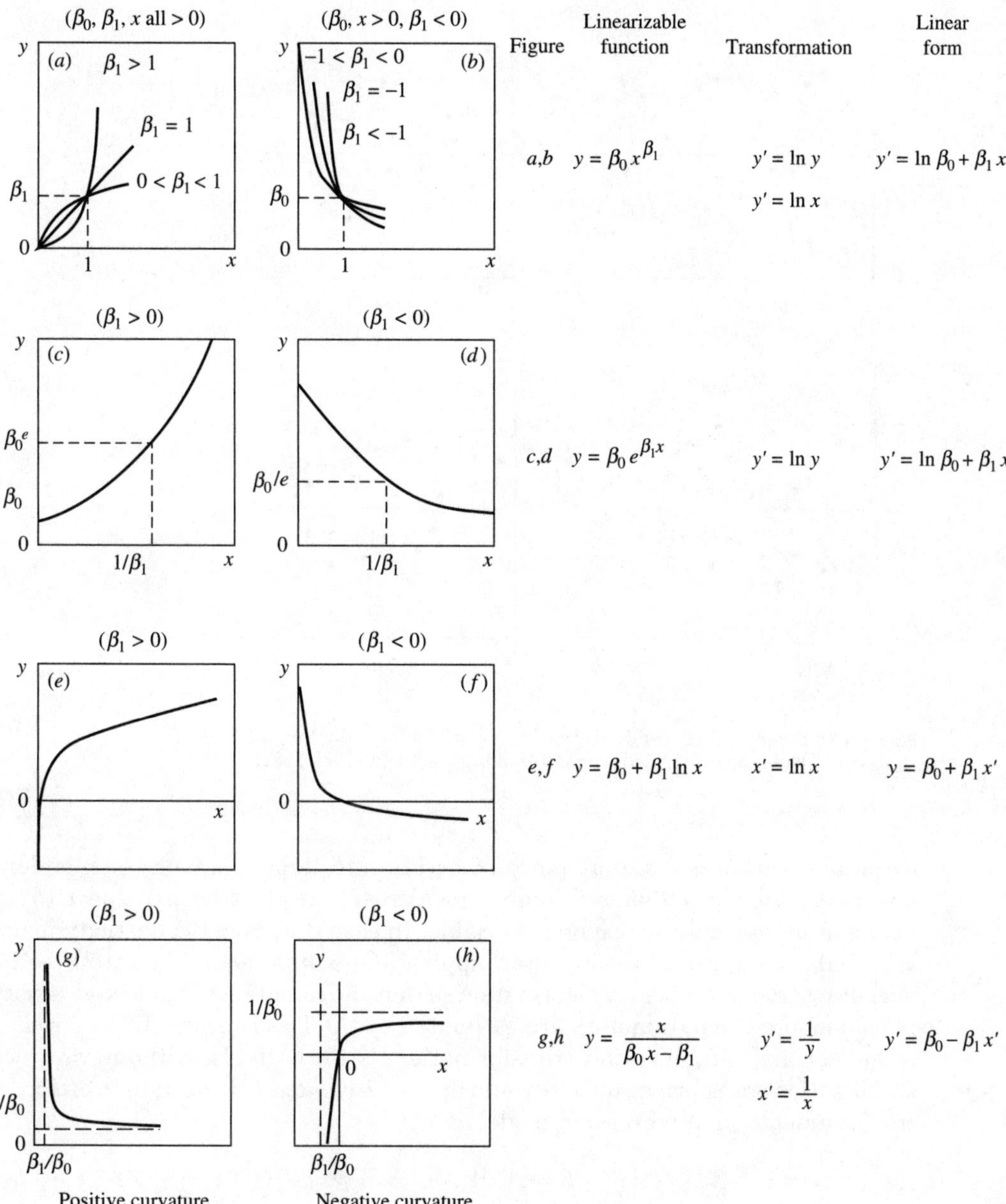

Figure 3.8   Some linearizable functions (after Danniel and Wood 1980).

the ordinary LS estimators of model parameters β's in a linear regression model are the best linearly unbiased estimators when the assumptions of the model outlined in Sec. 3.5 are held. When the assumption of constant variance (called *homoscedasticity*), that is, $\text{Var}(Y\,|\,x) = \text{Var}(\varepsilon\,|\,x) = \sigma_\varepsilon^2$, is not satisfied, the ordinary LS estimators are no longer efficient estimators. Homoscedasticity can be

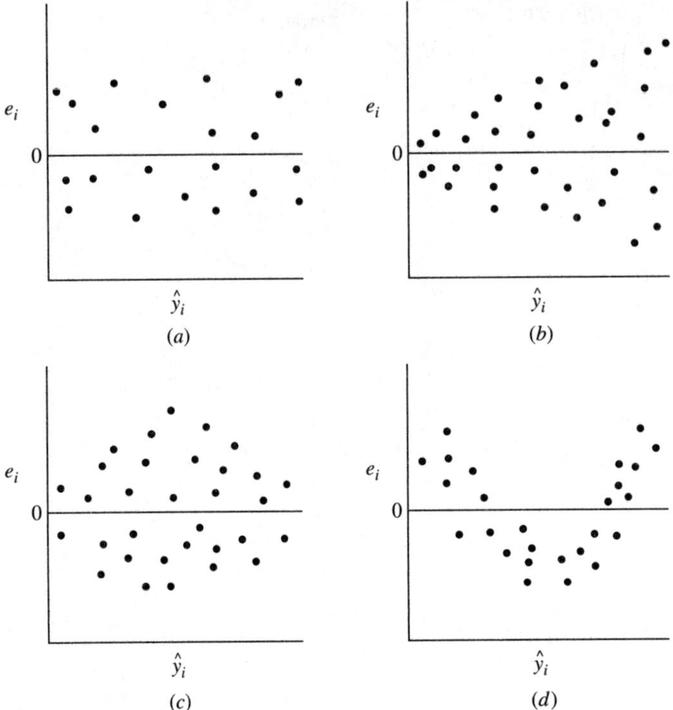

Figure 3.9 Some typical residual plots. (a) Satisfactory; (b) Funnel shaped; (c) Double bow; (d) Nonlinear. (after Montgomery and Peck 1982).

examined through residual plots of errors and dependent or independent variables. Figure 3.9 shows some typical residual plots with respect to an explanatory variable or response variable. In case that homoscedasticity is not satisfied, two approaches are often applied. One approach is to stabilize the variance through proper transformation so that the condition of homoscedasticity is held before the parameters are estimated by the LS criterion. Table 3.4 lists a few variance stabilization transformations. Alternatively, without variance stabilization transformation, one could apply a *weighted LS* criterion to estimate the parameters in a regression model as

$$\hat{\boldsymbol{\beta}} = (\boldsymbol{X}^t \boldsymbol{W} \boldsymbol{X})^{-1} \boldsymbol{X} \boldsymbol{W} \boldsymbol{y} \tag{3.30}$$

in which $\boldsymbol{W}$ is an $n \times n$ diagonal matrix of weights, $\boldsymbol{W} = \mathrm{diag}(w_1, w_2, ..., w_n)$, for each observation with $w_i \propto 1/\mathrm{Var}(Y_i)$.

Normality and normal transformation.   Compliance with the normality assumption is needed to ensure the validity of statistical inferences made for the estimated regression model and model parameters described in Secs. 3.6 and 3.7. The degree of linearity of errors plotted on a normal probability paper would provide a visual inspection about its compliance. Alternatively, formal statistical

TABLE 3.4   Appropriate Variance Stabilization Transformation When $\sigma_y = f(\eta)$ (after Draper and Smith 1981)

Nature of dependence $\sigma_y = f(\eta)^*$	Range for Y	Variance stabilizing transformation
$\sigma_y \propto \eta^k$ and in particular	$Y \geq 0$	Y^{1-k}
$\sigma_y \propto \eta^{1/2}$ poisson	$Y \geq 0$	$Y^{1/2}$
$\sigma_y \propto \eta$	$Y \geq 0$	$\ln(Y)$
$\sigma_y \propto \eta^2$	$Y \geq 0$	Y^{-1}
$\sigma_y \propto \eta^{1/2}(1-\eta)^{1/2}$ binomial	$0 \leq Y \leq 1$	$\sin^{-1}(Y^{1/2})$
$\sigma_y \propto (1-\eta)^{1/2}/\eta$ negative binomial	$0 \leq Y \leq 1$	$(1-Y)^{1/2} - (1-Y)^{3/2}/3$
$\sigma_y \propto (1-\eta^2)^{-2}$	$-1 \leq Y \leq 1$	$\ln\{(1+Y)/(1-Y)\}$

NOTE: $*\eta = E(Y)$.

goodness-of-fit tests, such as the Komolgorov-Smirnov test and others (Ang and Tang 1975; D'Agostino and Stephens 1986), can be conducted. In the case that the normality assumption is violated, a normal transformation can be applied to the dependent variable. There are many normal transformation schemes and a simple approach is the power transform proposed by Box and Cox (1964)

$$y(\lambda) = \frac{(y^\lambda - 1)}{\lambda} \quad \lambda \neq 0$$

$$y(\lambda) = \ln(y) \qquad \lambda = 0$$

(3.31)

in which λ is the parameter to be determined in such a way that the resulting $y(\lambda)$ approximately is normally distributed. For a dependent variable Y, which could be negative-valued, a constant y_0 can be added so that the value of $(y + y_0)$ is always positive.

3.9.3   Multicollinearity and outliers

Multicollinearity usually exists in regression analysis of practical problems due to the presence of linear dependence among explanatory variables. Explanatory variables might be intrinsically related or sometimes human introduced, such as adding polynomial terms. Refer to the term $X^t X$ involved in the estimation of regression model parameters and their statistical features. It is clear that a strong presence of multicollinearity would produce poor LS estimators of β's, result in large variance and covariance for them, and make selection of explanatory variables and assessment of their variance contribution more troublesome. For the treatment of multicollinearity, one can refer to any textbook on regression analysis.

In regression analysis, outliers are those data whose errors have much larger value in absolute magnitude than all the others. A practical way to label a residual to be an outlier is when its standardized value is larger than 3. One should carefully examine the cause of an outlier before dismissing or eliminating it from the data set.

3.9.4 Implementation of regression analysis

There are many well-developed computer software for conducting comprehensive regression analyses. Examples of those well-known statistical packages, just to name a few, are statistical Analysis System (SAS), Statistical Package for Social Sciences (SPSS), Biomedical Package (BMDP), MINITAB, and S-Plus. These statistical packages are capable of handling large data sets, estimating regression parameters along with their hypothesis tests, computing confidence and prediction intervals, selecting the best subset of independent variables by various schemes, residuals analysis, and many other functions. Although the application of statistical software greatly enhances the ability of data analysis, one should remain vigilant about the appropriateness of the techniques and the interpretation of the results.

Problems

3.1 Based on the water consumption data given in Table 3.1, perform the regression analysis to establish the relationship between annual water use (Q), population (POP), and unit water price ($PRICE$) in the following form:

$$Q = \beta_0 \times POP^{\beta_1} \times PRICE^{\beta_2}$$

a. Determine the model parameters by the least squares method.
b. Calculate the coefficient of determination and standard error of estimate of the model.
c. Compute the variances of the LS-estimators, $\hat{\beta}_1$, $\hat{\beta}_2$ and their correlation coefficient.
d. Determine the 95 percent confidence interval and prediction interval for the annual water use under $POP = 450{,}000$ and $PRICE = \$1.50$.

3.2 The following table contains data of the rainfall excess hyetograph and direct runoff hydrograph for a watershed having a drainage area of 242 mi.^{2}

Time (h)	1	2	3	4	5	6	7	8	9	10	11	12
Rainfall excess (in)	0.10	0.20	0.89	2.90								
Direct runoff (in/h)	0.01	0.03	0.1	0.3	1.18	1.00	0.64	0.37	0.21	0.12	0.07	0.03

a. Determine the LS-based 1h unit hydrograph (UH) for the watershed.
b. Determine variance-covariance matrix of the 1h UH ordinates.
c. Given the following effective rainfall hyetograph with $\Delta t = 1$h.

$$\mathbf{p}_* = (0.2 \text{ in}, 1.0 \text{ in}, 0.5 \text{ in})$$

determine the mean and variance of the direct runoff discharge at $t = 4$h based on the UH model.
d. Compute the probability that discharge$_t$ at $t = 4$ h would exceed 65,000 ft^3/s.

3.3 Consider the following data from studies on clear water scour around circular piers (Jain 1981). From the dimensional analysis of scour parameters, it was identified that the maximum scour depth is related to flow and pier properties as

$$\frac{d_s}{b} = g\left(\frac{y}{b}, F_c\right)$$

where d_s = maximum scour depth; b = pier diameter; y = flow depth; $F_c = V_c / \sqrt{gy}$ = threshold Froude number, with V_c = threshold flow velocity that initiates sediment movement. Under the assumption of the following model form

$$\frac{d_s}{b} = \beta_0 \left(\frac{y}{b}\right)^{\beta_1} (F_c)^{\beta_2} \varepsilon$$

with ε = model error term,

a. Determine the model parameters by the LS method.

b. Calculate the standard error of estimate and coefficient of determination of the model.

c. Compute the variance-covariance matrix of the LS-estimators.

d. Determine the 95 percent confidence interval and prediction interval of the maximum scour depth for the condition of $b = 1.5$ m, $y = 3.0$ m, $V = 1.8$ m/s, and $F_c = 0.50$.

e. Determine the probability that the actual maximum scour depth would exceed 30 cm under the condition stated in Part (d).

V (m/s)	b (cm)	y (cm)	d_s (mm)	F_c
0.82	5.1	24.7	8.7	0.46
0.40	15.2	21.9	18.0	0.21
0.32	15.2	11.6	13.4	0.27
0.36	15.2	15.6	15.8	0.24
0.38	15.2	20.6	17.1	0.22
0.44	15.2	21.0	21.0	0.22
0.41	15.2	26.3	18.6	0.20
0.38	15.2	17.6	16.6	0.25
0.50	91.4	61.0	54.9	0.15
0.85	5.0	20.0	7.0	0.56
0.85	10.0	20.0	12.5	0.56
0.85	15.0	20.0	18.5	0.56
0.76	5.0	10.0	8.7	0.71
0.76	10.0	10.0	13.1	0.71
0.76	15.0	10.0	17.5	0.71
0.66	5.0	20.0	9.8	0.39
0.66	10.0	20.0	17.0	0.39
0.66	15.0	20.0	20.3	0.39
0.40	5.0	19.7	9.5	0.25
0.40	10.0	19.7	12.2	0.25
0.40	15.0	19.7	14.9	0.25
0.42	5.0	35.0	9.0	0.20
0.42	10.0	35.0	12.0	0.20
0.42	15.0	35.0	13.7	0.20
0.37	10.0	10.0	11.5	0.32

References

Ang, A. H. S., and W. H. Tang (1975). *Probability Concepts in Engineering Planning and Design,* Vol. I, John Wiley and Sons, New York.

Blank, L. (1980). *Statistical Procedures for Engineering, Management, and Science,* McGraw-Hill, New York.

Box, G. E. P., and D. R. Cox (1964). "An Analysis of Transformations," *Journal of the Royal Statistical Society B,* **26**: 211–252.

D'Agostino, R. B., and M. A. Stephens (1986). *Goodness-of-Fit Procedures,* Marcel Dekker, New York.

Danniel, C., and F. S. Wood (1980). *Fitting Equations to Data,* 2d ed., John Wiley and Sons, New York.

Devore, J. L. (1987). *Probability and Statistics for Engineering and Sciences,* 2d ed., Brooks/Cole, Monterey, CA.

Draper, N., and H. Smith (1981) *Applied Regression Analysis,* 2d ed, John Wiley and Sons, New York.

Espey, W. H. Jr., and D. G. Altman (1978). Nomographs for 10-Minute Unit Hydrographs for Small Urban Watersheds. U.S. Environmental Protection Agency, *Report* EPA-600/9-78-035.

Jain, S. C. (1981). Maximum Clear-Water Scour Around Circular Piers, *Journal of the Hydraulics Division,* ASCE, **107**(5):611–626.

Mays, L. W., and Y. K. Tung (1992). *Hydrosystems Engineering and Management,* McGraw-Hill Book Company, New York.

Miller, N., and R. Cronshey (1992). "Runoff Curve Number: The Next Step," In *Catchment Runoff and Rational Formula,* B. C. Yen (ed.), Water Resources Publications, Littleton, CO.

Montgomery, D. C., and E. A. Peck (1982). *Introduction to Linear Regression Analysis,* John Wiley and Sons, New York.

Morgali, J., and R. K. Linsley (1965). "Computer Analysis of Overland Flow," *Journal of Hydraulic Division,* ASCE, **91**(HY3):81–100.

Mosley, M. P., and A. I. McKerchar (1992). "Chapter 8: Streamflow," In *Handbook of Hydrology,* D. Maidment (ed.), McGraw-Hill, New York.

Neter, J., W. Wasserman, and M. H. Kutner (1983). *Applied Linear Regression Models.* Richard D. Irwin, Homewood, IL.

Seber, G. A. F. (1977). *Linear Regression Analysis,* John Wiley and Sons, New York.

Woods, J., and J. S. Gulliver (1991). "Economic and Financial Analysis," In *Hydropower Engineering Handbook,* J. S. Gulliver and R.E.A. Arndt (eds.), 9.1–9.37, McGraw-Hill, New York.

Zeller, M. E., and W. T. Fullerton (1983). "A Theoretically Derived Sediment Transport Equation for Sand-Bed Channels in Arid Regions," *Proceedings of D.B. Simons Symposium on Erosion and Sedimentation,* R. M. Li and P. F. Lagasse (eds.), Bookcrafters, Chelsea, MI.

Analytic Methods for Uncertainty Analysis

Referring to the previous discussions on uncertainty in modeling and analysis of hydrosystem infrastructures, it is not difficult to find that many quantities of interest are functionally related to several variables, some or all of which are subject to uncertainty. For example, hydraulic engineers frequently apply the weir flow equation $Q = CLH^{\alpha}$ to estimate the spillway capacity in which the coefficients C and α, as well as head H are often subject to uncertainty. As a result, discharge over the spillway is not certain. Another example is the use of the Thiem equation (Prob. 4.39) to estimate the pressure drawdown in a confined aquifer due to pumpage in which the hydraulic conductivity and aquifer thickness are not completely known. Hence, the pressure drawdown cannot be predicted with certainty.

The main concern regarding the subject of algebra of random variables is to derive the probability distribution function (PDF) or statistical properties of a random variable, which is a function of other random variables. The types of methods applicable for uncertainty analysis are, in general, dictated by the information available with regard to the stochastic basic variables and the functional relationship among the variables. In principle, it would be most ideal to derive the exact PDF of the model output as a function of the involved stochastic basic variables. In this chapter, several analytical methods are discussed that would allow analytical derivations of an exact PDF and/or statistical moments of a random variable as a function of several other stochastic basic variables. In theory, the techniques described in this chapter are straightforward. However, the success of implementing these procedures largely depends on the functional relation, forms of the PDFs involved, and the analyst's mathematical skills. Commonly, situations arise in which analytical derivations are virtually impossible. For this reason, it is often practical to find an approximation of the statistical moments of the random variable of interest in terms of those of stochastic basic variables. Several useful techniques for approximation are described in Chap. 5.

This chapter starts with a discussion of the derived distribution method (Sec. 4.1) that allows derivation of the analytical expression of the joint PDF of several random variables as functions of several other random variables. Then, it is followed by two sections discussing the basic theories and applications of integral transforms to uncertainty analysis. Integral transform techniques originally found their roles in univariate statistical analysis. The well-known integral transforms are Fourier, Laplace, and exponential transforms (Sec. 4.2). Section 4.3 describes the Mellin transform, which is useful, but less used in hydrosystem engineering applications. Tung (1989; 1990) demonstrated the applications of the Mellin transform to uncertainty analysis of hydrologic and hydraulic problems. One major advantage of integral transforms is that, if such transforms of a PDF exist, the relationship between the PDF and its integral transform is unique (Kendall, Stuart and Ord 1987). In dealing with a multivariate problem in which a random variable is a function of several random variables, the convolution property of these integral transforms becomes analytically powerful, especially when the stochastic variables in the model are independent. Fourier and exponential transforms are powerful in treating the sum and difference of random variables while the Mellin transform is applicable to the quotient and product of random variables. In Sec. 4.4, several techniques are presented to construct a probability distribution based on a limited number of statistical moments without making parametric assumptions for the distributions. Finally, Sec. 4.5 summarizes the advantages and limitations of each method discussed in this chapter.

4.1 Derived Distribution Method

Suppose that a random variable W is related to another random variable X through the functional relationship $W = g(X)$. Furthermore, the PDF and cumulative distribution function (CDF) of X are known. The CDF of W can be obtained as

$$H_w(w) = P[W \le w] = P[g(X) \le w]$$

$$= P[X \le g^{-1}(w)]$$

$$= F_x[g^{-1}(w)] \tag{4.1}$$

where $g^{-1}(.)$ represents the inverse function of $g(.)$. As can be seen from Eq. (4.1), the CDF of W, $H_w(w)$, can be expressed in terms of the CDF of X, $F_x(x)$. The PDF of W, $h_w(w)$, then, can be obtained by taking the derivative of $H_w(w)$ with respect to w, that is, $h_w(w) = d[H_w(w)]/dw$.

Example 4.1 Consider the relationship between random variables X and W as $X = \ln(W)$. It is known that X is a normal random variable with the PDF as

$$f_x(x) = \frac{1}{\sqrt{2\pi}\, \sigma_x} \exp\left[-\frac{1}{2}\left(\frac{x - \mu_x}{\sigma_x} \right)^2 \right] \qquad \text{for} -\infty < x < \infty$$

in which μ_x and σ_x are the mean and standard deviation of X, respectively. Derive the PDF of W.

Solution   By Eq. (4.1), the CDF of the dependent variable $W = \exp(X)$ can be expressed as

$$H_w(w) = P(W \le w) = P[\exp(X) \le w] = P[X \le \ln(w)] = F_x[\ln(w)]$$

in which $F_x(.)$ is the CDF of the normal random variable X. The PDF of random variable W can then be obtained as

$$h_w(w) = \frac{dH_w(w)}{dw} = \frac{d}{dw}\ [F_x(\ln(w))]$$

$$= f_x\ [\ln(w)]\ \frac{d[\ln(w)]}{dw} = f_x\ [\ln(w)]\left(\frac{1}{w}\right)$$

Substituting $\ln(w)$ for x in the PDF of X, that is, $f_x[\ln(w)]$, the PDF of random variable W is

$$h_w(w) = \frac{1}{\sqrt{2\pi}\ \sigma_{\ln w} w}\ \exp\left[-\frac{1}{2}\left(\frac{\ln(w) - \mu_{\ln w}}{\sigma_{\ln w}}\right)^2\right]\qquad \text{for } w > 0$$

which is the PDF of a lognormal random variable as Eq. (2.57). Since $W = \exp(X)$ and $-\infty < x < \infty$, the valid range for the random variable W is $w > 0$.

In the case where the functional relationship between X and W is strictly increasing or strictly decreasing, the PDF of W can be derived directly from the PDF of X as

$$h_w(w) = f_x(x)\ |dx/dw| \qquad (4.2)$$

in which $|dx/dw|$ is called the *Jacobian*. This derived distribution method is also called *transformation of variables technique*. From Eq. (4.2), one realizes that when the function $W = g(X)$ is either monotonically increasing or decreasing, the PDF of the dependent variable W can be obtained from the known PDF of the independent variable X multiplied by the absolute-valued Jacobian.

Example 4.2   Referring to Example 4.1, derive the PDF of W from that of X using Eq. (4.2).

Solution   Since the relationship between X and W, $X = \ln W$, is a one-to-one strictly increasing function relationship, Eq. (4.2) can be applied to derive the PDF of W as

$$h_w(w) = f_x(x)\left|\frac{dx}{dw}\right| = \frac{1}{\sqrt{2\pi}\ \sigma_x}\ \exp\left[-\frac{1}{2}\left(\frac{x - \mu_x}{\sigma_x}\right)^2\right]\left|\frac{dx}{dw}\right|$$

where the Jacobian can be determined as $|dx/dw| = |d(\ln w)/dw| = 1/w$. Then, the PDF of W can be obtained by substituting the Jacobian and expressing x in terms of w, that is, $x = \ln(w)$, in the above equation as

$$h_w(w) = \frac{1}{\sqrt{2\pi}\ \sigma_{\ln w}\ w} \exp\left[-\frac{1}{2}\left(\frac{\ln(w) - \mu_{\ln w}}{\sigma_{\ln w}}\right)^2\right] \quad \text{for } w > 0$$

Referring to Fig. 4.1, functional relationships sometimes are not single-valued, i.e., the inverse $g^{-1}(w)$ may correspond to multiple values of x for a

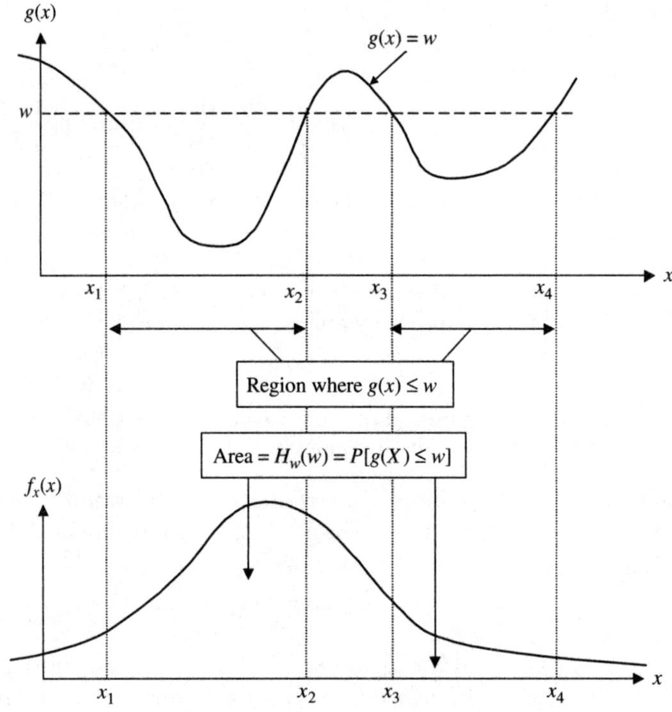

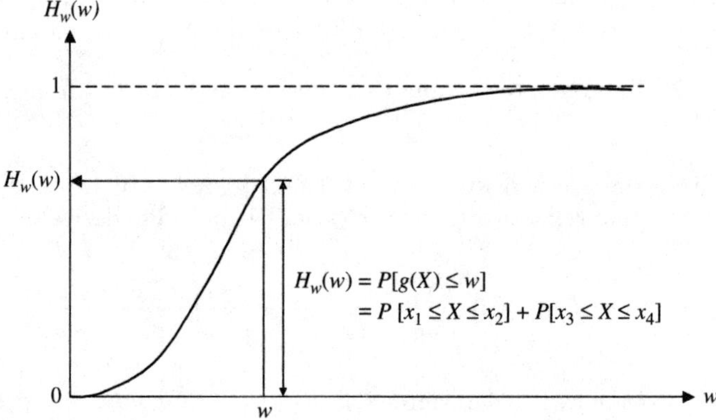

Figure 4.1   A general case of function of a random variable.

given value of w. For example, if $g^{-1}(w) = x_1, x_2,..., x_m$, then Eq. (4.2) can be extended to

$$h_w(w) = \sum_{i=1}^{m} f_x(x_i) \left| \frac{dx_i}{dw} \right| \tag{4.3}$$

Example 4.3   The kinetic energy of the turbulence (E) of fluid flow in a given direction can be computed as

$$E = \frac{\rho U^2}{2}$$

in which ρ is fluid density and U is the fluctuation of fluid velocity around its mean along the specified direction. Assume that U has a standard normal distribution with the following PDF

$$f_u(u) = \frac{1}{\sqrt{2\pi}} e^{-u^2/2} \qquad \text{for} -\infty < u < \infty$$

Derive the PDF of the kinetic energy E.

Solution   Note that the functional relationship between E and U is not one-to-one. To derive the PDF of E, two approaches can be used. One approach is to start with Eq. (4.1) by which the CDF of the kinetic energy E is

$$H_e(e) = P(E \le e) = P\left(U^2 \le \frac{2e}{\rho} \right) = P\left(-\sqrt{\frac{2e}{\rho}} \le U \le \sqrt{\frac{2e}{\rho}} \right)$$

$$= P\left(U \le \sqrt{\frac{2e}{\rho}} \right) - P\left(U \le -\sqrt{\frac{2e}{\rho}} \right)$$

$$= F_u\left(\sqrt{\frac{2e}{\rho}} \right) - F_u\left(-\sqrt{\frac{2e}{\rho}} \right)$$

Hence, the PDF of the kinetic energy E can be derived, for $e \ge 0$, as

$$h_e(e) = \frac{dH_e(e)}{de} = \frac{1}{\sqrt{2\rho e}} \left[f_u\left(\sqrt{\frac{2e}{\rho}} \right) + f_u\left(-\sqrt{\frac{2e}{\rho}} \right) \right] = \sqrt{\frac{2}{\pi}} e^{-e/\rho}$$

The second method is to apply Eq. (4.3) by which the original random variable U, in terms of the new random variable E, is obtained as

$$u_+ = +\sqrt{\frac{2e}{\rho}} \qquad u_- = -\sqrt{\frac{2e}{\rho}}$$

The Jacobian is

$$\left| \frac{du_+}{de} \right| = \left| \frac{du_-}{de} \right| = \frac{1}{\sqrt{2\rho e}}$$

According to Eq. (4.3), the PDF of the turbulent kinetic energy (E) is

$$h_e(e) = f_u(u_+)\left|\frac{du_+}{de}\right| + f_u(u_-)\left|\frac{du_-}{de}\right|$$

$$= \frac{1}{\sqrt{2\rho e}}\left[f_u\left(\sqrt{\frac{2e}{\rho}}\right) + f_u\left(-\sqrt{\frac{2e}{\rho}}\right)\right] = \sqrt{\frac{2}{\pi}}\,e^{-e/\rho}$$

for $e \geq 0$.

Consider a multivariate case in which K random variables W_1, W_2,..., W_K are related to K and other random variables X_1, X_2,..., X_K through a system of K equations as

$$W_1 = g_1(X_1, X_2,..., X_K)$$

$$W_2 = g_2(X_1, X_2,..., X_K)$$

$$\vdots$$

$$W_K = g_K(X_1, X_2,..., X_K)$$

When the functions $g_1(.)$, $g_2(.)$,...,$g_K(.)$ satisfy a monotonic relationship, the joint PDF of random variables Ws can be directly obtained by

$$h_w(w_1, w_2,..., w_K) = f_x(x_1, x_2,..., x_K)\,|\boldsymbol{J}| \tag{4.4}$$

where $f_x(x_1, x_2,..., x_K)$ and $h_w(w_1, w_2,..., w_K)$ are the joint PDFs of Xs and Ws, respectively; $|\boldsymbol{J}|$ is the absolute value of the determinant of the $K \times K$ *Jacobian matrix*

$$|\boldsymbol{J}| = \begin{vmatrix} \dfrac{\partial x_1}{\partial w_1} & \dfrac{\partial x_1}{\partial w_2} & \cdots & \dfrac{\partial x_1}{\partial w_K} \\ \vdots & \vdots & \vdots & \vdots \\ \dfrac{\partial x_K}{\partial w_1} & \dfrac{\partial x_K}{\partial w_2} & \cdots & \dfrac{\partial x_K}{\partial w_K} \end{vmatrix} \tag{4.5}$$

When one is interested in the PDF of a random variable as a function of several random variables, the joint PDF $h_w(w_1, w_2,..., w_K)$ can be used to find the marginal PDF of the random variable of interest (see Sec. 2.2.2).

Example 4.4   Given that the joint PDF of X and Y is

$$f_{x,y}(x, y) = \frac{1}{14}\left(5 - \frac{y}{2} + x^2\right) \qquad \text{for } 0 \leq x, y \leq 2$$

(a) What is the PDF for the sum of these two random variables?
(b) What are the mean and standard deviation of the sum?

Solution

(a) Let $W = X + Y$ be the new random variable whose PDF is to be derived. In order to apply Eq. (4.4), a new random variable, say $V = Y$, is artificially introduced so that there are two equations with two unknowns. To obtain the Jacobian matrix, the old random variables X and Y are expressed in terms of the new random variables W and V as

$$X = W - V \qquad Y = V$$

The absolute value of the determinant of the Jacobian matrix can then be computed as

$$|\boldsymbol{J}| = \begin{vmatrix} \dfrac{\partial x}{\partial w} & \dfrac{\partial x}{\partial v} \\[2mm] \dfrac{\partial y}{\partial w} & \dfrac{\partial y}{\partial v} \end{vmatrix} = \begin{vmatrix} 1 & 1 \\ 0 & 1 \end{vmatrix} = 1$$

Therefore, the joint PDF of W and V can be expressed as

$$h_{w,v}(w, v) = \frac{3}{70}\left(5 - \frac{v}{2} + (w - v)^2\right) \qquad \text{for } 0 \le v \le 2 \qquad v \le w \le v + 2$$

The valid domain of w and v is shown in Fig. 4.2(a).

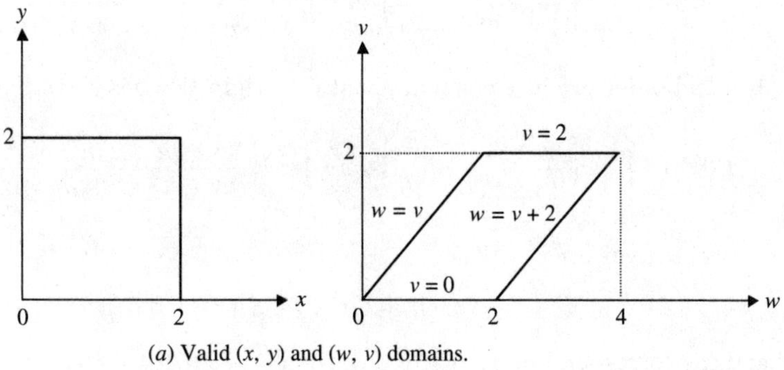

(a) Valid (x, y) and (w, v) domains.

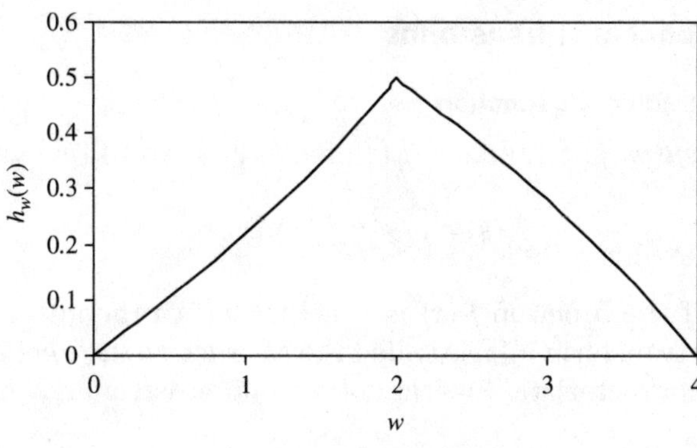

(b) PDF of random variable W.

Figure 4.2   Valid domains of random variables and the PDF of W of Example 4.4.

Since the problem is only interested in the PDF of W rather than the joint PDF, the marginal PDF of W can be derived by applying Eq. (2.15). Referring to Fig. 4.2(a), integration of $h_{w,v}(w, v)$ over $0 \leq v \leq 2$ has to be broken into two parts; one part for $0 \leq w \leq 2$ and the other for $2 \leq w \leq 4$, as follows:

(i) $0 \leq w \leq 2$,

$$h_w(w) = \int_0^w h_{w,v}(w, v)\, dv = \frac{3}{70} \int_0^w \left(5 - \frac{v}{2} + (w-v)^2 \right) dv = \frac{3w}{14} - \frac{3w^2}{280} + \frac{w^3}{70}$$

(ii) $2 \leq w \leq 4$,

$$h_w(w) = \frac{3}{70} \int_{w-2}^2 \left(5 - \frac{v}{2} + (w-v)^2 \right) dv = \frac{38}{35} - \frac{3w}{7} + \frac{27w^2}{280} - \frac{w^3}{70}$$

The resulting PDF of W is shown in Fig. 4.2(b).

(b) The mean of W can be obtained by

$$E(W) = \int_0^2 w \left(\frac{3w}{14} - \frac{3w^2}{280} + \frac{w^3}{70} \right) dw + \int_2^4 w \left(\frac{38}{35} - \frac{3w}{7} + \frac{27w^2}{280} - \frac{w^3}{70} \right) dw = 2.086$$

The 2nd-order product-moment about the origin $W = 0$ is

$$E(W^2) = \int_0^w w^2 \left(\frac{3w}{14} - \frac{3w^2}{280} + \frac{w^3}{70} \right) dw + \int_2^4 w^2 \left(\frac{38}{35} - \frac{3w}{7} + \frac{27w^2}{280} - \frac{w^3}{70} \right) dw = 5.025$$

Hence, the variance of W is

$$\text{Var}(W) = E(W^2) - E^2(W) = 0.675$$

and the corresponding standard deviation is $\sqrt{0.675} = 0.821$.

4.2 Fourier, Laplace, and Exponential Transforms

4.2.1 Fourier transform and characteristic function

The *Fourier transform* of a function, $f_x(x)$, is defined for all real values of s as

$$\mathbb{F}_x(s) = \int_{-\infty}^{\infty} e^{isx} f_x(x)\, dx \tag{4.6}$$

where $i = \sqrt{-1}$. If the function $f_x(x)$ is the PDF of a random variable X, the resulting Fourier transform $\mathbb{F}_x(s)$ is called the *characteristic function*. According to Eq. (4.6), the characteristic function of a random variable X having a PDF $f_x(x)$ is

$$\mathbb{F}_x(s) = E\left[e^{isX} \right] \tag{4.7}$$

The characteristic function of a random variable always exists for all values of the argument s. Furthermore, the characteristic function for a random variable under consideration is unique. In other words, two distribution functions are identical if and only if the corresponding characteristic functions are identical (Patel, Kapadia, and Owen 1976). Therefore, given a characteristic function of a random variable, its PDF can be uniquely determined through the inverse transform as

$$f_x(x) = \frac{1}{2\pi} \int_{-\infty}^{\infty} e^{-isx}\, F_x(s)\,ds \qquad (4.8)$$

Some useful operational properties of Fourier transforms on a PDF are given in Table 4.1. Furthermore, the characteristic functions for some commonly used PDFs are shown in the third column of Table 4.2.

Using the characteristic function, the rth-order moment about the origin of the random variable X can be obtained as

$$E[X^r] = \mu_r' = \frac{1}{i^r}\left[\frac{d^r F_x(s)}{ds^r}\right]_{s=0} \qquad (4.9)$$

The formulas for computing $E[X^r]$ or $E[(X - \mu_x)^r]$ of various commonly used distributions are shown in the last column of Table 4.2. One could easily obtain one type of moments from the other type by using Eq. (2.21) or (2.22).

The characteristic function of a random variable can be expanded in a power series, in terms of its moments, as

$$F_x(s) = \sum_{r=0}^{\infty} \mu_r' \frac{(is)^r}{r!} \qquad (4.10)$$

In case where the moments of a random variable are known, Eq. (4.10) can be applied along with Eq. (4.8), to derive the corresponding PDF.

Example 4.5 Consider that the recreational benefit from a proposed multipurpose reservoir is uncertain, having a uniform distribution with a pessimistic value of a dollars and an optimistic value of b dollars. Derive the characteristic function of the

TABLE 4.1 Operation Properties of the Fourier Transform on a PDF (after Springer 1978)

Property	PDF	Random variable	Fourier transform	Laplace Transform
Standard	$f_x(x)$	X	$F_x(s)$	$L_x(s)$
Scaling	$f_x(ax)$	X	$a^{-1}F_x(s/a)$	$a^{-1}L_x(s/a)$
Linear	$af_x(x)$	X	$aF_x(s)$	$aL_x(s)$
Translation 1	$e^{ax}f_x(x)$	X	$F_x(s - ia)$	$L_x(s + a)$
Translation 2	$f_x(x - a)$	X	$e^{ias}F_x(s)$	$e^{as}L_x(s),\ x > a$

TABLE 4.2 Characteristic Functions and Moment Generating Functions of Some Commonly Used Distribution Functions

Distribution	PDF	Characteristic function	Moment generating function	Product moments
Uniform	Eq. (2.92)	$\dfrac{e^{ibs} - e^{ias}}{i(b-a)s}$	$\dfrac{e^{bs} - e^{as}}{(b-a)s}$	$\mu_r' = \dfrac{b^{r+1} - a^{r+1}}{(b-a)(r+1)}$
Normal	Eq. (2.50)	$\exp\{i\mu s - 0.5s^2\sigma^2\}$	$\exp\{\mu s - 0.5s^2\sigma^2\}$	$\mu_{2r} = \dfrac{(2r)!}{2^r r!}\,\sigma^{2r} \qquad r = 1,\,2,\ldots$
Lognormal	Eq. (2.57)	—	—	$\mu_r' = \exp\!\left(r\mu_{\ln x} + \tfrac{1}{2}r^2\sigma_{\ln x}^2\right)$
Gamma	Eq. (2.64)	$\left(\dfrac{1/\beta}{(1/\beta) - is}\right)^{\alpha}$	$\left(\dfrac{1/\beta}{(1/\beta) - s}\right)^{\alpha}$	$\mu_r' = \dfrac{\Gamma(r+\alpha)}{\Gamma(\alpha)}\,\beta^r$
Exponential	Eq. (2.71)	$\dfrac{1/\beta}{(1/\beta) - is}$	$\dfrac{1/\beta}{(1/\beta) - s}$	$\mu_r' = \Gamma(r+1)\,\beta^r$
Standard beta	Eq. (2.90)	—	—	$\mu_r' = \dfrac{\Gamma(\alpha+r)\,\Gamma(\alpha+\beta)}{\Gamma(\alpha)\,\Gamma(\alpha+\beta+r)}$
Extreme-value I (max)	Eq. (2.77)	$e^{i\xi s}\Gamma(1 - i\beta s)$	$e^{\xi s}\Gamma(1 - \beta s)$	—
Weibull	Eq. (2.81)	—	—	$E[(X-\xi)^r] = \beta^r\,\Gamma\!\left(\dfrac{r}{\alpha} + 1\right)$
Chi-square	Eq. (2.94)	$(1 - 2is)^{-K/2}$	$(1 - 2s)^{-K/2}$	$\mu_r' = \dfrac{2^r\,\Gamma(K/2+r)}{\Gamma(K/2)}$

PDF for the recreational benefit. Furthermore, determine the mean and variance of the random recreational benefit.

Solution   The pessimistic and optimistic values of the recreation benefit, i.e., a and b, can be regarded as the lower and upper bounds of the random recreation benefit as Fig. P2.2. The PDF of the recreational benefit, according to Eq. (2.92), is

$$f_x(x) = \frac{1}{(b-a)} \qquad \text{for } a \le x \le b$$

According to Eq. (4.7), the characteristic function of the random recreation benefit can be derived by

$$\mathbb{F}_x(s) = E(e^{isX}) = \frac{1}{b-a} \int_a^b e^{isx}\, dx = \frac{e^{ibs} - e^{ias}}{is(b-a)}$$

The 1st- and 2nd-order derivatives of the characteristic function can be obtained as

$$\frac{d\mathbb{F}_x(s)}{ds} = \frac{d}{ds}\left[\frac{e^{ibs} - e^{ias}}{i(b-a)s} \right] = \frac{(ibs\, e^{ibs} - ias\, e^{ias}) - (e^{ibs} - e^{ias})}{i(b-a)s^2}$$

$$\frac{d^2 \mathbb{F}_x(s)}{ds^2} = \frac{d^2}{ds^2}\left[\frac{e^{ibs} - e^{ias}}{i(b-a)s} \right] = \frac{d}{ds}\left[\frac{(ibs\, e^{ibs} - ias\, e^{ias}) - (e^{ibs} - e^{ias})}{i(b-a)s^2} \right]$$

$$= \frac{(a^2 e^{ias} - b^2 e^{ibs})s^2 - 2(be^{ibs} - ae^{ias})is + 2(e^{ibs} - e^{ias})}{i(b-a)s^3}$$

Hence, the mean of the random recreation benefit can be obtained by

$$\mu_x' = \mu_1' = \left[\frac{1}{i}\frac{d\mathbb{F}_x(s)}{ds} \right]_{s=0} = \left[\frac{(ibs\, e^{ibs} - ias\, e^{ias}) - (e^{ibs} - e^{ias})}{i^2(b-a)s^2} \right]_{s=0}$$

$$= \left[\frac{a^2 s e^{ias} - b^2 s e^{ibs}}{2s(b-a)} \right]_{s=0} \qquad (by\ L'\,Hospital's\ rule)$$

$$= \frac{a+b}{2}$$

Similarly, the 2nd-order product-moment about the origin can be obtained by

$$\mu_2' = \left[\frac{1}{i^2}\frac{d^2\mathbb{F}_x(s)}{ds^2} \right]_{s=0}$$

$$= \left[\frac{(a^2\, e^{ias} - b^2\, e^{ibs})s^2 - 2(be^{ibs} - ae^{ias})is + 2(e^{ibs} - e^{ias})}{i^3(b-a)s^3} \right]_{s=0}$$

By using L'Hospital's rule again, the result for μ_2' is

$$\mu_2' = \sigma_x^2 = \frac{a^2 + ab + b^2}{3}$$

Consequently, the variance of the recreation benefit is

$$\sigma_x^2 = \mu_2' - \mu_x^2 = \frac{(b-a)^2}{12}$$

4.2.2 Convolution properties of characteristic functions

The Fourier transform is particularly useful when random variables are independent and related linearly. In such cases, the convolution property of the Fourier transform can be applied to derive the characteristic function of the resulting random variable. For example, the PDF of $W = X_1 + X_2$, in which X_1 and X_2 are independent random variables with the PDFs $f_1(x_1)$ and $f_2(x_2)$, respectively, can be expressed as the following convolution

$$h_w(w) = \int_{-\infty}^{\infty} f_1(w - x_2)\, f_2(x_2)\, dx_2 = \int_{-\infty}^{\infty} f_1(x_1)\, f_2(w - x_1)\, dx_1 \tag{4.11}$$

More specifically, consider that $W = X_1 + X_2 + \cdots + X_K$ and all Xs are independent random variables with known PDF, $f_k(x_k)$, $k = 1, 2,..., K$. The characteristic function of W can then be obtained as

$$\mathbb{F}_w(s) = \mathbb{F}_1(s) \times \mathbb{F}_2(s) \times \cdots \times \mathbb{F}_K(s) = \prod_{k=1}^{K} \mathbb{F}_k(s) \tag{4.12}$$

which is the product of the characteristic functions of each individual random variable. The resulting characteristic function for W can be used in Eq. (4.9) to obtain the statistical moments of any order for the random variable W. The *inverse Fourier transform* of $\mathbb{F}_w(s)$, according to Eq. (4.8), can be made to derive the PDF of W as

$$h_w(w) = \frac{1}{2\pi} \int_{-\infty}^{\infty} e^{isw} \left[\prod_{k=1}^{K} \mathbb{F}_k(s) \right] ds \tag{4.13}$$

if it is analytically tractable. Otherwise, numerical algorithms for inverse transform have to be applied.

Example 4.6   A water resource project has two benefit components that are subject to uncertainty. The two benefit components are assumed to be independent and their distributional properties are listed below. Derive the characteristic function for the total benefit and determine its mean and standard deviation.

Benefit item	Distribution	Statistical properties
Flood control	Uniform	Lower bound = $1 million Upper bound = $2 millions
Water supply	Normal	Mean = $1.5 millions Standard deviation = $0.5 millions

Solution   Let random variables X_1 and X_2, respectively, be the benefits from the flood control and water supply. Based on the problem statements, the total benefit $T = X_1 + X_2$ is also a random variable. Since X_1 and X_2 are independent, according to Eq. (4.12), the characteristic function of the random total benefit can be expressed as

$$\mathbb{F}_T(s) = \mathbb{F}_1(s)\,\mathbb{F}_2(s)$$

in which $\mathbb{F}_1(s)$ and $\mathbb{F}_2(s)$ are the characteristic functions of the random benefits from flood control and water supply, respectively.

From Table 4.2, the characteristic functions of the benefits from flood control and water supply, for this example, could respectively be expressed as

$$\mathbb{F}_1(s) = \frac{e^{2is} - e^{is}}{is} \qquad \mathbb{F}_2(s) = e^{1.5is - 0.25s^2/2}$$

Hence, the characteristic function of the total benefit can be expressed as

$$\mathbb{F}_T(s) = \left(\frac{e^{2is} - e^{is}}{is}\right)\left(e^{1.5is - 0.25s^2/2}\right)$$

The mean and 2nd-order product-moment about the origin of the random total benefit could be obtained, according to Eq. (4.9), as the following:

$$E(T) = \left[\frac{1}{i}\frac{d\mathbb{F}_T(s)}{ds}\right]_{s=0}$$

$$= \left[\frac{0.25\,e^{0.125s(-s+20i)}\left[(14is - 4 - s^2)e^{is} - (10is - 4 - s^2)\right]}{i^2 s^2}\right]_{s=0}$$

$$= \$3\,\mathrm{M}$$

$$E(T^2) = \frac{1}{i^2}\left[\frac{d^2\mathbb{F}_T(s)}{ds^2}\right]_{s=0} = \frac{1}{12}\left(\frac{109i^2 - 3}{i^2}\right) = \$9.333\ \mathrm{million}^2$$

Hence, the variance of the total random benefit is

$$\mathrm{Var}(T) = 9.3333 - 3^2 = 0.3333\ \mathrm{million}^2$$

which can alternatively be calculated as $\mathrm{Var}(T) = \mathrm{Var}(X_1) + \mathrm{Var}(X_2)$. The corresponding standard deviation is $\sigma_T = \sqrt{0.3333} = \$0.577$ millions.

4.2.3   Laplace and exponential transforms and moment generating functions

The *Laplace and exponential transforms* of a function, $f_x(x)$, are defined, respectively, as

$$\mathbb{L}_x(s) = \int_0^\infty e^{sx}\, f_x(x)\,dx \tag{4.14}$$

and

$$\mathbb{E}_x(s) = \int_{-\infty}^{\infty} e^{sx} f_x(x)\, dx \tag{4.15}$$

As can be seen, the Laplace and exponential transforms are practically identical except that the former is applicable to functions with a nonnegative argument. In the case when $f_x(x)$ is the PDF of a random variable, the Laplace and exponential transforms, defined in Eqs. (4.14) and (4.15), respectively, can be stated as

$$\begin{aligned} \mathbb{L}_x(s) &= E\,[e^{sX}] && \text{for } x \geq 0 \\ \mathbb{E}_x(s) &= E\,[e^{sX}] && \text{for } -\infty < x < \infty \end{aligned} \tag{4.16}$$

Useful operational properties of the Laplace transform on a PDF are given in the last column of Table 4.1. The transformed function given by Eq. (4.14) or (4.15) of a PDF is called the *moment generating function (MGF)* as shown in Table 4.2 for some commonly used distribution functions.

Similar to the characteristic function, statistical moments of a random variable X can be derived from its moment generating function as

$$E[X^r] = \mu_r' = \left[\frac{d^r \mathbb{L}_x(s)}{ds^r}\right]_{s=0} \tag{4.17}$$

The MGF has convolution properties similar to those of the characteristic function as described in Sec. 4.2.2. Some useful operational rules relevant to the MGF are shown in Table 4.3. It can also be expressed in power series, in terms of statistical moments, as

$$\mathbb{L}_x(s) = \sum_{r=0}^{\infty} \mu_r' \frac{s^r}{r!} \tag{4.18}$$

There are two deficiencies associated with moment generating functions: (1) the MGF of a random variable may not always exist for all distribution functions and all values of s and (2) the correspondence between a PDF and the MGF may not necessarily be unique. Springer (1978) stated three theorems describing the conditions under which unique correspondence between a PDF and a MGF exists. These conditions, however, are generally satisfied in most situations.

TABLE 4.3   Operational Rules for the Moment Generating Function (after Springer 1978)

$W = cX$	$\mathbb{L}_w(s) = \mathbb{L}_x(cs)$, $c = $ constant
$W = c + X$	$\mathbb{L}_w(s) = e^{cs}\,\mathbb{L}_x(s)$, $c = $ constant
$W = \sum_i X_i$	$\mathbb{L}_w(s) = \Pi_i\, \mathbb{L}_i(s)$, when all X_i are independent
$W = \sum_i c_i X_i$	$\mathbb{L}_w(s) = \Pi_i\, \mathbb{L}_i(c_i s)$, when all X_i are independent

Example 4.7　In many economic analyses, the discrete cash flow pattern is replaced by its continuous equivalence such as

$$\text{PVR} = \int_0^T R(t)e^{-rt}\,dt$$

in which PVR is the present value of return; r is the *nominal continuous interest rate* that is related to the *discrete annual effective interest rate* i as $i = e^r - 1$; $R(t)$ is the continuous economic return; and T is the project life. Consider that the continuous cash flow pattern $R(t) = R_o$ and both R_o and project life T are independent random variables. Derive the expressions for the first three moments about the origin of the PVR.

Solution　The present value of return with $R(t) = R_o$ can be obtained by carrying out the integration as

$$\text{PVR} = \int_0^T R_o\,e^{-rt}\,dt = R_o\left[\frac{e^{-rt}}{-r}\right]_0^T = \frac{R_o}{r}(1 - e^{-rT})$$

The mth-order product-moment about the origin for the PVR can be expressed as

$$E(\text{PVR}^m) = E\left[\frac{R_o^m}{r^m}(1 - e^{-rT})^m\right]$$

Because R_o and T are statistically independent of each other, the above equation can be rewritten as

$$E(\text{PVR}^m) = \frac{1}{r^m}\,E\!\left(R_o^m\right)E[(1 - e^{-rT})^m]$$

The terms $(1 - e^{-rT})^m$ in the above equation can be expressed, using the binomial expansion, as

$$(1 - e^{-rT})^m = \sum_{j=0}^m \binom{m}{j} 1^j(-e^{-rT})^{m-j} = \sum_{j=0}^m \binom{m}{j}(-1)^{m-j}e^{-r(m-j)T}$$

Applying the expectation operator to the binomial expansion of $(1 - e^{-rT})^m$ results in

$$E[(1 - e^{-rT})^m] = E\left[\sum_{j=0}^m \binom{m}{j}(-1)^{m-j}e^{-r(m-j)T}\right]$$

$$= \sum_{j=0}^m \binom{m}{j}(-1)^{m-j}\,E[e^{-r(m-j)T}] = \sum_{j=0}^m \binom{m}{j}(-1)^{m-j}\mathbb{L}_T[-(m-j)r]$$

in which $\mathbb{L}_T[.]$ is the MGF of the random project life T. Putting together the equations, the resulting expression for the mth-order product-moment about the origin of the PVR is

$$E(\text{PVR}^m) = \frac{1}{r^m}\,E\!\left(R_o^m\right)\left[\sum_{j=0}^m \binom{m}{j}(-1)^{m-j}\mathbb{L}_T[-(m-j)r]\right]$$

Hence, the first three moments of the PVR about the origin can be obtained as follows:

$m = 1$

$$E(\text{PVR}) = \frac{1}{r} E(R_o) \left[\sum_{j=0}^{m=1} \binom{1}{j} (-1)^{1-j} \mathbb{L}_T[-(1-j)r] \right]$$

$$= \frac{E(R_o)}{r} [-\mathbb{L}_T(-r) + \mathbb{L}_T(0)]$$

$m = 2$

$$E(\text{PVR}^2) = \frac{1}{r^2} E\left(R_o^2\right) \left[\sum_{j=0}^{2} \binom{2}{j} (-1)^{2-j} \mathbb{L}_T[-(2-j)r] \right]$$

$$= \frac{E\left(R_o^2\right)}{r^2} [\mathbb{L}_T(-2r) - 2\mathbb{L}_T(-r) + \mathbb{L}_T(0)]$$

$m = 3$

$$E(\text{PVR}^3) = \frac{1}{r^3} E\left(R_o^3\right) \left[\sum_{j=0}^{3} \binom{3}{j} (-1)^{3-j} \mathbb{L}_T[-(3-j)r] \right]$$

$$= \frac{E\left(R_o^3\right)}{r^3} [-\mathbb{L}_T(-3r) + 3\mathbb{L}_T(-2r) - 3\mathbb{L}_T(-r) + \mathbb{L}_T(0)]$$

From the knowledge of the first three moments about the origin, the corresponding central moments can be obtained according to Eq. (2.21).

Example 4.8   Solve Example 4.7 numerically, considering that R_o and T are independent uniform random variables with bounds (r_{o1}, r_{o2}) and (t_1, t_2), respectively. Calculate the mean, standard deviation, and skewness coefficient of the PVR, based on the following data: $i = 5$ percent; $(r_{o1}, r_{o2}) = (\$2000, \$3000)$; $(t_1, t_2) = (45$ year, 55 year).

Solution   From Example 4.5, $E(R_o)$ and $E(R_o^2)$ can numerically be computed as

$$E(R_o) = \frac{r_{o1} + r_{o2}}{2} = 2500$$

$$E\left(R_o^2\right) = \frac{r_{o1}^2 + r_{o1}\, r_{o2} + r_{o2}^2}{3} = 6.3333 \times 10^6$$

The moment generating function of the random project life, according to Table 4.2, is

$$\mathbb{L}_T(s) = \frac{e^{t_2 s} - e^{t_1 s}}{(t_2 - t_1)s} = \frac{e^{55s} - e^{45s}}{10s}$$

The nominal interest rate $r = \ln(1 + i) = \ln(1.05) = 0.0488$. Substituting the value of the nominal interest rate r in $\mathbb{L}_T(-r)$, $\mathbb{L}_T(-2r)$, and $\mathbb{L}_T(-3r)$, one obtains $\mathbb{L}_T(0) = 1$, $L_T(-0.0488) = 0.088$, $\mathbb{L}_T(-0.0976) = 0.007902$, and $\mathbb{L}_T(-0.1464) = 0.0007229$. Hence, based on the formulas derived in Example 4.7, the first three moments of the PVR about the origin can be obtained as

$$E(\text{PVR}) = \frac{2500}{0.0488}[-0.088 + 1] = \$46720$$

$$E(\text{PVR}^2) = \frac{6.3333 \times 10^6}{0.0488^2}[0.007902 - 2(0.088) + 1] = 2.2123 \times 10^9$$

$$E(\text{PVR}^3) = \frac{1.6041 \times 10^{10}}{0.0488^3}[-0.0007229 + 3(0.007902) - 3(0.088) + 1] = 1.0475 \times 10^{14}$$

The variance of the PVR then can be computed as

$$\text{Var(PVR)} = E(\text{PVR}^2) - E^2(\text{PVR}) = 2.2123 \times 10^9 - 46720^2 = 2.9511 \times 10^7$$

and the corresponding standard deviation is $\sqrt{\text{Var (PVR)}} = \$5432.37$.

To compute the skewness coefficient of the PVR, the 3rd-order central moment $\mu_{3,\text{PVR}}$ can be computed, according to Eq. (2.21), as

$$\mu_{3,\text{PVR}} = E(\text{PVR}^3) - 3E(\text{PVR})E(\text{PVR}^2) + 2E^3(\text{PVR})$$

$$= 1.359 \times 10^{14}$$

Then, skewness coefficient of the PVR can be calculated, based on Eq. (2.39), as

$$\gamma_{\text{PVR}} = \frac{\mu_{3,\text{PVR}}}{[\text{Var(PVR)}]^{1.5}} = 8.48$$

Example 4.9 In hydrological rainfall-runoff modeling, the *geomorphological instantaneous unit hydrograph* (GIUH) is used to relate effective rainfall with the geomorphological features of a basin, such as overland flow and channel flow. It is defined as the PDF of the total travel time for a droplet in rainfall excess to travel across an overland surface of a certain order to a channel of the same order, followed by a sequence of channels of higher orders and eventually reach the outlet of the basin (Bras 1990). Hence, the total travel time (T) for a rain droplet along a particular flow path can be written as

$$T = T_{oi} + \sum_{j=i}^{N} T_{cj}$$

where T_{oi} = travel time in the ith-order overland surface
$\quad\quad T_{cj}$ = travel time in the jth-order channel
$\quad\quad N$ = the order of the basin (i.e., highest order of channel)

The travel time in different states (namely, overland surfaces and channel reaches) are affected by geomorphological features of the watershed (e.g., slope and roughness of overland surfaces and channels), which are random. Therefore, the component travel time and total travel time are random. Assuming independency of the involved component travel time with known PDFs, the PDF of the total travel time can be derived through an inverse Laplace transform of its MGF, which can be obtained by the convolution integral as

$$\mathbb{L}_T(s) = f_{oi}(t_{oi}) * f_{ci}(t_{ci}) * \cdots * f_{cN}(t_{cN})$$
$$= \mathbb{L}_{T_{oi}}(s) \times \mathbb{L}_{T_{ci}}(s) \times \cdots \times \mathbb{L}_{T_{cN}}(s)$$

where $*$ is a convolution operator. The GIUH of a watershed, $u(t)$, is obtained as

$$u(t) = \sum_{\omega=1}^{\Omega} f_{T_\omega}(t) \cdot p(\omega)$$

where $f_{T_\omega}(t)$ = PDF of total travel time for rain droplet following a particular flow path ω

$p(\omega)$ = probability that rain droplet would follow flow path ω

Ω = total number of possible flow paths

For illustration, consider a 2nd-order watershed with the channel network shown in Fig. 4.3. There are two possible flow paths, i.e.,

Path 1: $O_1 \rightarrow C_1 \rightarrow C_2 \rightarrow$ outlet

Path 2: $O_2 \rightarrow C_2 \rightarrow$ outlet

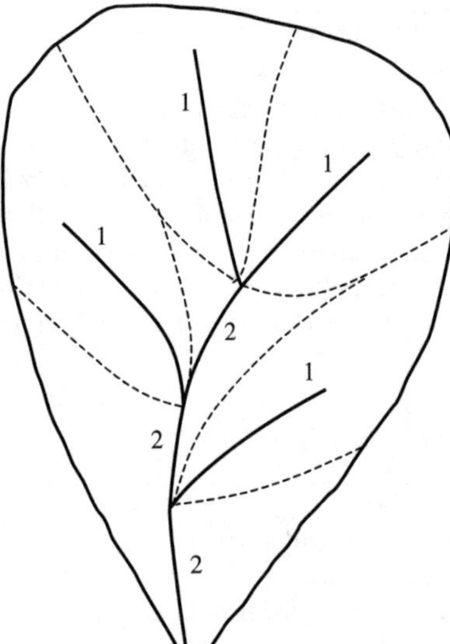

Figure 4.3 Example 2nd-order watershed.

The total travel time associated with each flow path, respectively, are

$$T_1 = T_{o1} + T_{c1} + T_{c2}$$
$$T_2 = T_{o2} + T_{c2}$$

Assuming the component travel time are independent and exponentially distributed random variables with respective means μ_j; then, the MGF corresponding to each component travel time is

$$\mathbb{L}_j(s) = \frac{1/\mu_j}{1/\mu_j - s} \qquad \text{for } j = o1, o2, c1, \text{ and } c2$$

,

The MGF for the total travel time associated with the two paths can be expressed as

$$\mathbb{L}_{T1}(s) = \left(\frac{1/\mu_{o1}}{1/\mu_{o1} - s} \right) \times \left(\frac{1/\mu_{c1}}{1/\mu_{c1} - s} \right) \times \left(\frac{1/\mu_{c2}}{1/\mu_{c2} - s} \right)$$

$$\mathbb{L}_{T2}(s) = \left(\frac{1/\mu_{o2}}{1/\mu_{o2} - s} \right) \times \left(\frac{1/\mu_{c2}}{1/\mu_{c2} - s} \right)$$

Performing an inverse Laplace transform on $\mathbb{L}_{T1}(s)$ and $\mathbb{L}_{T2}(s)$ yields the PDFs for T_1 and T_2, respectively, as

$$f_{T1}(t) = \frac{\mu_{o1}\, e^{-t/\mu_{o1}}}{(\mu_{o1} - \mu_{c1})(\mu_{o1} - \mu_{c2})} + \frac{\mu_{c1}\, e^{-t/\mu_{c1}}}{(\mu_{c1} - \mu_{o1})(\mu_{c1} - \mu_{c2})} + \frac{\mu_{c2}\, e^{-t/\mu_{c2}}}{(\mu_{c2} - \mu_{o1})(\mu_{c2} - \mu_{c1})}$$

$$f_{T2}(t) = \frac{e^{-t/\mu_{o2}} - e^{-t/\mu_{c2}}}{(\mu_{o2} - \mu_{c2})}$$

Suppose that the areas of the 1st- and 2nd-order overland surfaces are 12.57 and 4.93 km^2, respectively, the probabilities associated with path 1 and path 2 can be estimated as $p_1 = 12.57/(12.57 + 4.93) = 0.72$ and $p_2 = 0.28$, respectively. Then, the resulting GIUH for the watershed can be defined as

$$u(t) = 0.72 f_{T1}(t) + 0.28 f_{T2}(t)$$

With $\mu_{o1} = 8035s$, $\mu_{o2} = 9233s$, $\mu_{c1} = 257s$, $\mu_{c2} = 491s$, the resulting GIUH is shown in Fig. 4.4. For more detailed derivations of the GIUH under different distributional assumptions of component travel time, readers can refer to Cheng (1982).

4.3   Mellin Transform

When the functional relation $W = g(X)$ satisfies two conditions, the exact moments for W of any order can be derived analytically as functions of the statistical moments of several stochastic variables Xs by the Mellin transform without

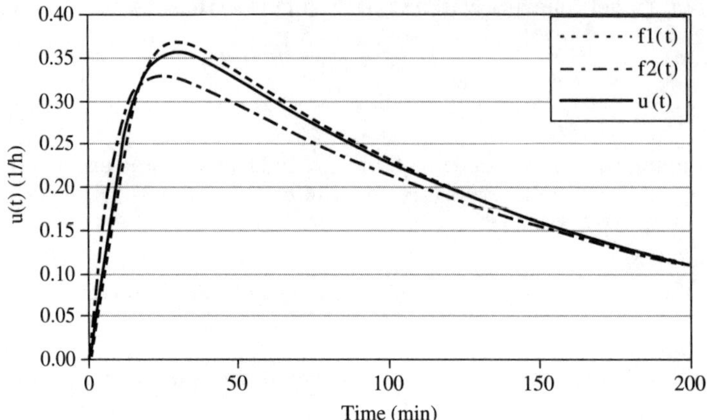

Figure 4.4   The geomorphological instantaneous unit hydrograph (GIUH) and contributing instantaneous unit hydrographs (IUHs) for the example watershed.

extensive simulation or using any approximation methods. The two conditions are:

1. The function $g(\boldsymbol{X})$ has a multiplicative form as

$$W = g(\boldsymbol{X}) = a_o \prod_{k=1}^{K} X_k^{a_k} \tag{4.19}$$

where a_k are constants.

2. The stochastic basic variables, Xs, are independent and nonnegative.

The Mellin transform is particularly attractive in the uncertainty analysis of hydrosystems engineering problems because many models used and the involved stochastic basic variables satisfy the above two conditions (Tung 1989; 1990). In general, the nonnegativity condition on the Xs is not strictly required by the Mellin transform, but it would require some mathematical manipulations to find the Mellin transform of a function involving random variables that can take negative values (Epstein 1948; Springer 1978).

4.3.1   Statistical moments and the Mellin transform

Mellin transform of a function $f_x(x)$, where x is positive, is defined as (Giffin 1975; Springer 1978)

$$M_x(s) = M[f_x(x)] = \int_0^\infty x^{s-1} f_x(x)\,dx \qquad \text{for } x > 0 \tag{4.20}$$

where $M_x(s)$ is the Mellin transform of the function $f_x(x)$. Like Fourier and Laplace transforms, a one-to-one correspondence between $M_x(s)$ and $f_x(x)$ exists.

When $f_x(x)$ is a PDF, one can immediately recognize that the relationship between the Mellin transform of a PDF and the statistical moments about the origin is

$$\mu'_{s-1} = E(X^{s-1}) = M_x(s) \tag{4.21}$$

for $s = 1, 2, \ldots$ As can be seen, the Mellin transform provides an alternative way to find the moments of any order for nonnegative random variables.

Example 4.10　Consider that Manning's roughness coefficient is a random variable with a uniform PDF

$$f_n(n) = \frac{1}{(n_b - n_a)} \qquad \text{for } n_a \leq n \leq n_b$$

in which n_a and n_b are the lower and upper bounds of the roughness coefficient, respectively. Determine the Mellin transform of the above uniform PDF and determine the mean and variance of the roughness coefficient.

Solution　According to Eq. (4.20), the Mellin transform of a uniform PDF is

$$M_n(s) = M[f_n(n)] = \int_{n_a}^{n_b} n^{s-1} f_n(n)\, dn$$

$$= \frac{1}{n_b - n_a} \int_{n_a}^{n_b} n^{s-1} dn = \frac{n_b^s - n_a^s}{s(n_b - n_a)}$$

The mean of the roughness coefficient can be obtained, according to Eq. (4.21), by setting $s = 2$, as

$$\mu_n = E(n) = M_n(s = 2) = \frac{n_b^2 - n_a^2}{2(n_b - n_a)} = \frac{n_b + n_a}{2}$$

The variance of the roughness coefficient can be obtained by first computing $E(n^2)$ as

$$E(n^2) = M_n(s = 3) = \frac{n_b^3 - n_a^3}{3(n_b - n_a)} = \frac{n_b^2 + n_b n_a + n_a^3}{3}$$

Then, $\text{Var}(n)$ can be obtained, according to Eq. (2.3.19), as

$$\sigma_n^2 = \text{Var}(n) = E(n^2) - \mu_n^2$$

$$= \left(\frac{n_b^2 + n_b n_a + n_a^2}{3} \right) - \left(\frac{n_b + n_a}{2} \right)^2 = \frac{(n_b - n_a)^2}{12}$$

The results are identical to those of Example 4.5 using the characteristic function approach.

4.3.2 Operational properties of the Mellin transform

Consider that a random variable W is the product of two independent non-negative random variables, that is, $W = XY$. The PDF of W, $h_w(w)$, can be obtained as

$$h_w(w) = \int_0^\infty \frac{1}{y} f_x\left(\frac{w}{y}\right) g_y(y)\, dy \tag{4.22}$$

where $f_x(\cdot)$ and $g_y(\cdot)$ are the PDFs of X and Y, respectively. In fact, Eq. (4.22) is actually the definition of the *Mellin convolution* (Springer 1978). Therefore, similar to the convolution property of the Laplace and Fourier transforms, the Mellin transform of $h_w(w)$ can be obtained as

$$M_w(s) = M[h_w(w)] = M[f_x(x) * g_y(y)] = M_x(s) \times M_y(s) \tag{4.23}$$

in which * is the convolution operator. From Eq. (4.23), the Mellin transform of the convolution of the PDFs associated with two independent random variables in a product form is simply equal to the product of the Mellin transforms of two individual PDFs. Equation (4.23) can be extended to the general case involving more than two independent random variables.

From this convolution property of the Mellin transform and its relationship with statistical moments, one can immediately see the advantage of the Mellin transform as a tool for obtaining the moments of a random variable that is related to the other independent random variables in a multiplicative fashion. In addition to the convolution property, which is of primary importance, the Mellin transform has several useful operational properties that are summarized in Table 4.4 (Bateman 1954; Park 1987). These properties of the Mellin transform can be derived from the basic definition given in Eq.(4.20).

Applying the definition of the Mellin transform and its basic operational properties, along with the convolution properties, the Mellin transform of random variables in the form of products and quotients can be derived. Some useful results are summarized in Table 4.5.

TABLE 4.4 Operational Properties of the Mellin Transform on a PDF (after Park 1987)

Property	PDF	Random variable	Mellin transform
Standard	$f_x(x)$	X	$M_x(s)$
Scaling	$f_x(ax)$	X	$a^{-s} M_x(s)$
Linear	$a f_x(x)$	X	$a M_x(s)$
Translation	$x^a f_x(x)$	X	$M_x(a + s)$
Exponentiation	$f_x(x^a)$	X	$a^{-1} M_x(s/a)$

**TABLE 4.5   The Mellin Transform of Products and Quotients
of Random Variables (after Park 1987)**

Random variable	PDF given	$M_w(s)$
$W = X$	$f_x(x)$	$M_x(s)$
$W = X^b$	$f_x(x)$	$M_x(bs - b + 1)$
$W = 1/X$	$f_x(x)$	$M_x(2 - s)$
$W = XY$	$f_x(x), g_y(y)$	$M_x(s)\, M_y(s)$
$W = X/Y$	$f_x(x), g_y(y)$	$M_x(s)\, M_y(-2s)$
$W = aX^b Y^c$	$f_x(x), g_y(y)$	$a^{s-1} M_x(bs - b + 1)\, M_y(cs - c + 1)$

NOTE: a, b, c constants; X, Y, W: random variables.

Example 4.11   Manning's formula is frequently used for determining the flow capacity
of a storm sewer by

$$Q = 0.463\, n^{-1}\, D^{2.67}\, S^{0.5}$$

in which Q is the flow rate (in ft^3/s); n is the roughness coefficient; D is the sewer
diameter (in feet); and S is the pipe slope (in ft/ft). Assume that the three model
parameters are independent random variables. Derive the expression of the Mellin
transform for the sewer flow capacity.

Solution   By definition, the Mellin transform of the sewer flow capacity is

$$M_Q(s) = E[Q^{s-1}] = E[(0.463 n^{-1} D^{2.67}\, S^{0.5})^{s-1}]$$
$$= 0.463^{s-1}\, E[n^{-s+1}\, D^{2.67s-2.67}\, S^{0.5s-0.5}]$$

Because of the independence of stochastic model parameters, the expectation of the
above expression can be decomposed as

$$M_Q(s) = 0.463^{s-1} E[n^{-s+1}] E[D^{2.67s-2.67}] E[S^{0.5s-0.5}]$$
$$= 0.463^{s-1}\, E[n^{(-s+1+1)-1}] E[D^{(2.67s-2.67+1)-1}] E[S^{(0.5s-0.5+1)-1}]$$

By the definition of the Mellin transform, each expectation term in the above expression
can be written as

$$E[n^{(-s+1+1)-1}] = M_n(-s + 1 + 1) = M_n(-s + 2)$$
$$E[D^{(2.67s-2.67+1)-1}] = M_D(2.67s - 2.67 + 1) = M_D(2.67s - 1.67)$$
$$E[S^{(0.5s-0.5+1)-1}] = M_S(0.5s - 0.5 + 1) = M_S(0.5s + 0.5)$$

And the resulting $M_Q(s)$ can be written as

$$M_Q(s) = 0.463^{s-1} M_n(-s + 2)\, M_D(2.67s - 1.67)\, M_S(0.5s + 0.5)$$

4.3.3   Mellin transform of some probability density functions

In uncertainty analysis, model parameters with uncertainty are treated as
random variables associated with a PDF. Given the functional relationship as
Eq. (4.19), the statistical moments of W can be obtained by the Mellin transform

TABLE 4.6   Mellin Transforms for Some Commonly Used Probability Density Functions

Probability	Density function, $f_x(x)$	Mellin transform, $M_x(s)$
Uniform	Eq. (2.92)	$\dfrac{b^s - a^s}{s(b-a)}$
Standard normal	Eq. (2.51)	$\dfrac{2^{(s-1)/2}\,\Gamma\!\left(\frac{s}{2}\right)}{\sqrt{\pi}}$ for $s=$ odd; $\quad 0$ for $s=$ even
Lognormal	Eq. (2.57)	$\exp\!\left[(s-1)\mu_{\ln x} + \frac{1}{2}(s-1)^2\sigma_{\ln x}^2\right]$
Exponential	Eq. (2.71)	$\beta^{1-s}\,\Gamma(s)$
Gamma	Eq. (2.64)	$\dfrac{\beta^{1-s}\,\Gamma(\alpha+s-1)}{\Gamma(\alpha)}$
Triangular	$f_x(x)=\begin{cases}\dfrac{2(x-a)}{(b-a)(m-a)} & a \le x \le m \\[2mm] \dfrac{2(b-x)}{(b-a)(b-m)} & m \le x \le b\end{cases}$	$\dfrac{2}{s(s+1)(b-a)}\left[\dfrac{b(b^s - m^s)}{b-m} - \dfrac{a(m^s - a^s)}{m-a}\right]$
Weibull	Eq. (2.81)	$\displaystyle\sum_{j=0}^{s-1}\binom{s-1}{j}\beta^j\,\xi^{s-1-j}\,\Gamma\!\left(\frac{j}{\kappa}+1\right)$
Standard beta	Eq. (2.90)	$\dfrac{\Gamma(\alpha+\beta)\,\Gamma(\alpha+s-1)}{\Gamma(\alpha)\,\Gamma(\alpha+\beta+s-1)}$
Nonstandard beta	Eq. (2.88)	$\displaystyle\sum_{j=0}^{s-1}\binom{s-1}{j}a^{s-1-j}\,(b-a)^j M_x(j)$ where $M_x(j)$ for standard beta

of the PDFs of random variables. From the previous studies (Epstein 1948; Park 1987), the Mellin transforms of some commonly used PDFs are tabulated in Table 4.6 that can be easily obtained from the expression for μ'_r shown in the last column of Table 4.2. For models involving independent random variables related in multiplicative, linear, and combination of the two, Tyagi and Haan (2001) reexpress $E(X^r)$ for some commonly used distribution functions in terms of the mean and coefficient of variation, rather than the distribution parameters. Using the results in Tables 4.5 and 4.6, one can derive the exact moments of the random model output W.

Although the Mellin transform is useful for uncertainty analysis under the conditions stated previously, it possesses one drawback: namely, under some combinations of the distribution and functional form, the resulting transform may not be defined for all values of s. This could occur, especially, when quotients or variables with negative exponents are involved. For example, if the random variable W is related to the inverse of X, that is, $W = 1/X$, and X has a uniform distribution in $(0,1)$, then $M_w(s) = M_x(2-s) = 1/(2-s)$. In this case, the expected value of W, $E(W)$, which can be calculated, by $M_w(s=2)$, does not exist because $M_w(s=2) = 1/0$, which is not defined. Under such circumstances, other transforms, such as the Laplace or Fourier transform, could be used.

Example 4.12   Referring to Example 4.11, derive the expression for the Mellin transform for the sewer flow capacity assuming the following distributional properties for the three stochastic model parameters.

Parameter	Distribution
n	Uniform distribution with lower bound n_a and upper bound n_b
D	Triangular distribution with lower bound d_a, mode d_m, and upper bound d_b
S	Uniform distribution with bounds (S_a, S_b)

Solution   From Example 4.11, the general expression of the Mellin transform of the sewer flow capacity is

$$M_Q(s) = 0.463^{s-1} M_n(-s+2)\, M_D(2.67s - 1.67)\, M_S(0.5s + 0.5)$$

For the Manning roughness coefficient n having a uniform distribution, from Table 4.6, one obtains

$$M_n(-s+2) = \frac{n_b^{-s+2} - n_a^{-s+2}}{(-s+2)\,(n_b - n_a)}$$

For sewer diameter D with a triangular distribution, one obtains

$$M_D(2.67s - 1.67) = \frac{2}{(d_b - d_a)\,(2.67s - 1.67)\,(2.67s - 0.67)}$$
$$\times \left[\frac{d_b\left(d_b^{2.67s-1.67} - d_m^{2.67s-1.67}\right)}{d_b - d_m} - \frac{d_a\left(d_m^{2.67s-1.67} - d_a^{2.67s-1.67}\right)}{d_m - d_a} \right]$$

and for sewer slope S with a uniform distribution,

$$M_s(0.5s + 0.5) = \frac{(S_b)^{0.5s+0.5} - (S_a)^{0.5s+0.5}}{(0.5s + 0.5)(S_b - S_a)}$$

Substituting individual terms in $M_Q(s)$ results in the expression of the Mellin transform of sewer flow capacity specifically for the distributions associated with the three stochastic model parameters.

Example 4.13   Referring to Example 4.12, numerically solve for the mean and variance of the storm sewer capacity using the Mellin transform. It is known that the roughness coefficient has a uniform distribution with a lower bound and upper bound of 0.0137 and 0.0163, respectively; sewer diameter has a triangular distribution with lower bound, mode, and upper bound being 2.853, 3.00, and 3.147 ft, respectively; and sewer slope has a uniform distribution with lower and upper bounds being 0.00457 and 0.00543, respectively.

Solution   To compute the mean and variance of sewer flow capacity, Q, the 1st- and 2nd-order moments about the origin for Q are computed. Based on the information given, the Mellin transforms of each stochastic model parameter can be expressed as

$$M_n(s) = \frac{0.0163^s - 0.0137^s}{0.0026s}$$

$$M_D(s) = \frac{2}{(0.294)\, s \,(s+1)} \left[\frac{3.147\,(3.147^s - 3.00^s)}{0.147} - \frac{2.853\,(3.00^s - 2.853^s)}{0.147} \right]$$

$$M_s(s) = \frac{0.00543^s - 0.00457^s}{0.00086s}$$

The computations are shown in the following table:

	$s = 2$	$s = 3$
0.463^{s-1}	0.463	0.2144
$M_n(-s+2)$	$M_n(0) = 66.834$	$M_n(-1) = 4478.080$
$M_D(2.67s - 1.67)$	$M_D(3.67) = 18.806$	$M_D(6.34) = 354.681$
$M_n(0.5s + 0.5)$	$M_S(1.50) = 0.0707$	$M_S(2.00) = 0.005$

Therefore, the mean sewer flow capacity can be determined as

$$E(Q) = M_Q(s = 2) = 0.463\, M_n(0)\, M_D(3.67)\, M_S(1.50)$$
$$= 0.463(66.834)(18.806)(0.0707) = 41.14 \text{ ft}^3/\text{s}$$

The 2nd-order product-moment about the origin of sewer flow capacity is

$$E(Q^2) = M_Q(s = 3) = 0.463^2\, M_n(-1)\, M_D(6.34)\, M_S(2.00)$$
$$= 0.463^2\,(4478.08)\,(354.681)\,(0.005) = 1702.40 \text{ (ft}^3/\text{s)}^2$$

Then, the variance of sewer flow capacity can be determined as

$$\text{Var}(Q) = E(Q^2) - E^2(Q) = 1702.40 - 41.137^2 = 10.15 \text{ (ft}^3/\text{s)}^2$$

The corresponding standard deviation of sewer flow capacity is

$$\sigma_Q = \sqrt{10.147} = 3.19 \text{ ft}^3/\text{s}$$

4.3.4   Sensitivity of component uncertainty on overall uncertainty

In engineering designs, sensitivity analysis is commonly used when the designs are performed under uncertainty. In uncertainty analysis, investigating the impact of component uncertainty on the overall output uncertainty provides important information regarding the relative contribution of component uncertainty to the overall uncertainty of model output. In the framework of the Mellin transform, the sensitivity analysis can be performed as the following (Tung 1990).

Refer to the multiplicative model involving independent, nonnegative random variables, Eq. (4.19). The first two moments about the origin of the model output W, using Table 4.5, can be obtained, as

$$E(W) = M_w(2) = a_0 \prod_{k=1}^{K} M_k(1 + a_k) \tag{4.24}$$

$$E(W^2) = M_w(3) = a_0^2 \prod_{k=1}^{K} M_k(1 + 2a_k) \tag{4.25}$$

where $M_k(1 + a_k)$ and $M_k(1 + 2a_k)$ are the first two product-moments about the origin for the kth-term, $W_k = X_k^{a_k}$, in Eq. (4.19). The variance of the model output W can be expressed as

$$\mathrm{Var}(W) = a_0^2 \left[\prod_{k=1}^{K} M_k(1 + 2a_k) - \prod_{k=1}^{K} M_k^2(1 + a_k) \right] \tag{4.26}$$

and the corresponding coefficient of variation (Ω_w) as

$$\Omega_w^2 = \prod_{k=1}^{K} \left[\Omega_{w_k}^2 + 1 \right] - 1 \tag{4.27}$$

where Ω_{w_k} is the coefficient of variation of $W_k = X_k^{a_k}$.

To examine the impact of component uncertainty on the overall uncertainty of the model output W, it is necessary to express the coefficient of variation of W in terms of the coefficients of variation of stochastic basic variables, Xs. Since $W_k = X_k^{a_k}$, the relationship between the coefficients of variation of W_k and X_k can be similarly derived as

$$\Omega_{w_k}^2 = \eta_k^2 \, \Omega_{x_k}^2 \tag{4.28}$$

where

$$\eta_k^2 = \left(\frac{M_{x_k}^2(2)}{M_{x_k}^2(1 + a_k)} \right) \left(\frac{M_{x_k}(1 + 2a_k) - M_{x_k}^2(1 + a_k)}{M_{x_k}(3) - M_{x_k}^2(2)} \right) \tag{4.29}$$

and Ω_{x_k} is the coefficient of variation of the stochastic variable X_k which is computed as

$$\Omega_{x_k} = \frac{\sqrt{M_{x_k}(3) - M_{x_k}^2(2)}}{M_{x_k}(2)} \tag{4.30}$$

Substituting Eq. (4.28) in Eq.(4.27), one obtains the following relationship

$$\Omega_w^2 = \prod_{k=1}^{K}\left[\eta_k^2\Omega_{x_k}^2 + 1\right] - 1 \tag{4.31}$$

The sensitivity of the model output uncertainty with respect to the uncertainty of the kth stochastic basic variable, X_k, can be obtained as

$$\frac{\partial\Omega_w}{\partial\Omega_{x_k}} = \frac{\eta_k^2\,\Omega_{x_k}\left[\Omega_w^2 + 1\right]}{\Omega_w\left[\eta_k^2\,\Omega_{x_k}^2 + 1\right]} \tag{4.32}$$

The sensitivity coefficients computed by Eq. (4.32) represent the rate of change in model output uncertainty resulting from a unit change in the uncertainty of the kth stochastic basic variable. Such information could be used as an important guide for future data collection program design in an attempt to reduce the total model output uncertainty.

Referring to Eq. (4.31), it is seen that, under Eq. (4.19), the relationship between the model output uncertainty and those of the stochastic basic parameters is essentially multiplicative. Therefore, isolation of the exact impact of individual component uncertainty is difficult. Under the condition that $\prod_{j=1}^{K\geq 2}\Omega_j^2 \approx 0$, Eq. (4.31) reduces to

$$\Omega_w^2 \approx \sum_{k=1}^{K}\eta_k^2\,\Omega_{x_k}^2 \tag{4.33}$$

From Eq. (4.33), the percentage of contribution of each individual stochastic model variable to the overall output uncertainty can be estimated.

As a further approximation, the 1st-order variance estimation method (see Sec. 5.1.3) leads to

$$\Omega_w^2 \approx \sum_{k=1}^{K}a_k^2\,\Omega_{x_k}^2 \tag{4.34}$$

From Eqs. (4.33) and (4.34), the approximated sensitivity coefficients, with respect to the individual component uncertainty, based on Eq. (4.33), can be derived as

$$\frac{\partial\Omega_w}{\partial\Omega_{x_k}} \approx \frac{\eta_k^2\,\Omega_{x_k}}{\Omega_w} \approx \frac{a_k^2\Omega_{x_k}}{\Omega_w} \tag{4.35}$$

It should be emphasized that Eqs. (4.33) and (4.34) are approximations of the true relationships given in Eq. (4.31) while Eq. (4.35) is an approximation of Eq. (4.32). In fact, Eq. (4.34) is the result from the first-order approximation of Eq. (4.19) (Prob. 5.10).

Example 4.14 Referring to Example 4.13, determine the sensitivity of total uncertainty of the sewer capacity with respect to the uncertainty of the individual stochastic basic variable.

Solution Based on Example 4.13, the coefficient of variation of the sewer flow capacity is $\Omega_Q = 3.186/41.14 = 0.0774$. By Eq. (4.32), the sensitivity of Ω_Q with respect to the coefficient of variation of Manning's roughness, Ω_n, can be computed as

$$\frac{\partial \Omega_Q}{\partial \Omega_n} = \frac{\eta_n^2 \Omega_n [\Omega_Q^2 + 1]}{\Omega_Q [\eta_n^2 \Omega_n^2 + 1]}$$

In the above equation, η_n^2 is computed, according to Eq. (4.29), with $a_n = -1$ as

$$\eta_n^2 = \left(\frac{M_n^2(2)}{M_n^2(0)} \right) \left(\frac{M_n(-1) - M_n^2(0)}{M_n(3) - M_n^2(2)} \right)$$

$$= \left(\frac{0.015^2}{66.834^2} \right) \left[\frac{4478.08 - 6.6834^2}{0.000225563 - 0.015^2} \right]$$

$$= 1.006055$$

and Ω_n, according to the result in Example 4.5 for uniform random variable or by Eq. (4.30), is

$$\Omega_n = \left(\frac{0.0163 - 0.0137}{2\sqrt{3}} \right) \Big/ \left(\frac{0.0163 + 0.0137}{2} \right) = 0.050037$$

Hence, the sensitivity of Ω_Q with respect to Ω_n is

$$\frac{\partial \Omega_Q}{\partial \Omega_n} = \frac{(1.006055)(0.050037)\,(0.0774^2 + 1)}{(0.0774)[(1.006055)(0.050037)^2 + 1]} = 0.65224$$

Similar computations can be performed for the stochastic basic variables D and S, with the results summarized in the following table.

Stochastic parameter	Ω_k Eq. (4.30)	η_k Eq. (4.29)	$\partial \Omega_Q / \partial \Omega_k$ Eq. (4.32)	$\partial \Omega_Q / \partial \Omega_k$ Eq. (4.35)
n	0.050037	1.003023	0.65224	0.64607
D	0.020004	2.668640	1.84522	1.84132
S	0.049652	0.500460	0.16143	0.16027

From the above table, the uncertainty of the sewer flow capacity is most sensitive to that of the sewer diameter, followed by Manning's roughness coefficient. Comparing the last two columns, the approximated sensitivity coefficients computed by Eq. (4.35) are very close to that of the exact values provided by Eq. (4.32).

4.4   Estimations of Probability and Quantile Using Moments

Although it is generally difficult to analytically derive a PDF from the results of the integral transform techniques described in the previous sections of this chapter; it is, however, rather straightforward to obtain or estimate the statistical moments of the random variable one is interested in. Based on the computed statistical moments, one is able to estimate the distribution and quantile of the random variable. This section describes two such approaches: one is based on the asymptotic expansion about the normal distribution for calculating the values of CDF and quantile and the other is based on the maximum entropy concept.

4.4.1   Edgeworth asymptotic expansion of PDF and CDF

In terms of the statistical moments and standard normal distribution, the general *Edgeworth asymptotic expansion* for the PDF and CDF of any standardized random variable, $X' = (X - \mu_x)/\sigma_x$, can be found in Abramowitz and Stegun (1972) and Kendall, Stuart, and Ord (1987). For practical applications, considering that the first four product-moments are available or estimated, the Edgeworth asymptotic expansion of the PDF of standardized variable, $f_{x'}(x')$, can be approximated as

$$f_{x'}(\xi) = \phi(\xi) - \left[\left(\frac{\gamma_x}{6}\right)\phi^{(3)}(\xi)\right] + \left[\left(\frac{\kappa_x - 3}{24}\right)\phi^{(4)}(\xi) + \left(\frac{\gamma_x^2}{72}\right)\phi^{(6)}(\xi)\right]$$

$$= \phi(\xi) - [c_3\phi^{(3)}(\xi)] + \left[c_4\phi^{(4)}(\xi) + c_6\phi^{(6)}(\xi)\right] \qquad (4.36)$$

in which $\phi(\xi)$ is the standard normal PDF; γ_x and κ_x are, respectively, the skewness coefficient and kurtosis of the random variable X; $\phi^{(r)}(\xi)$ is the rth-order derivative of the standard normal PDF; $c_3 = \gamma_x/6$; $c_4 = (\kappa_x - 3)/24$; and $c_6 = \gamma_x^2/72$. Hence, the CDF, $F_{x'}(\xi)$, can be obtained from integrating Eq. (4.36) as

$$F_{x'}(\xi) \approx \Phi(\xi) - c_3\phi^{(2)}(\xi) + c_4\phi^{(4)}(\xi) + c_6\phi^{(5)}(\xi) \qquad (4.37)$$

in which $\Phi(\xi)$ is the standard normal CDF.

The rth-order derivative of the standard normal PDF can be computed as

$$\phi^{(r)}(\xi) = \frac{d^r[\phi(\xi)]}{d\xi^r} = (-1)^r H_r(\xi)\,\phi(\xi) \qquad (4.38)$$

in which $H_r(\xi)$ is the rth-order *Hermite polynomial*, which can be computed by (Abramowitz and Stegun 1972)

$$H_r(\xi) = \xi^r - \frac{r^2}{2 \times 1!}\,\xi^{r-2} + \frac{r^4}{2^2 \times 2!}\xi^{r-4} - \frac{r^6}{2^2 \times 3!}\,\xi^{r-6} + \cdots \qquad (4.39)$$

or more specifically as

$$H_1(\xi) = \xi$$

$$H_2(\xi) = \xi^2 - 1$$

$$H_3(\xi) = \xi^3 - 3\xi$$

$$H_4(\xi) = \xi^4 - 6\xi^2 + 3$$

$$H_5(\xi) = \xi^5 - 10\xi^3 + 15\xi$$

$$H_6(\xi) = \xi^6 - 15\xi^4 + 45\xi^2 - 15$$

$$H_7(\xi) = \xi^7 - 21\xi^5 + 105\xi^3 - 105\xi$$

In terms of Hermite polynomials, Eqs. (4.36) and (4.37) can be expressed as

$$f_{x'}(\xi) \approx \phi(\xi)\,[1 + c_3 H_3(\xi) + c_4 H_4(\xi) + c_6 H_6(\xi)] \tag{4.40}$$

$$F_{x'}(\xi) \approx \Phi(\xi) - \phi(\xi)\,[c_3 H_2(\xi) + c_4 H_3(\xi) + c_6 H_5(\xi)] \tag{4.41}$$

It should be pointed out that, using finite terms in the Edgeworth series expansion, there is a possibility to produce negative values for the PDF and CDF toward the tail portions of a distribution. Figure 4.5 shows the effect of the skewness coefficient and kurtosis on the PDF of a standardized random variable using the three-term Edgeworth expansion.

Example 4.13   Referring to Example 4.12, use the first two terms of the Edgeworth expansion to delineate the PDF of the sewer flow capacity.

Solution   The first two product-moments, i.e., the mean and standard deviation, of the sewer flow capacity have been computed in Example 4.13 and they are $\mu_Q = 41.14$ ft^3/s and $\sigma_Q = 3.19$ ft^3/s. To use the first two terms in the Edgeworth expansion, the skewness coefficient of the sewer flow capacity must be computed. According to the Mellin transform, as illustrated in Example 4.13, the 3rd-order moment about the origin for the sewer flow capacity can be calculated, using $s = 4$, as

$$E(Q^3) = (0.463)^3\,M_n(-2)\,M_D(9.01)\,M_S(2.5)$$

$$= (0.463)^3\,(300797.98)\,(6708.18)\,(0.0003539)$$

$$= 70872.73\ (\text{ft}^3/\text{s})^3$$

Then, the 3rd-order central moment of the sewer flow capacity, according to Eq. (2.21) or Prob. 2.5, can be computed as

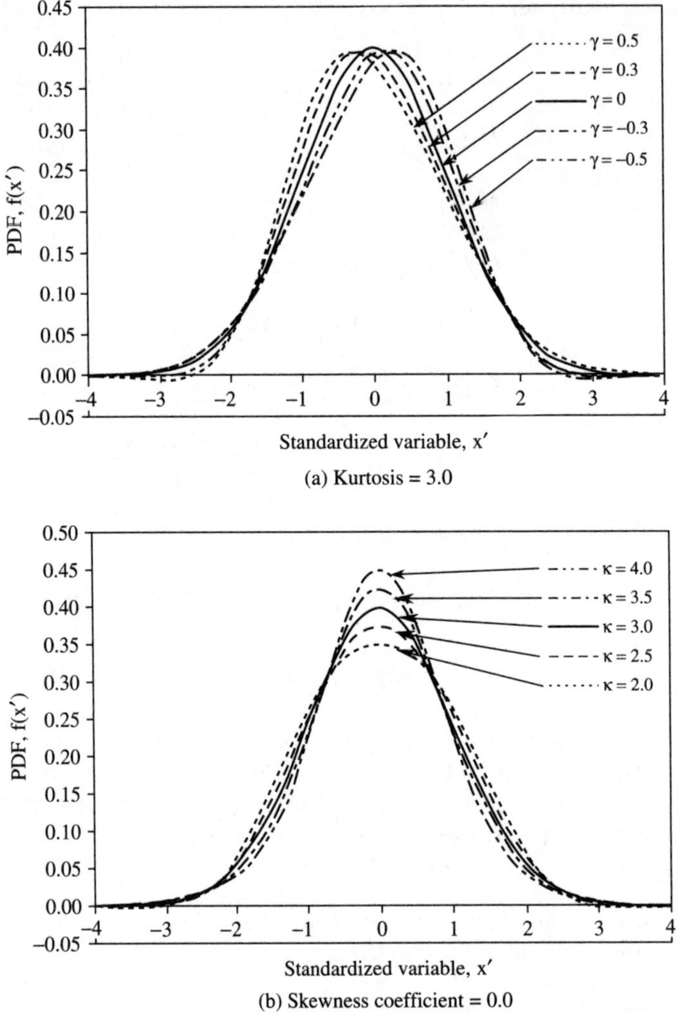

(a) Kurtosis = 3.0

(b) Skewness coefficient = 0.0

Figure 4.5 Effects of skewness coefficient and kurtosis on the three-term Edgeworth expansion.

$$\mu_3(Q) = E(Q^3) - 3\mu_Q\, E(Q^2) + 2\mu_Q{}^3$$

$$= 70872.73 - 3(41.14)(1702.4) + 2(41.14)^3$$

$$= 6.174\ (\text{ft}^3/\text{s})^3$$

The skewness coefficient of the sewer flow capacity, according to Eq. (2.39), is

$$\gamma_Q = \frac{\mu_3(Q)}{\sigma_Q^3} = \frac{6.174}{(3.186)^3} = 0.191$$

Using Eq. (4.40), the PDF of the standardized sewer capacity, $Q' = (Q - \mu_Q)/\sigma_Q$, can be expressed as

$$f_{Q'}(q') \approx \phi(q')\,[1 + 0.0318\,H_3(q')] = \phi(q')\,[1 - 0.0955\,q' + 0.0318\,q'^3]$$

The ordinates of the PDF for sewer flow capacity are calculated and shown in column (4) of the following table.

q' (1)	q (ft^3/s) (2)	$\phi(q')$ (3)	$f_{Q'}(q')$ (4)
−4.0	28.384	0.000133	−0.000088
−3.0	31.573	0.004432	0.001892
−2.0	34.762	0.053991	0.050554
−1.5	36.356	0.129518	0.134156
−1.0	37.951	0.241971	0.257376
−0.5	39.545	0.352065	0.367476
0.0	41.140	0.398942	0.398942
0.5	42.734	0.352065	0.336655
1.0	44.329	0.241971	0.226565
1.5	45.923	0.129518	0.124879
2.0	47.518	0.053991	0.057428
3.0	50.707	0.004432	0.006971
4.0	53.896	0.000134	0.000355

Note that when $f_{Q'}(q') = -4.0$ or $q = 28.38$ ft^3/s, the two-term Edgeworth expansion results in a negative value for the PDF.

4.4.2 Fisher-Cornish asymptotic expansion of quantile

Inversely, to estimate the quantile x'_p in which $P(X' \leq x'_p) = p$, the *Fisher-Cornish asymptotic expansion* (Fisher and Cornish 1960; Kendall, Stuart, and Ord 1987), considering the first four moments, can be expressed as

$$x'_p = z_p + c_2 H_2(z_p) + c_4 H_3(z_p) - c_6[2H_3(z_p) + H_1(z_p)]$$
$$= -c_3 + (1 - 3c_4 + 10c_6)z_p + c_3\,z_p^2 + (c_4 - 2c_6)z_p^3 \tag{4.42}$$

in which $z_p = \Phi^{-1}(p)$ and $H_r(z_p)$ are hermite polynomials defined in Eq. (4.39). The quantile of the original scale can be easily computed as $x_p = \mu_x + x'_p\,\sigma_x$. For more complete expansion series, which would require higher-order moments, readers are referred to Kendall, Stuart, and Ord (1987). As can be seen from Eqs. (4.36) and (4.42), if only the first two moments are available, the two asymptotic expansions reduce to the case of the normal distribution.

Winterstein (1986) used an improved Fisher-Cornish expression without considering the last expansion terms in Eq. (4.42), i.e., without c_6

$$x'_p = -\tilde{k}\tilde{c}_3 + \tilde{k}(1 - 3\tilde{c}_4)z_p + \tilde{k}\tilde{c}_3\,z_p^2 + \tilde{k}\tilde{c}_4\,z_p^3 \tag{4.43}$$

where

$$\tilde{c}_3' = \gamma_x \Big/ \left(4 + 2\sqrt{1 + 1.5(\kappa_x - 3)}\right) \tag{4.44a}$$

$$\tilde{c}_4' = \left(\sqrt{1 + 1.5(\kappa_x - 3)} - 1\right) \Big/ 18 \tag{4.44b}$$

$$\tilde{k} = 1 \Big/ \sqrt{1 + 2\tilde{c}_3^2 + 6\tilde{c}_4^2} \tag{4.44c}$$

Note that the improved Fisher-Cornish expansion requires that $\kappa_x > 7/3$ to ensure positiveness of the square root terms in Eqs. (4.44a) and (4.44b).

4.4.3 Maximum entropy distribution

The use of the *entropy* concept for measuring the amount of uncertainty in a statistical experiment was originated by Shannon (1948). It is based on Boltzmann's entropy from statistical physics, which has been used as an indicator of disorder in a physical system. Shannon's entropy has been used in a wide variety of areas including information and communication, economics, physics, ecology, reliability, and so on. The entropy concept has been applied to model velocity distribution in open channel (Chiu 1987; 1988; 1989; Chiu and Said 1995) and in pipe flow (Chiu, Lin, and Lu 1993), hydrology (Armorocho and Espildora 1973; Singh, Rajagopal, and Singh 1986; Singh and Krstanovic 1987; Singh and Rajagopal 1987), and water-quality monitoring (Kusmulyono and Goulter 1994).

The *Shannon entropy* is defined, for a discrete case, as

$$H(X) = -\sum_{i=1}^{n} \ln(p_i)\, p_i \tag{4.45a}$$

and, for a continuous case, as

$$H(X) = -\int_{x_{\min}}^{x_{\max}} \ln[f_x(x)]\, f_x(x)\, dx \tag{4.45b}$$

in which $p_i = P(X = x_i)$ and $f_x(x)$ is the PDF of the continuous random variable X. The degree of uncertainty (or information) associated with the realization of a random variable is measured by

$$I(x_i) = -\ln(p_i) \qquad \text{for a discrete case}$$
$$I(x) = -\ln[f_x(x)] \qquad \text{for a continuous case}$$

As can be seen, the entropy is the expected information content associated with a random variable X over its entire range, that is, $H(X) = E[I(X)]$. It should be pointed out that the value of entropy for discrete random variables is nonnegative, whereas, for continuous random variables, the entropy value could be negative. For detailed discussions on the properties of entropy, readers are referred to Guiasu (1977) and Jumarie (1990).

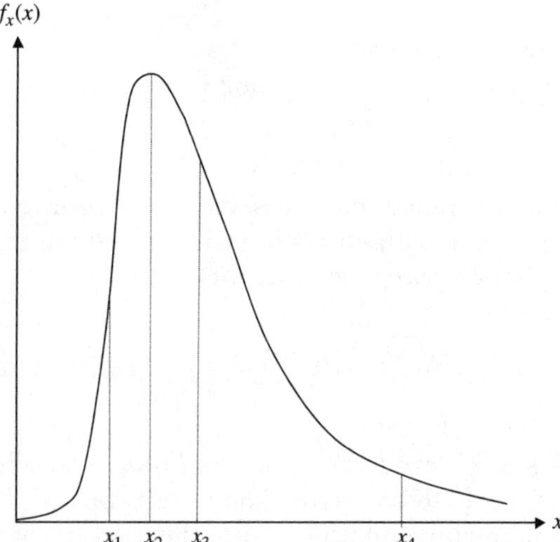

Figure 4.6　Illustration of information content associated with the observed values of a random variable.

To see why the information about a random variable is dependent on the probability of occurrence of a value of the random variable, refer to Fig. 4.6, in which four values of a random process are observed. One realizes that it is more likely to obtain x_1, x_2, and x_3, than x_4. However, x_1, x_2, and x_3 provide little information regarding the distributional characteristics of the random process. A rare event, such as x_4, provides more useful information about the tail parts of the distribution in which most engineering problems are interested. Therefore, it is meaningful to think of the information content of a random process as being inversely proportional to the likelihood of the event occurring. An extreme case is for a deterministic condition, where there is no uncertainty, for which the entropy is zero. The use of a logarithmic scale in the definition of entropy is only a fictitious one for providing consistent and desirable properties of the entropy function, such as (1) as the likelihood of occurrence of a random event decreases, the information $I(x)$ monotonically increases; (2) the greater the uncertainty, the larger the entropy value will be; and (3) the entropy of the joint occurrence of two independent events is simply the addition of the entropy of the individual events.

The *maximum entropy principle* was proposed by Jaynes (1957), stating that of all the distributions that satisfy the constraints supplied by the known information is the one that has the largest entropy for the random variables. Consider a continuous random variable, X, for which some of its properties are known a priori. Using the maximum entropy principle, the PDF of the random variable, X, can be derived by solving the following optimization problem (Cover and Thomas 1991).

$$\text{Maximize}\quad H(X) = -\int_{x_{\min}}^{x_{\max}} \ln[f_x(x)]\, f_x(x)\, dx \qquad (4.46a)$$

subject to

$$\int_{x_{min}}^{x_{max}} h_j(x)\, f_x(x)\, dx = a_j \qquad \text{for } j = 1, \ldots, m \tag{4.46b}$$

where $h_j(x)$ is a function of random variable, X.

The above maximization problem can be solved by *Lagrangian multiplier method*, which converts a constrained optimization problem into an unconstrained problem through a *Lagrangian function* as

$$\text{Maximize } L(f_x, \lambda) = -\int_{x_{min}}^{x_{max}} \ln(f_x(x))\, f_x(x)\, dx - \sum_{j=1}^{m} \lambda_j \left[\int_{x_{min}}^{x_{max}} h_j(x) f_x(x)\, dx - a_j \right] \tag{4.47}$$

in which $f_x(x)$ is the PDF and λ_j's are *Lagrangian multipliers*, which could be positive, zero, or negative. The PDF satisfying the constraint Eq. (4.46b) can be obtained by solving the following equation, using the calculus of variation,

$$\begin{aligned}
\frac{\partial L(f_x, \lambda)}{\partial f_x} &= \int [-1 - \ln(f_x(x))]\, dx - \sum_{j=1}^{m} \lambda_j \int h_j(x)\, dx \\
&= \int \left[-1 - \ln(f_x(x)) - \sum_{j=1}^{m} \lambda_j h_j(x) \right] dx = 0
\end{aligned} \tag{4.48}$$

The entropy-based PDF from Eq. (4.48) then is obtained as

$$f_x(x) = \exp\left[-1 - \sum_{j=1}^{m} \lambda_j h_j(x) \right] \tag{4.49}$$

A special case of Eq. (4.49) is that the constraint equations are related to the moments of the random variable. In other words, suppose that the first m product-moments about the origin are available from the appropriate integral transformation techniques described in the previous three sections. The constrained maximization problem of Eqs. (4.46a) and (4.46b), then, can be written as

$$\text{Maximize } H(X) = -\int_{x_{min}}^{x_{max}} \ln[f_x(x)]\, f_x(x)\, dx \tag{4.50a}$$

subject to

$$\int x^j f_x(x)\, dx = \mu_j' \qquad \text{for } j = 0, 1, \ldots, m \tag{4.50b}$$

The entropy-based distribution satisfying Eqs. (4.50b) is

$$f_x(x) = \exp\left[-1 - \lambda_0 - \sum_{j=1}^{m} \lambda_j x^j \right] \tag{4.51}$$

To obtain the entropy-based distribution as given in Eq. (4.51), the values of Lagrangian multipliers λ's must be determined. A system of $(m+1)$ nonlinear equations containing $(m+1)$ unknown λ's can be established by substituting Eq. (4.51) into the constraint Eq. (4.50b) and the results are

$$-\lambda_0 + \ln\left[\int x^r \exp\left(-\sum_{j=1}^{m} \lambda_j x^j\right) dx\right] = 1 + \ln(\mu_r') \qquad \text{for } r = 0, 1, \ldots, m \quad (4.52)$$

The above system of nonlinear equations can be solved by using appropriate numerical techniques. Alternatively, values of the λ's can be obtained by solving the following nonlinear optimization model

$$\text{Minimize} \sum_{r=0}^{m} \left(e_r^+ + e_r^-\right) \tag{4.53a}$$

subject to

$$-\lambda_0 + \ln\left[\int x^r \exp\left(\sum_{j=1}^{m} \lambda_j x^j\right) dx\right] - e_r^+ + e_r^- = 1 + \ln(\mu_r') \qquad \text{for } r = 0, 1, \ldots, m \quad (4.53b)$$

in which e_r^+ and e_r^- are nonnegative variables representing, respectively, the errors of over and underestimating the right-hand side values of the rth constraint. To solve for λ's, one should be cautious about the possibility of numerical overflow associated with the higher-order moments.

Example 4.14   Given the mean of zero and variance of σ_x^2 of a random variable, X, find its PDF maximizing the entropy function.

Solution   The PDF of the random variable, X, satisfying the maximum entropy principle can be obtained by solving the following problem:

$$\text{Max } H(X) = -\int \ln(f_x) f_x \, dx$$

s.t.

$$\int f_x dx = 1 \tag{a}$$

$$\int x f_x \, dx = 0 \tag{b}$$

$$\int x^2 f_x \, dx = \sigma_x^2 \tag{c}$$

The PDF satisfying the above three constraints, according to Eq. (4.51), is

$$f_x(x) = \exp\left(-1 - \lambda_0 - \lambda_1 x - \lambda_2 x^2\right) \tag{d}$$

To solve for λ_0, λ_1, and λ_2, (d) is substituted into (a) to (c) resulting in

$$e^{-1-\lambda_0} \int_{-\infty}^{\infty} e^{-\lambda_1 x - \lambda_2 x^2} dx = e^A \sqrt{\frac{\pi}{\lambda_2}} = 1 \tag{e}$$

$$e^{-1-\lambda_0} \int_{-\infty}^{\infty} x\, e^{-\lambda_1 x - \lambda_2 x^2}\, dx = -\frac{\lambda_1 \sqrt{\pi}\, e^A}{2\lambda_2^{1.5}} = 0 \qquad\qquad \text{(f)}$$

To ensure that the $f_x(x)$ in (e) behaves properly, it is necessary that $\lambda_2 > 0$; otherwise, the value of $\int f_x\, dx$ will be infinity. With $\lambda_2 > 0$, from (f), one has $\lambda_1 = 0$ and $e^A = e^{-1-\lambda_0}$. From (e), $\exp(-1 - \lambda_0) = \sqrt{\lambda_2/\pi}$ and the PDF can be expressed as

$$f_x(x) = \sqrt{\frac{\lambda_2}{\pi}}\; e^{-\lambda_2 x^2} \qquad\qquad \text{(g)}$$

Substituting (g) into (c), one obtains $\lambda_2 = 1/(2\sigma_x^2)$. The final form of the PDF is

$$f_x(x) = \frac{1}{\sqrt{2\pi}\,\sigma_x^2}\; e^{-x^2/2\sigma_x^2} \qquad \text{for} -\infty < x < \infty$$

which is exactly the PDF for the normal variable with zero mean and variance σ_x^2.

From this example, one learns an interesting fact—if the mean and variance of some random processes are known a priori, the normal distribution is a minimally prejudiced probability distribution that contains the largest amount of information among all possible competitive distributions. This fact, as stated in Sec. 2.4.1, provides a rather strong argument and justification for the use of the normal distribution in situations where only the first two moments are known or given.

4.5 Concluding Remarks

As discussed previously, applications of models in hydrosystem engineering infrastructural design and analysis often involve quantities subject to uncertainties. One of the main objectives of uncertainty analysis is to assess uncertainty features of system model outputs as affected by the presence of various uncertainties involved in model, parameters, data, and other factors. In this chapter, several analytical approaches applicable for uncertainty analysis of models are described. Each uncertainty analysis technique described has different levels of mathematical sophistication, computational complexity, and data requirements. Also, each technique possesses limitations and advantages regarding their applicability. The analytical methods described in this chapter are powerful—as well as mathematically elegant—tools for problems that are not too complex. Although this may be too much to expect when dealing with real-life problems, examples and problems of this chapter offer plenty of illustrations that such circumstances do exist. For those situations, analytical techniques could be applied to obtain exact uncertainty features of model outputs without approximation or extensive simulation. When using analytical techniques for uncertainty analysis, knowledge about the PDFs of stochastic parameters in the model is required.

The derived distribution method, if successfully implemented, would provide engineers with complete information regarding the uncertainty features of model outputs, that is, their PDFs. From the PDF, a complete description of statistical characteristics of the model output subject to uncertainty could be obtained. The main concern that dictates its application is the complexity of mathematical manipulations required of engineers, especially in multivariate problems.

When stochastic parameters in a model are statistically independent and are strictly multiplicative or additive, the convolutional property of an appropriate integral transform can be applied to obtain the exact statistical moments of the model output. Tables of various integral transforms of functions can be found in many mathematical handbooks (e.g., Abramowitz and Stegun 1972 and specialized books (Bateman 1954). In a general case, where stochastic model parameters are correlated and the model functional form is more complicated, integral transforms would become analytically and computationally difficult, if not impossible. This appears to be the main factor that severely restricts the practical usefulness of integral transforms for uncertainty analysis in hydrosystem engineering problems. Another potential shortcoming of using integral transforms for uncertainty analysis is that integral transforms do not always exist analytically under some conditions.

Problems

4.1 Growth forecasts are used to provide information regarding the dynamic change in a system and organization caused by an increase of certain state variables of the system, such as population, income, or production. One such model is a simple exponential growth model described by

$$dP_t/dt = RP_t$$

in which P_t is the variable to be projected and R is the growth rate. The solution to the above equation is

$$P_t = P_0\, e^{-Rt}$$

where P_0 is the initial condition of the system. Suppose that the growth rate R is a uniform random variable within the bound $(r - b, r + b)$ and P_0 is a constant. Derive the PDF of P_t, its mean, and variance.

4.2 Referring to Example 4.3, note that, from Sec. 2.4.6, U^2 has a Chi-square distribution with one degree of freedom. Use this fact to derive the PDF for the kinetic energy, E.

4.3 Refer to Prob. 4.2. Derive the PDF for the total kinetic energy, E, which is the sum the of kinetic energy resulting from fluctuations in all three directions, that is,

$$E = (\rho/2)\,(U_x^2 + U_y^2 + U_z^2)$$

in which U_x, U_y, and U_z are velocity fluctuations in an x-, y-, and z-direction, respectively, and they are independent normal random variables with a zero mean and unit standard deviation.

4.4 Referring to Prob. 4.2, derive the PDF for the kinetic energy, E, if the random variable U has a normal distribution with mean 0 and standard deviation σ.

4.5 The height of earth dams must allow sufficient freeboard above the maximum reservoir level to prevent waves from washing over the top. The determination of the height would include the considerations of wind tide and wave height. The wind tide T (in feet) above the still-water level can be estimated by

$$T = \frac{F}{1400\,d}V^2$$

where V = wind speed (in mi/h) blowing toward the direction of the dam
F = fetch or length of water surface over which the wind blows (in ft)
d = average depth of the lake along the fetch (in ft).

If the wind speed has an exponential distribution with a mean speed of v_o; that is,

$$f_v(v) = \frac{1}{v_o}e^{-v/v_o} \qquad \text{for } v > 0$$

Determine the distribution for the wind tide, T (Ang and Tang 1975).

4.6 Suppose that the economic benefit and cost of a water resource project are independent log-normal random variables having the PDFs shown by Eq. (2.57). Derive the PDF for the benefit-cost ratio. Justify your result from the properties of lognormal random variables described in Sec. 2.4.2.

4.7 Given the joint PDF of two random variables, X and Y, as $f_{x,y}(x, y)$, derive (a) the PDF for the sum, $U = X + Y$; and (b) the PDF of the difference, $V = X - Y$.

4.8 Show that, from the results obtained in Prob. 4.7, the PDF of the sum and difference of two independent random variables, X and Y, can be expressed, respectively, as

$$h_u(u) = \int_{-\infty}^{\infty} f_x(u-y)f_y(y)\,dy = \int_{-\infty}^{\infty} f_x(x)f_y(u-x)\,dx$$

$$h_v(v) = \int_{-\infty}^{\infty} f_x(v+y)f_y(y)\,dy = \int_{-\infty}^{\infty} f_x(x)\,f_y(x-v)\,dx$$

4.9 The amount of pollutant build-up on a street surface can be estimated by (Roesner 1982)

$$P_b = \frac{aT}{b+T}$$

in which P_b is the mass of the pollutant build-up per unit area; T is the elapsed time between two consecutive storm events; and a and b are constants. Derive the probability density function for P_b under the condition that the random T is a uniform random variable bounded between $[t_a, t_b]$. The two model constants are $a = 5 \times 10^{-3}$ kg/m^2/day and $b = 2.2$ days. Furthermore, with $[t_a, t_b] = [3$ days, 7 days], compute the probability that the pollutant build-up over a 1-km highway with a road width of 10 m would exceed 100 kg.

4.10 Referring to the pollutant build-up model in Prob. 4.9, derive the probability density function for P_b under the condition that random T is a lognormal random variable. Furthermore, with $\mu_T = 5$ days and $\sigma_T = 2$ days, compute the probability that the pollutant build-up over a 1-km highway with a road width of 10 m would exceed 100 kg.

4.11 Verify your derivation of the mean and variance in Prob. 4.1 using the characteristic function of random growth rate.

4.12 Refer to Prob. 4.1. Suppose that the initial condition $P_0 \sim N(\mu_{P_0}, \sigma_{P_0})$ and the growth rate $R \sim N(\mu_R, \sigma_R)$ are two independent normal random variables with their respective mean and variances. Derive the expressions for the mean and variance of P_t.

4.13 Repeat Prob. 4.12 by considering that P_0 and R are independent uniform random variables where P_0 is bounded within $(p_0 - a, p_0 + a)$ and R is bounded within $(r - b, r + b)$.

4.14 Compare the variances derived in Probs. 4.11 and 4.12 and discuss the effect of the presence of uncertainty in P_0.

4.15 Referring to Figure P2.1, derive the moment generating function of the triangular distribution and the expression of the first four moments about the origin.

4.16 Suppose that the benefit associated with a hydropower operation has a triangular distribution with $a = \$1M$, $m = \$3M$, and $b = \$6M$. Numerically compute the mean, standard deviation, skewness coefficient, and kurtosis.

4.17 In water resource project evaluations, the present value of the net benefit is frequently used to indicate the economic merit of a project. Consider a pumping station that is to be operated over a 3-year period. The present value of the net benefit of the pumping station can be calculated as

$$PV = -I_0 + \sum_{n=1}^{3} (1+i)^n R_n + S$$

where PV = present value of the net benefit
$\quad\quad I_o$ = initial investment cost
$\quad\quad\quad i$ = interest rate
$\quad\quad R_n$ = return of the nth year
$\quad\quad\quad S$ = salvage value of the pumping station

Assume that interest rate is deterministic and all benefit and cost items are independent normal random variables. Derive the characteristic function of the random PV and show that the PDF of PV also is a normal distribution. Furthermore, find the mean and variance of PV.

4.18 Referring to Example 4.7, examine the effect of uncertainty in the project life on PVR as compared to life without considering such uncertainty.

4.19 Refer to Example 4.7. The total present value of a project can be expressed as

$$\text{PV} = -I_0 + \int_0^T e^{-rt} R(t)\, dt + S_T e^{-rT}$$

where r = nominal continuous interest rate
$\quad R(t)$ = continuous economic return
$\quad\quad T$ = project life
$\quad\quad S_T$ = salvage value at the end of the project life

Assume that the economic return profile $R(t)$ can be expressed as $R(t) = R_o$, with R_o being the fixed, but random return. Derive the moment generating function for the random PV, considering that I, R_o, and S_T are independent normal random variables.

4.20 Refer to Example 4.7. Consider, further, that project life T also is a random variable having a uniform distribution in an interval of $[t_a, t_b]$. Assuming that the random project life is independent of R_o, derive the expression for the mean and variance of the total return, excluding salvage value, and compare it to those without considering random project life. Discuss the implication of the random project life on the mean and variance of the present value evaluation.

4.21 Refer to Example 4.7. Derive the expressions for the mean, variance, and skewness coefficient for the random PV.

4.22 Repeat Prob. 4.21 for the following three cash-flow patterns:

$a.\ \ R(t) = R_o + r_2\, t, \qquad 0 \le t \le T$
$b.\ \ R(t) = R_o + r_2\, (T - t), \qquad 0 \le t \le T$
$c.\ \ R(t) = R_o\, e^{-gt}, \qquad 0 \le t \le T$

in which R_o is the random initial return and r_2 and g are the constants.

4.23 Refer to Prob. 4.21. Consider further that project life T also is a random variable. Assuming that the random project life is independent of I_o, R_o, and S_T, derive the expression for the mean and variance of the total return, excluding salvage value, and compare it to those without considering random project life. Discuss the implication of the random project life on the mean and variance of the present value evaluation.

4.24 Solve Prob. 4.21 numerically based on the following data:

$T = 50$ years, $i = 5\%$
I_o, normal distribution with mean = \$12,000, standard deviation = \$3000
R_o, uniform distribution with the bounds [\$2000, \$3000]
S_T, triangular distribution with $a = \$5000$, $m = \$6000$, and $b = \$8000$

4.25 Solve Prob. 4.23 numerically based on the following data:

$i = 5\%$
T, uniform distribution with [45, 55] years;
I_o, normal distribution with mean = \$12,000, standard deviation = \$3000;
R_o, uniform distribution with the bounds [\$2000, \$3000];
S_T, triangular distribution with $a = \$5000$, $m = \$6000$, and $b = \$8000$.

4.26 Derive Eqs. (2.58a) and (2.58b) using the moment generating function.

4.27 Derive Eq. (2.110) using the moment generating function.

4.28 A frequently used surface pollutant washoff model is the 1st-order decay function (Sartor and Boyd 1972)

$$P_t = P_0\, e^{-kRt}$$

where P_o = initial pollutant mass at time $t = 0$
$\quad R$ = runoff intensity
$\quad k$ = washoff coefficient
$\quad P_t$ = mass of pollutant remaining on the street surface
$\quad t$ = time elapsed since the beginning of the storm

The above model does not consider pollutant build-up and is generally appropriate for within-storm event analysis.

Suppose that P_o and k are constants and R is a uniform random variable in $[r_a, r_b]$. Derive the mean, standard deviation, and skewness coefficient for (a) P_t; (b) pollutant amount washed-off after t; and (c) percentage of pollutant remaining on the street surface; and (d) derive the correlation between P_t and P_s at different times $s \neq t$.

4.29 Referring to Prob. 4.28, suppose that P_o is also a normal random variable with mean μ_{P_o} and standard deviation σ_{P_o}. Furthermore, P_o and R are independent of each other. Resolve Prob. 4.28.

4.30 Refer to Prob. 4.29. For a given stretch of highway, the washoff coefficient is 1.84/cm, and the mean and standard deviation of the initial pollutant mass are 10,000 kg and 2000 kg, respectively. The uniformly distributed random runoff intensity R is bounded between [8 cm/h, 12 cm/h]. Numerically compute the mean, coefficient of variation, skewness coefficient, and correlation matrix of P_t for $t = 1, 2, 3, 4, 5$ hour and plot them with respect to time.

4.31 Referring to the surface pollutant model in Prob. 4.28, its discrete version can be written as (Patry and Kennedy 1989)

$$P_n = P_{n-1}e^{-kR_n\,\Delta t_n}$$

or

$$\Delta P_n = P_n - P_{n-1}\,(1 - e^{-kR_n\,\Delta t_n})$$

where R_n = runoff intensity during time step n from t_{n-1} to t_n

$\Delta t_n = t_n - t_{n-1}$

ΔP_n = amount of pollutant washed-off during the nth time interval

Consider that $\Delta t_n = \Delta t$ and $R_n = R$ for all n. Furthermore, suppose that P_o and R are independent of each other with P_o being a normal random variable having a mean of μ_{P_o} and a standard deviation of σ_{P_o}, and R being a uniform random variable in range $[r_a, r_b]$. Derive the following quantities: (a) mean of ΔP_n; (b) standard deviation of ΔP_n; (c) skewness coefficient of ΔP_n; and (d) correlation of ΔP_n and ΔP_m for $m \neq n$.

4.32 Referring to Prob. 4.31, numerically compute the mean, coefficient of variation, skew coefficient, and correlation matrix of ΔP_n for $n = 1, 2, 3, 4, 5$ based on the data provided in Prob. 4.30.

4.33 Considering K independently exponentially distributed random variables having the marginal PDF $f_k(x) = e^{-x/\mu_k}/\mu_k$, $k = 1,2,\ldots, K$, show that the PDF of sum of such K random variables, $X = X_1 + X_2 + \cdots + X_K$, is

$$f_x(x) = \sum_{k=1}^{K} \frac{(\mu_k)^{K-1}}{\Pi_{j\neq k}^{K}(\mu_j - \mu_k)}\, e^{-x/\mu_k}$$

4.34 Consider a 3rd-order basin with an area of 32.56 km^2 in which there are following four possible paths are possible for rain water to travel until it reaches the basin outlet:

Path no.	Flow path	Probability
1	$O_1 \rightarrow C_1 \rightarrow C_2 \rightarrow C_3 \rightarrow$ outlet	0.322
2	$O_1 \rightarrow C_1 \rightarrow C_3 \rightarrow$ outlet	0.064
3	$O_2 \rightarrow C_2 \rightarrow C_3 \rightarrow$ outlet	0.185
4	$O_3 \rightarrow C_3 \rightarrow$ outlet	0.429

NOTE: O_i = ith-order overland surface; C_j = jth-order channel.

Assume that travel time in different states are independently distributed exponential random variables with known mean values. (a) Derive the PDF for total travel time in each flow path. (b) Derive the expression for GIUH of the basin.

4.35 Referring to Prob. 4.34, numerically find the GIUH based on the following component travel time information.

Overland surface	Mean travel time (h)	Channel reach	Mean travel time (h)
O_1	2.23	C_1	0.06
O_2	1.26	C_2	0.22
O_3	4.54	C_3	0.24

4.36 Referring to Table 4.4, prove scaling, translation, and exponentiation properties of the Mellin transform.

4.37 Referring to Table 4.5, show that (i) $M_w(s) = a^{s-1} M_x(s)$ when $W = aX$ and (ii) $M_w(s) = M_x(bs - b + 1)$ when $W = X^b$.

4.38 Based on the Mellin transform of the standard normal PDF given in Table 4.6, use the binomial expansion to derive the expression for the Mellin transform of integer-valued argument s of general normal PDF with mean μ_x and standard deviation σ_x.

4.39 In groundwater study, the Theim equation is frequently used to estimate the drawdown in a confined aquifer system. The drawdown of the groundwater table at some distance away from the production well can be estimated as

$$s = \frac{Q\ln(r_o/r)}{2\pi Kb}$$

where s = drawdown
$\quad Q$ = pumpage at the production well
$\quad K$ = aquifer conductivity
$\quad b$ = aquifer thickness
$\quad r_o$ = radius of influence of the production well
$\quad r$ = distance away from the production well where drawdown is estimated.

Consider that, in a homogenous aquifer, the conductivity and aquifer thickness are two independent random variables with the following statistical properties:

$\quad K$: lognormal distribution with mean 10 m/day and coefficient of variation 0.3

$\quad b$: triangular distribution with 90, 100, 120 m as the lower bound, mode, and upper bound, respectively.

Use the Mellin transform to determine the first three moments of drawdown under the following conditions: $Q = 1000$ m^3/day, $r = 200$ m, and $r_o = 500$ m.

4.40 Suppose that there are two pumping wells, A and B, in operation in the same stochastically homogenous confined aquifer. The steady-state drawdown at any location in the aquifer can be estimated by the Theim equation described in Prob. 4.39. Let s_A and s_B be the drawdowns at the location X resulting from pumpage at production wells, A and B, respectively. (1) Explain your reasons why s_A and s_B should be correlated. (2) Use the Mellin transform to determine the covariance and correlation coefficient between s_A and s_B.

4.41 Refer to the same groundwater aquifer in Prob. 4.39. Suppose that there are three production wells A, B, and C in operation. The effect of these three production wells on the drawdown at any point in the aquifer is additive. Namely, the total drawdown at location X (s_X) in the aquifer could be estimated as

$$s_X = s_{AX} + s_{BX} + s_{CX}$$

in which s_{iX} is the drawdown incurred by production well i, for $i = A, B, C$, at location X, and s_{iX} can be calculated by the Theim equation, for a steady-state condition, as

$$s_{iX} = \frac{Q_i\ln(r_{io}/r_{iX})}{2\pi Kb} \qquad i = A, B, C$$

with Q_i and r_{io} being the pumpage and radius of influence of the ith production well, respectively; and r_{iX} being the distance between the ith production well and the point of interest X. Assuming that the following conditions are known, (1) use the procedure developed in Prob. 4.40 to determine the mean and variance of the total drawdown at the point of interest X and (2) estimate the probability that the total drawdown at location X would exceed 1 m.

$$Q_A = 800 \text{ m}^3/\text{day} \qquad r_{Ao} = 400 \text{ m} \qquad r_{AX} = 150 \text{ m}$$

$$Q_B = 1000 \text{ m}^3/\text{day} \qquad r_{Bo} = 500 \text{ m} \qquad r_{BX} = 200 \text{ m}$$

$$Q_C = 1200 \text{ m}^3/\text{day} \qquad r_{Co} = 500 \text{ m} \qquad r_{CX} = 150 \text{ m}$$

4.42 Uncertainty analyses of hydraulic computations in channel flood routing are mainly concerned with the assessment of the uncertainty features of the computation results. In channel flood routing, the results of primary interest are the travel time of the floodwater, the magnitude of the peak, and the corresponding water surface profile, along with the area of inundation. Using Manning's formula, the travel time, T, of a kinematic wave in a wide rectangular channel carrying a flow of Q can be determined by (Chow, Maidment, and Mays 1988)

$$T = \frac{3}{5} \left(\frac{n B^{2/3}}{1.49 \ S_o^{1/2}} \right)^{3/5} Q^{-2/5} L$$

where B = channel width
n = Manning's roughness coefficient
L = length of the channel reach

Based on the information given in the following table, determine the mean, standard deviation, and skewness coefficient of the travel time.

Variable	Distribution	Lower bound	Mode	Upper bound
n	Triangular	0.030	0.045	0.055
B (ft)	Triangular	180	200	220
S_o (ft/ft)	Triangular	0.00025	0.00035	0.00045
Q (cfs)	Triangular	9800	10,000	12,000
L (mi)	Triangular	99	100	101

4.43 Consider the design of a storm sewer system. The sewer flow carrying capacity, Q_C, is determined by Manning's formula

$$Q_C = \frac{0.463}{n} \lambda_m D^{8/3} S_o^{1/2}$$

where n = Manning's roughness coefficient
λ_m = model correction factor to account for the model uncertainty
D = actual pipe diameter
S_o = pipe slope

The inflow, Q_L, to the sewer is the surface runoff whose peak discharge is estimated by the rational formula

$$Q_L = \lambda_L CiA$$

where λ_L = correction factor for the model uncertainty
$\quad C$ = runoff coefficient
$\quad\ i$ = rainfall intensity
$\quad A$ = runoff contribution area

In practice, it is reasonable to assume that all the parameters on the right-hand side of the two equations are subject to uncertainty. The sewer capacity, Q_C, and peak inflow, Q_L, from the surface runoff, consequently, cannot be quantified with absolute certainty. Assume that all the stochastic parameters in the models for Q_C and Q_L are statistically independent, use the Mellin transform to estimate the first four moments of Q_C and Q_L. Furthermore, estimate the reliability that the sewer capacity could accommodate peak runoff based on the following data.

Stochastic parameter	Distribution	Parameters
λ_L	Triangular	$(a, m, b) = (0.85, 1.00, 1.15)$
C	Triangular	$(a, m, b) = (0.708, 0.825, 0.942)$
i (in/h)	Triangular	$(a, m, b) = (3.60, 4.00, 4.40)$
A (acres)	Triangular	$(a, m, b) = (9.5, 10.0, 10.5)$
λ_m	Triangular	$(a, m, b) = (0.98, 1.10, 1.22)$
n	Gamma	$(\mu, \sigma) = (0.015, 0.00083)$
D (ft)	Triangular	$(a, m, b) = (2.97, 3.0, 3.03)$
S	Triangular	$(a, m, b) = (0.002, 0.005, 0.007)$

4.44 The Hazen-Williams equation is frequently applied to compute friction losses in pipe flow analysis. In terms of flow rate (Q), it can be expressed as

$$h_L = \frac{4.728\, LQ^{1.852}}{C^{1.852}\, D^{4.87}}$$

where h_L = head loss (in ft)
$\quad L$ = pipe length (in ft)
$\quad Q$ = flow rate (in ft^3/s)
$\quad C$ = Hazen-Williams coefficient
$\quad D$ = pipe diameter (in ft)

Suppose that Q, C, and D are independent random variables with the following statistical properties, use the Mellin transform technique to quantify the first three moments of the friction loss in a 2000-ft long cast iron pipe.

Variable	Distribution type and the associated parameters
Q	Uniform distribution with $a = 8$ ft^3/s, $b = 12$ ft^3/s.
C	Triangular distribution with $a = 125$, $m = 130$, $b = 135$.
D	Normal distribution with $\mu = 6$ in., $\sigma = 0.3$ in.

4.45 Referring to Example 4.13, use the three-term Edgeworth expansion to delineate the PDF of the sewer flow capacity.

4.46 Based on the information about the moments of the sewer flow capacity in Example 4.13 and Prob. 4.45, develop the curves for the CDF using the one-term, two-term, and three-term Edgeworth expansion. Also, compare the results.

4.47 Use the Fisher-Cornish asymptotic expansion to determine quantiles of the sewer flow capacity with $p = 0.01, 0.05, 0.10, 0.90, 0.95,$ and 0.99 based on (a) the first three moments and (b) the first four moments.

4.48 Show that the entropy distribution for a random variable bounded between the interval of $[a, b]$ is a uniform one.

4.49 Show that, for a nonnegative valued random variable X with a known mean, μ_x, the corresponding entropy distribution is exponential.

4.50 Find the entropy distribution for a random variable bounded between $[5, 20]$ with a mean equal to 10.

4.51 Show that, for a nonnegative valued random variable with a known median, x_{md}, the corresponding entropy distribution is exponential. Plot the PDFs for $x_{md} = 1$ and 10.

4.52 Find the entropy distribution for a random variable bounded between $[5, 20]$ with a median equal to 10.

4.53 Find the entropy distribution for a nonnegative random variable with a known mean of 10 and a coefficient of variation of 0.3.

References

Abramowitz, M., and I. A. Stegun (ed.) (1972). *Handbook of Mathematical Functions With Formulas, Graphs, and Mathematical Tables*, 9th ed., Dover Publications, New York, pp. 1019–1030.

Amorocho, J., and B. Espildora (1973). "Entropy in the Assessment of Uncertainty of Hydrologic Systems and Models," *Water Resources Research*, **9**(6):1515–1522.

Ang, A. H. S., and W. H. Tang (1975). *Probability Concepts in Engineering Planning and Design*, Vol. I, John Wiley and Sons, New York.

Bateman, H. (1954). *Tables of Integral Transforms*, Vol. I, McGraw-Hill, New York.

Bras, R. (1990). *Hydrology: An Introduction to Hydrological Science*, Addison Wesley, New York, Chap. 19, Sec. 12.3.1.

Chen, C. N., and T. S. W. Wang (1989). "Re-evaluation of Rational Method Using Kinematic Wave Approach," *Proceedings of International Conference on Channel Flow and Catchment Runoff: Centennial of Manning's Formula and Kuichling's Rational Formula*, University of Virginia, Charlottesville, VA, pp. 61–70.

Cheng, B. L. M. (1982). "A Study of Geomorphological Unit Hydrograph," Ph.D. Thesis, Department of Civil and Environmental Engineering, University of Illinois at Urbana-Champaign, IL.

Chiu, C. L. (1987). "Entropy and Probability Concepts in Hydraulics," *Journal of Hydraulic Engineering*, ASCE, **113**(5):583–600.

Chiu, C. L. (1988). "Entropy and 2D Velocity Distribution in Open Channels," *Journal of Hydraulic Engineering*, ASCE, **114**(7):738–756.

Chiu, C. L. (1989). "Velocity Distribution of Open Channel Flow," *Journal of Hydraulic Engineering*, ASCE, **115**(5):576–594.

Chiu, C. L., G. F. Lin, and J. M. Lu (1993). "Application of Probability and Entropy Concepts in Pipe-Flow Study," *Journal of Hydraulic Engineering*, ASCE, **119**(6):742–757.

Chiu, C. L., and C. A. A. Said (1995). "Maximum and Mean Velocities and Entropy in Open Channel Flow," *Journal of Hydraulic Engineering*, ASCE, **121**(1):26–35.

Chow, V. T., D. Maidment, and L. W. Mays (1988). *Applied Hydrology*, McGraw-Hill, New York.

Cover, T. M., and J. A. Thomas, (1991). *Elements of Information Theory*, John Wiley and Sons, New York.

Epstein, B. (1948). "Some Application of the Mellin Transform in Statistics," *Annals of Mathematical Statistics*, **19**:370–379.

Fisher, R. A., and E. A. Cornish (1960). "The Percentile Points of Distributions Having Known Cumulants," *Technometrics*, **2**(2):209–225.

Giffin, W. C. (1975). *Transform Techniques for Probability Modeling*, Academic Press, San Diego, CA.

Guiasu, S. (1977). *Information Theory with Applications*. McGraw-Hill, New York.

Jaynes, E. T. (1957). "Information Theory and Statistical Mechanics," *Physics Review*, **106**:620–630; **108**:171–182.

Jumarie, G. (1990). *Relative Information: Theories and Applications*. Springer-Verlag, New York.

Kendall, M., A. Stuart, and J. K. Ord (1987). *Kendall's Advanced Theory of Statistics,* Vol. 1: *Distribution Theory*, 5th ed, Oxford University Press, New York.

Kusmulyono, A., and I. Goulter (1994). "Entropy Principles in the Prediction of Water Quality Values at Discontinued Monitoring Stations," *Journal of Stochastic Hydrology and Hydraulics*, **8**(4):301–317.

Park, C. S. (1987). "The Mellin Transform in Probabilistic Cash Flow Modeling," *The Engineering Economist*, **32**(2):115–134.

Patel, J. K., C. H. Kapadia, and D. B. Owen (1976). *Handbook of Statistical Distributions*, Marcel Deckker, New York.

Patry, G. G., and A. Kennedy (1989). "Pollutant Washoff Under Noise-Corrupted Runoff Conditions," *Journal of Water Resources Planning and Management*, ASCE, **115**(5):646–657.

Roesner, L. A. (1982). "Quality of Urban Runoff," In: *Urban Stormwater Hydrology*, D.F. Kibler (ed.), American Geophysical Union, Washington, DC, p. 161–187.

Sartor, J. D., and G. B. Boyd, (1972). "Water Pollution Aspects of Street Surface Contaminants," *Report* No. EPA-R2-72-081, U.S. Environmental Protection Agency, Washington, DC.

Shannon, C. E. (1948). "A Mathematical Theory of Communication," *Bell System Technical Journal*, **27**:379–423;623–656.

Singh, V. P., A. K. Rajagopal, and K. Singh (1986). "Derivation of Some Frequency Distributions Using the Principle of Maximum Entropy (POME)," *Advances in Water Resources*, **9**:91–106.

Singh, V. P., and P. F Krstanovic (1987). "A Stochastic Model for Sediment Yield Using the Principle of Maximum Entropy," *Water Resources Research*, **23**(5):781–793.

Singh, V. P., and A. K. Rajagopal (1987). "Some Recent Advances in the Application of the Principle of Maximum Entropy (POME) in Hydrology," *Water for Future: Hydrology in Perspective*, IAHS Publication No. **164**:353–364.

Springer, M. D. (1978). *The Algebra of Random Variables*, John Wiley and Sons, New York, pp. 80, 81, 470.

Tung, Y. K. (1989). "Uncertainty on Travel Time in Kinematic Wave Channel Routing," *Proceedings of International Conference on Channel Flow and Catchment Runoff*, University of Virginia, Charlottesville, VA, pp. 767–781.

Tung, Y. K. (1990). "Mellin Transform Applied to Uncertainty Analysis in Hydrology/Hydraulics," *Journal of Hydraulic Engineering*, ASCE, **116**(5):659–674.

Tyagi, A., and C. T. Haan (2001a). "Reliability, Risk, and Uncertainty Analysis Using Generic Expectation Functions," *Journal of Environmental Engineering*, ASCE, **127**(10):938–945.

Winterstein, S. R. (1986). "Nonlinear Vibration Models for Extremes and Fatigue," *Journal of Engineering Mechanics*, ASCE, **114**(10):1772–1790.

Approximation Methods for Uncertainty Analysis

Methods for performing uncertainty analysis vary in degrees of sophistication. They are also dictated by the information available about the stochastic basic variables, complexity of the problem, and resource constraints. In principle, it would be ideal to derive the exact probability distribution of the model output as a function of the stochastic basic variables. In Chap. 4, methods that allow one to analytically derive the probability density function (PDF) and/or statistical moments of functions of random variables are described. However, many analytical methods are restrictive in practical applications because they require simple functional relationships and independence of stochastic basic variables that might not be satisfied in real-life problems. Most of the models or design procedures used in hydrosystems infrastructural engineering and analysis are nonlinear and highly complex. This basically prohibits analytical derivation of the probability distributions of the model outputs. As a practical alternative, engineers frequently resort to methods that yield approximations to the statistical properties of model outputs subject to uncertainty. In this chapter, several methods useful for the uncertainty analysis of the general hydrosystems engineering infrastructural design are described. They include the 1st-order variance estimation method and several probabilistic point estimation procedures. Monte Carlo simulation techniques are another type of approximation methods and they are discussed separately in Chap. 6.

The value of these approximate methods was concisely explained by Cornell (1972) as follows:

> An approach based on means and variances may be all that is justified when one appreciates: (1) that data and physical arguments are often insufficient to establish the full probability law of a variable; (2) that most engineering analyses include an important component of real, but difficult to measure professional uncertainty; and (3) that the final output, namely the decision or design parameters, is often not sensitive to moments higher than the mean and variance.

Furthermore, Cornell (1972) pointed out the most important consideration of any reliability analysis:

> It is important to engineering applications that we avoid the tendency to model only those probabilistic aspects that we think we know how to analyze. It is far better to have an approximate model of the whole problem than an exact model of only a portion of it.

In hydrosystems engineering, the latter quote is aimed at the tremendous effort required for determining the frequency of floods or rainfall while ignoring all other uncertainties in the hydraulic design and analysis processes. These quotes were originally made with respect to the 1st-order variance estimation method, but the concepts expressed are equally valid for the other methods described in this chapter.

5.1   First-Order Variance Estimation Method

The *first-order variance estimation* (FOVE) method, also called the *variance propagation method* (Berthouex 1975), estimates the uncertainty features of a model output based on the statistical properties of the model's stochastic basic variables. The basic idea of the method is to approximate a model involving stochastic basic variables by a Taylor series expansion.

5.1.1   Univariate FOVE method

Consider a model output W that is related to a single stochastic basic variable X in a functional form as $W = g(X)$, with $g(X)$ being a general expression for the functional relationship representing a system or a model. Because the quantity W is a function of the stochastic basic variable X, it is also a random variable subject to uncertainty (see Fig. 5.1). The problem is to estimate the mean and variance of W from the knowledge of the stochastic basic variable X. It is sometimes possible to estimate the higher-order moment of $g(X)$, provided that the higher-order moments of X are known.

By the definition of statistical expectation described in Sec. 2.3, the mean and variance of $g(X)$ can theoretically be expressed, respectively, as

$$E[W] = E[g(X)] = \mu_w = \int_{-\infty}^{\infty} g(x) f_x(x)\, dx \tag{5.1}$$

and

$$\text{Var}[W] = \text{Var}[g(X)] = \sigma_w^2 = \int_{-\infty}^{\infty} [g(x) - \mu_w]^2\, f_x(x)\, dx \tag{5.2}$$

in which $f_x(x)$ is the PDF of the stochastic basic variable X. As can be seen, the mean and variance of $W = g(X)$ are, strictly speaking, functions of the PDF of random variable X. It should be made clear that knowing the PDF of a random

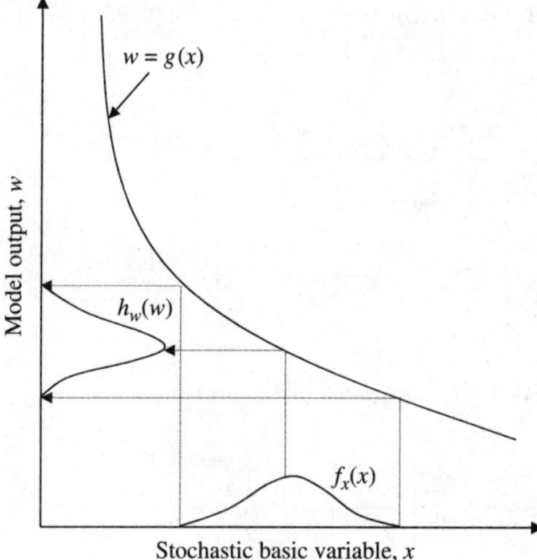

Figure 5.1   Transmission of uncertainty from a stochastic basic variable to model output.

variable means that complete information about the random variable, including all its moments and statistical properties, is known. The mean and variance provide only partial information about the uncertainty features of the random variable of interest.

In many practical engineering analyses and designs, the PDF of a random variable X, $f_x(x)$, might not be exactly known and, sometimes, the functional forms of $f_x(x)$ and/or $g(X)$ are too complex for an analytical solution of Eqs. (5.1) and (5.2). Alternative methods then have to be used to approximate the mean and variance of $g(X)$ on the basis of the known mean and variance of X.

A practical approximation is to expand the function $g(X)$ in the Taylor series about a selected point x_o in the sample space of the random variable X as

$$W = w_o + \sum_{r=1}^{\infty} \left[\frac{\partial^r W}{\partial X^r} \right]_{x_o} \frac{(X - x_o)^r}{r!}$$

$$= w_o + \left[\frac{\partial W}{\partial X} \right]_{x_o} (X - x_o) + \left[\frac{\partial^2 W}{\partial X^2} \right]_{x_o} \frac{(X - x_o)^2}{2} + \cdots \qquad (5.3)$$

in which $w_o = g(x_o)$; $[\partial W / \partial X]_{x_o}$ is called the *1st-order sensitivity coefficient* (or simply *sensitivity coefficient*) indicating the rate of change of the function value $g(X)$ with respect to its parameter at $X = x_o$; and $[\partial^2 W / \partial X^2]_{x_o}$ is the *2nd-order sensitivity coefficient* representing the curvature of the function $g(X)$ at $X = x_o$. If the function $W = g(X)$ is linear, all 2nd- and higher-order terms in Eq. (5.3) vanish and the approximation, then, is exact.

Applying the expectation operator to Eq. (5.3), the mean of $W = g(X)$ can be obtained as

$$E[W] = w_o + E\left(\sum_{r=1}^{\infty}\left[\frac{\partial^r W}{\partial X^r}\right]_{x_o}\frac{(X-x_o)^r}{r!}\right)$$

$$= w_o + \left[\frac{\partial W}{\partial X}\right]_{x_o}E[(X-x_o)] + \left[\frac{\partial^2 W}{\partial X^2}\right]_{x_o}\frac{E[(X-x_o)^2]}{2} + \cdots \qquad (5.4)$$

Similarly, the variance of $W = g(X)$ can be obtained as

$$\text{Var}\,[W] = \text{Var}[w_o] + \text{Var}\left(\sum_{r=1}^{\infty}\left[\frac{\partial^r W}{\partial X^r}\right]_{x_o}\frac{(X-x_o)^r}{r!}\right)$$

$$= \left[\frac{\partial W}{\partial X}\right]_{x_o}^2\text{Var}[(X-x_o)] + \left[\frac{\partial^2 W}{\partial X^2}\right]_{x_o}^2\frac{\text{Var}\,[(X-x_o)^2]}{4} + \cdots \qquad (5.5)$$

As can be seen, the mean and variance of $W = g(X)$ depend, among others, on the expansion point x_o. Furthermore, attempts to enhance the accuracy of the approximation by incorporating higher-order terms would render increasing algebraic complexity and require more statistical information, such as the third and fourth moments of the stochastic basic variables that may not be reliably obtained in practical situations. A practical approach is to adopt the 2nd-order approximation for estimating the mean of $W = g(X)$ as

$$E[W] \approx w_o + \left[\frac{\partial W}{\partial X}\right]_{x_o}(\mu_x - x_o) + \left[\frac{\partial^2 W}{\partial X^2}\right]_{x_o}\frac{\sigma_x^2 + (x_o - \mu_x)^2}{2} \qquad (5.6)$$

or simply by the 1st-order approximation as

$$E[W] \approx w_o + \left[\frac{\partial W}{\partial X}\right]_{x_o}(\mu_x - x_o) \qquad (5.7)$$

For the variance of $W = g(X)$, the 1st-order approximation is generally used resulting in

$$\text{Var}\,[W] \approx \left[\frac{\partial W}{\partial X}\right]_{x_o}^2\sigma_x^2 \qquad (5.8)$$

Note that the variance of $W = g(X)$ depends not only on the variance of X, but also on the value of the sensitivity coefficient. This can be graphically shown in Fig. 5.2 in which the degree of uncertainty of the model output (indicated by the

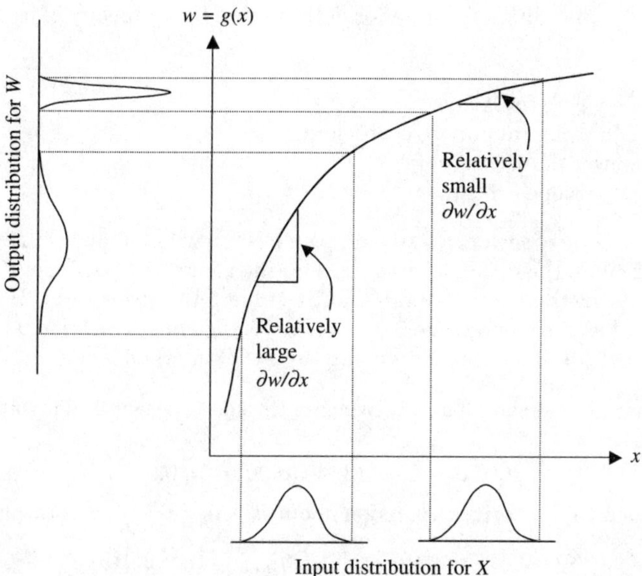

Figure 5.2   Effect of sensitivity and parameter uncertainty on model output uncertainty (after Verdeman, 1994).

width of the PDF, $h_w(w)$) clearly depends on the sensitivity of the model output with respect to the model parameter.

The general practice of the FOVE method is to take the mean of the stochastic basic variable as the expansion point, that is, $x_o = \mu_x$. The mean and variance of W can then be approximated, respectively, as

$$E[W] \approx \overline{w} + \left[\frac{\partial^2 W}{\partial X^2} \right]_{\mu_x} \frac{\sigma_x^2}{2} \tag{5.9a}$$

for a 2nd-order approximation, or

$$E[W] \approx \overline{w} \tag{5.9b}$$

for a 1st-order approximation with $\overline{w} = g(\mu_x)$ and

$$\mathrm{Var}\,[W] \approx \left[\frac{\partial W}{\partial X} \right]_{\mu_x}^2 \sigma_x^2 \tag{5.10}$$

for a 1st-order approximation. In general, $E[g(X)] \neq g[\mu_x]$ and the equality holds only when $g(X)$ is a linear function of X. The FOVE method would provide an accurate estimation of the mean and variance when the function $W = g(X)$ is close to linear and/or the uncertainty of the stochastic basic variables is small.

Example 5.1   Manning's formula for determining flow capacity of a storm sewer is

$$Q = 0.463 \, n^{-1} \, D^{2.67} \, S^{0.5}$$

where Q = flow rate (in ft^3/s)

n = Manning roughness coefficient

D = sewer diameter (in ft)

S = pipe slope (in ft/ft)

Consider a section of sewer in a storm sewer system with a diameter $D = 3.0'$ and slope $S = 0.005$. Due to the difficulty in assessing the roughness coefficient accurately, the flow capacity of the sewer would also be subject to uncertainty. If the roughness coefficient n has the mean value 0.015 with a coefficient of variation 0.05, quantify the uncertainty of the sewer capacity using the FOVE method.

Solution   Manning's formula for the sewer of the specified size and layout can be rewritten as

$$Q = 0.463 \, n^{-1} \, (3)^{2.67} \, (0.005)^{0.5} = 0.615 \, n^{-1}$$

The 1st-order Taylor series expansion about $n_o = \mu_n = 0.015$, according to Eq. (5.3), is

$$Q \approx 0.615 \, (0.015)^{-1} + [\partial Q/\partial n]_{n=0.015} \, (n - 0.015)$$

$$= 0.615 \, (0.015)^{-1} + [-0.615 \, (0.015)^{-2}] \, (n - 0.015)$$

$$= 41.01 - 2733.99 \, (n - 0.015)$$

Therefore, based on Eqs. (5.9) and (5.10), the approximated mean and variance of the sewer flow capacity, respectively, are

$$\mu_Q \approx 41.01 \; \text{ft}^3/\text{s}$$

$$\sigma_Q^2 \approx (2733.99)^2 \, \text{Var}(n) = (2733.99)^2 \, (\Omega_n \, \mu_n)^2$$

$$= (2733.99)^2 \, (0.05 \times 0.015)^2 = (2.05 \; \text{ft}^3/\text{s})^2$$

Hence, the standard deviation of the sewer flow capacity is 2.05 ft^3/s, which is about 5 percent of the estimated mean sewer flow capacity.

5.1.2   Bivariate FOVE method

Consider that the model output W is a function of two stochastic basic variables X_1 and X_2 as $W = g(X_1, X_2)$. It is known that the means and standard deviations of X_1 and X_2 are, respectively, μ_1, μ_2, and σ_1, σ_2. The Taylor series expansion of $W = g(X_1, X_2)$ with respect to the specified expansion point $\boldsymbol{\mu}_x = (\mu_1, \mu_2)^t$ is

$$W = g(\mu_1, \mu_2) + \sum_{k=1}^{2} \left[\frac{\partial W}{\partial X_k}\right]_{\boldsymbol{\mu}_x} (X_k - \mu_k) + \frac{1}{2} \sum_{k=1}^{2} \sum_{k'=1}^{2} \left[\frac{\partial^2 W}{\partial X_k \partial X_{k'}}\right]_{\boldsymbol{\mu}_x} (X_k - \mu_k)(X_{k'} - \mu_{k'}) + \varepsilon$$

$$= \bar{w} + \sum_{k=1}^{2} \left[\frac{\partial W}{\partial X_k}\right]_{\boldsymbol{\mu}_x} (X_k - \mu_k) + \frac{1}{2} \sum_{k=1}^{2} \left[\frac{\partial^2 W}{\partial X_k^2}\right]_{\boldsymbol{\mu}_x} (X_k - \mu_k)^2$$

$$+ \left[\frac{\partial^2 W}{\partial X_1 \partial X_2}\right]_{\boldsymbol{\mu}_x} (X_1 - \mu_1)(X_2 - \mu_2) + \varepsilon \tag{5.11}$$

in which $\overline{w} = g(\mu_1, \mu_2)$ and $\boldsymbol{\mu_x} = (\mu_1, \mu_2)^t$. The 1st-order approximation of $W = g(X_1, X_2)$ can be obtained from Eq. (5.11) by dropping the 2nd- and higher-order terms as

$$W \approx g(\mu_1, \mu_2) + \sum_{k=1}^{2} \left[\frac{\partial W}{\partial X_k}\right]_{\boldsymbol{\mu_x}} (X_k - \mu_k) = \overline{w} + \left[\frac{\partial W}{\partial X_1}\right]_{\boldsymbol{\mu_x}} (X_1 - \mu_1) + \left[\frac{\partial W}{\partial X_2}\right]_{\boldsymbol{\mu_x}} (X_2 - \mu_2)$$

$$(5.12)$$

According to Eq. (2.29), the expectation of W by the FOVE method for a bivariate case can be obtained as

$$\mu_w \approx \overline{w} = g(\mu_1, \mu_2) \tag{5.13}$$

Since there are two stochastic basic variables involved, X_1 and X_2, the variance of W can be obtained by considering their possible correlation, according to Eq. (2.48), as

$$\sigma_w^2 \approx \left[\frac{\partial W}{\partial X_1}\right]_{\boldsymbol{\mu_x}}^2 \sigma_1^2 + \left[\frac{\partial W}{\partial X_2}\right]_{\boldsymbol{\mu_x}}^2 \sigma_2^2 + 2 \left[\frac{\partial W}{\partial X_1}\right]_{\boldsymbol{\mu_x}} \left[\frac{\partial W}{\partial X_2}\right]_{\boldsymbol{\mu_x}} \mathrm{Cov}(X_1, X_2) \tag{5.14}$$

in which $\mathrm{Cov}(X_1, X_2)$ is the covariance between the two stochastic basic variables.

Example 5.2   Referring to Example 5.1, consider that the sewer diameter D is also subject to uncertainty due to manufacturing imprecision. Consulting with the manufacturer, it is known that the manufacturing error associated with the pipe is about 5 percent of its nominal diameter. Determine the uncertainty of the sewer flow capacity using the FOVE method for a section of a sewer with a nominal diameter $D = 3.0$ ft and slope $S = 0.005$. The roughness coefficient n has the mean value of 0.015 with a coefficient of variation 0.05. Assume that the correlation coefficient between the roughness coefficient n and diameter D is −0.75.

Solution   Sewer capacity of the specified pipe size and layout can be calculated as

$$Q = 0.463\, n^{-1}\, D^{2.67}\, (0.005)^{0.5} = 0.0327\, n^{-1}\, D^{2.67}$$

The 1st-order Taylor series expansion about $n_o = \mu_n = 0.015$ and $D_o = \mu_D = 3.0$, according to Eq. (5.12), is

$$Q \approx 0.0327\, (0.015)^{-1}\, (3)^{2.67} + [\partial Q/\partial n]_{n=0.015, D=3.0}\, (n - 0.015)$$

$$+ [\partial Q/\partial D]_{n=0.015, D=3.0}\, (D - 3.0)$$

$$= 41.01 + [-0.0327\, (0.015)^{-2}\, (3.0)^{2.67}](n - 0.015)$$

$$+ [\, 0.0327(2.67)(0.015)^{-1}\, (3.0)^{1.67}](D - 3.0)$$

$$= 41.01 - 2733.99\, (n - 0.015) + 36.50(D - 3.0)$$

Therefore, based on Eq. (5.13), the approximated mean of the sewer flow capacity is

$$\mu_Q \approx 41.01\ \mathrm{ft}^3/\mathrm{s}$$

According to Eq. (5.14), the approximated variance of the sewer flow capacity is

$$\sigma_Q^2 \approx (2733.99)^2\, \text{Var}(n) + (36.50)^2\, \text{Var}(D) - 2(2733.99)(36.50)\, \text{Cov}(n, D)$$

The variances of roughness coefficient n and sewer diameter D and the covariance between the two stochastic basic variables can be computed as follows:

$$\text{Var}(n) = (\Omega_n\, \mu_n)^2 = (0.05 \times 0.015)^2 = (7.5 \times 10^{-4})^2$$

$$\text{Var}(D) = (\Omega_D\, \mu_D)^2 = (0.05 \times 3.0)^2 = (1.5 \times 10^{-1})^2$$

$$\text{Cov}(n, D) = \rho_{n,D}\, \sigma_n\, \sigma_D = (-0.75)(0.00075)(0.15) = -8.4375 \times 10^{-5}$$

Therefore, the variance of sewer flow capacity can be computed as

$$\sigma_Q^2 \approx (2733.99)^2\, (7.5 \times 10^{-4})^2 + (36.50)^2\, (1.5 \times 10^{-1})^2$$
$$- 2(2733.99)(36.50)(-8.4375 \times 10^{-5})$$
$$= 2.05^2 + 5.47^2 + 16.84 = 50.95\ (\text{ft}^3/\text{s})^2$$

Hence, the standard deviation of the sewer flow capacity is $\sqrt{50.95} = 7.14$ ft^3/s, which is about 17 percent of the estimated mean sewer flow capacity.

From the preceding calculations for the variance of the sewer flow capacity, one can also estimate the percentage contribution of uncertainties of n and D to the overall uncertainty of the sewer flow capacity. If n and D are uncorrelated, the variance of the sewer flow capacity is

$$\text{Var}(Q) = 2.05^2 + 5.47^2 = 34.12$$

Therefore, the contribution to the total variance of the sewer flow capacity due to the uncertainty in roughness coefficient n is

$$\eta_n = 2.05^2/\text{Var}(Q) = 12.3\ \text{percent}$$

By the same token, the uncertainty in sewer size contributes to $5.47^2/\text{Var}(Q) = 87.7$ percent of the total uncertainty in the sewer flow capacity.

As can be seen, not considering the correlation between the roughness coefficient and sewer size leads to underestimation in total variance of the sewer flow capacity by $(50.95 - 34.12)/50.95 = 33$ percent. It should be noted that the consideration of correlation resulting in an increase in total variance in this example is coincident. The effect of the correlation among stochastic basic variables on the total variability model output, in general, is problem specific, which depends on the strength and sign of the correlation as well as the functional form of the model.

5.1.3   Multivariate FOVE method

Consider that a hydraulic or hydrologic design quantity W is related to K stochastic basic variables $X_1, X_2, \ldots, X_K$ as

$$W = g(\boldsymbol{X}) = g(X_1, X_2, \ldots, X_K) \tag{5.15}$$

where $X = (X_1, X_2,..., X_K)^t$, a K-dimensional column vector of variables in which all X's are subject to uncertainty, the superscript t represents the transpose of a matrix or vector. To extend the previous results, the Taylor series expansion of the function $g(X)$ with respect to a selected point of stochastic basic variables $X = x_o$ in the parameter space can be expressed as

$$W = w_o + \sum_{k=1}^{K} \left(\frac{\partial W}{\partial X_k} \right)_{x_o} (X_k - x_{ko}) + \frac{1}{2} \sum_{i=1}^{K} \sum_{j=1}^{K} \left(\frac{\partial^2 W}{\partial X_i \, \partial X_j} \right)_{x_o} (X_i - x_{io})(X_j - x_{jo}) + \varepsilon \quad (5.16)$$

in which $w_o = g(x_o)$ and ε represents the higher-order terms. Again, the partial derivative terms are called sensitivity coefficients, each representing the rate of change in the model output W with respect to the unit change of the corresponding variable at x_o.

Dropping the higher-order terms represented by ε, Eq. (5.16) is a 2nd-order approximation of the model $g(X)$. Applying the expectation operator to Eq. (5.16), according to Eq. (2.30), results in

$$E[W] \approx w_o + \sum_{k=1}^{K} \left[\frac{\partial W}{\partial X_k} \right]_{x_o} (\mu_k - x_{ko})$$

$$+ \frac{1}{2} \sum_{i=1}^{K} \sum_{j=1}^{K} \left(\frac{\partial^2 W}{\partial X_i \, \partial X_j} \right)_{x_o} [\mathrm{Cov}(X_i, X_j) + (\mu_i - x_{io})(\mu_j - x_{jo})] \quad (5.17)$$

in which μ_k is the mean of the kth stochastic basic variable X_k and the variance for $W = g(X)$, according to Eq. (2.48), can be expressed as

$$\mathrm{Var}[W] \approx \sum_{i=1}^{K} \sum_{j=1}^{K} \left[\frac{\partial W}{\partial X_i} \right]_{x_o} \left[\frac{\partial W}{\partial X_j} \right]_{x_o} [\mathrm{Cov}(X_i, X_j) + (\mu_i - x_{io})(\mu_j - x_{jo})]$$

$$+ \frac{1}{2} \sum_{i=1}^{K} \sum_{j=1}^{K} \sum_{k=1}^{K} \left[\frac{\partial W}{\partial X_i} \right]_{x_o} \left[\frac{\partial^2 W}{\partial X_j \, \partial X_k} \right]_{x_o} E\left[(X_i - x_{io})(X_j - x_{jo})(X_k - x_{ko}) \right] \quad (5.18)$$

As can be seen from Eq. (5.18), when stochastic basic variables are correlated, the estimation of the variance of W using the 2nd-order approximation would require knowledge about the cross-product moments among the stochastic basic variables, which is rarely available in practice. When the stochastic basic variables are independent, Eqs. (5.17) and (5.18) can be simplified, respectively, to

$$E[W] \approx w_o + \sum_{k=1}^{K} \left(\frac{\partial W}{\partial X_k} \right)_{x_o} (\mu_k - x_{ko}) + \frac{1}{2} \sum_{k=1}^{K} \left[\frac{\partial^2 W}{\partial X_k^2} \right]_{x_o} (\mu_k - x_{ko})^2 \quad (5.19)$$

and

$$\text{Var}[W] \approx \sum_{k=1}^{K}\left[\frac{\partial W}{\partial X_k}\right]_{x_o}^2 \sigma_k^2 + \frac{1}{2}\sum_{k=1}^{K}\left[\frac{\partial W}{\partial X_k}\right]_{x_o}\left[\frac{\partial^2 W}{\partial X_k^2}\right]_{x_o} E[(X_k - x_{ko})^3] \qquad (5.20)$$

Referring to Eq. (5.20), the variance of W from the 2nd-order approximation, under the condition that all stochastic basic variables are statistically independent, would require knowing their 3rd moments. For most practical applications where higher-order moments and cross-product moments are not easily available, the 1st-order approximation is frequently adopted.

By truncating the 2nd- and higher-order terms of the Taylor series, the 1st-order approximation of W is reduced to

$$W = g(\mathbf{X}) \approx w_o + \sum_{k=1}^{K}\left(\frac{\partial W}{\partial X_k}\right)_{x_o}(X_k - x_{ko}) \qquad (5.21)$$

or in a matrix form as

$$W \approx g(\boldsymbol{x_o}) + \boldsymbol{s}_o^t \cdot (\boldsymbol{X} - \boldsymbol{x_o}) \qquad (5.22)$$

where $\boldsymbol{s}_o = \nabla_{\boldsymbol{x}}W(\boldsymbol{x_o})$ is the column vector of sensitivity coefficients with each element representing $\partial W/\partial X_k$ evaluated at $\boldsymbol{X} = \boldsymbol{x_o}$. The mean and variance of W by the 1st-order approximation can be expressed, respectively, as

$$E[W] \approx w_o + \sum_{k=1}^{K}\left(\frac{\partial W}{\partial X_k}\right)_{x_o}(\mu_k - x_{ko}) \qquad (5.23)$$

and

$$\text{Var}[W] \approx \text{Var}\left[w_o + \sum_{k=1}^{K}\left(\frac{\partial W}{\partial X_k}\right)_{x_o}(X_k - x_{ko})\right]$$

$$= \sum_{k=1}^{K}\sum_{j=1}^{K}\left(\frac{\partial W}{\partial X_k}\right)_{x_o}\left(\frac{\partial W}{\partial X_j}\right)_{x_o}\text{Cov}(X_k, X_j) \qquad (5.24)$$

In matrix forms, Eqs. (5.23) and (5.24) can be expressed as

$$E[W] \approx w_o + \boldsymbol{s}_o^t \cdot (\boldsymbol{\mu_x} - \boldsymbol{x_o}) \qquad (5.25)$$

and

$$\text{Var}[W] \approx \boldsymbol{s}_o^t \boldsymbol{C}_x \boldsymbol{s}_o \qquad (5.26)$$

in which $\boldsymbol{\mu_x}$ and $\boldsymbol{C}_x$ are the vector of the mean and covariance matrix of the stochastic basic variable $\boldsymbol{X}$, respectively.

Commonly, the 1st-order variance estimation method consists of taking the expansion point $x_o = \mu_x$ at which the mean and variance of W reduce to

$$E[W] \approx g(\mu_x) = \overline{w} \tag{5.27}$$

and

$$\text{Var}[W] \approx s^t \, C_x \, s \tag{5.28}$$

in which $s = \nabla_x W(\mu_x)$ is a K-dimensional vector of the sensitivity coefficients evaluated at $x_o = \mu_x$. When all the stochastic basic variables are independent, the variance of the model output W could be approximated as

$$\text{Var}[W] \approx \sum_{k=1}^{K} s_k^2 \, \sigma_k^2 = s^t D_x \, s \tag{5.29}$$

in which $D_x = \text{diag}(\sigma_1^2, \sigma_2^2, \ldots, \sigma_K^2)$ is a $K \times K$ diagonal matrix of variances of the involved stochastic basic variables. From Eq. (5.29), the ratio $s_k^2 \, \sigma_k^2 / \text{Var}[W]$ indicates the proportion of the overall uncertainty in the model output contributed by the uncertainty associated with the stochastic basic variables X_k.

Example 5.3   Refer to Example 5.2 and consider that all parameters, namely, roughness coefficient n, sewer diameter D, and sewer slope S, in Manning's formula are subject to uncertainty due to the manufacturing imprecision and construction error. Again, the uncertainties associated with the roughness coefficient and pipe diameter are 5 percent of their nominal values. Furthermore, the sewer slope has 5 percent installation error of its intended value 0.005. Determine the uncertainty of the sewer flow capacity using the FOVE method for a section of 3-ft sewer having roughness coefficient with the nominal value 0.015. Assume that the correlation coefficient between the roughness coefficient n and sewer diameter D is −0.75. The sewer slope S is uncorrelated with the other two stochastic basic variables.

Solution   The 1st-order Taylor series expansion of Manning's formula about $n_o = \mu_n = 0.015$, $D_o = \mu_D = 3.0$, and $S_o = \mu_S = 0.005$, according to Eq. (5.21), is

$$Q \approx 0.463(0.015)^{-1}(3)^{2.67}(0.005)^{0.5} + [\partial Q/\partial n](n - 0.015)$$

$$+ [\partial Q/\partial D](D - 3.0) + [\partial Q/\partial S](S - 0.0005)$$

$$= 41.01 + [0.463(-1)(0.015)^{-2}(3.0)^{2.67}(0.005)^{0.5}](n - 0.015)$$

$$+ [0.463(2.67)(0.015)^{-1}(3.0)^{1.67}(0.005)^{0.5}](D - 3.0)$$

$$+ [0.463(0.5)(0.015)^{-1}(3.0)^{2.67}(0.005)^{-0.5}](S - 0.005)$$

$$= 41.01 - 2733.99(n - 0.015) + 36.50(D - 3.0) + 4100.99(S - 0.005)$$

Again, based on Eq. (5.27), the approximated mean of the sewer flow capacity is

$$\mu_Q \approx 41.01 \text{ ft}^3/\text{s}$$

According to Eq. (5.28), the approximated variance of the sewer flow capacity is

$$\sigma_Q^2 \approx (2733.99)^2 \, \text{Var}(n) + (36.50)^2 \, \text{Var}(D) + (4100.99)^2 \, \text{Var}(S)$$
$$- 2(2733.99)(36.50) \, \text{Cov}(n, D) - 2(2733.99)(4100.99) \, \text{Cov}(n, S)$$
$$+ 2(36.50)(4100.99) \, \text{Cov}(D, S)$$

The above expression reduces to

$$\sigma_Q^2 \approx (2733.99)^2 \, \text{Var}(n) + (36.50)^2 \, \text{Var}(D) + (4100.99)^2 \, \text{Var}(S)$$
$$- 2(2733.99)(36.50) \, \text{Cov}(n, D)$$

because $\text{Cov}(n, S) = \text{Cov}(D, S) = 0$. The variances of the pipe slope is

$$\text{Var}(S) = (\Omega_S \, \mu_S)^2 = (0.05 \times 0.005)^2 = 0.00025^2 = 6.25 \times 10^{-8}$$

Using the information from Example 5.3 about $\text{Var}(n)$, $\text{Var}(D)$, and $\text{Cov}(n,D)$, the variance of the sewer flow capacity can be computed as

$$\sigma_Q^2 \approx (2733.99)^2 \, (7.5 \times 10^{-4})^2 + (36.50)^2 \, (1.5 \times 10^{-1})^2$$
$$+ (4100.99)^2 \, (2.5 \times 10^{-4})^2 - 2 \, (2733.99)(36.50)(-8.4375 \times 10^{-5})$$
$$= 2.05^2 + 5.47^2 + 1.03^2 + 16.84 = 52.02 \ (\text{ft}^3/\text{s})^2$$

Hence, the standard deviation of the sewer flow capacity is $\sqrt{52.02} = 7.21$ ft^3/s, which is 17.6 percent of the estimated mean sewer flow capacity.

Without considering the correlation between n and D, the percentage contributions of uncertainty of n, D, and S to the overall uncertainty of the sewer flow capacity are, respectively, 11.9, 85.1, and 3.0 percent. The uncertainty associated with the sewer slope contributes less significantly to the total sewer flow capacity uncertainty compared with the other two stochastic basic variables even though it has the highest sensitivity coefficient among the three. This is because the variance of S, $\text{Var}(S)$, is significantly smaller than the variances of the other two stochastic basic variables n and D.

In general, $E[g(X)] \neq g(\mu_x)$ unless the function $g(X)$ is a linear function of X. Improvement of the accuracy can be made by incorporating the higher-order terms in the Taylor expansion. However, from the preceding descriptions, one immediately realizes that as the higher-order terms are included, both the mathematical complexity and the required information increase rapidly. This is especially true for estimating the variance. The method can be expanded to include the 2nd-order term for improving the estimation of the mean to account for the presence of model nonlinearity and correlation between the stochastic basic variables. From the models involving independent stochastic basic variables in multiplicative and linear forms, Tyagi and Haan (2001) develop the correction factor for $E(X^r)$ estimated by the FOVE method based on the exact moments of some distributions shown in Tables 4.2 and 4.6. It should be noted that for practical problems the difference between the 1st- and 2nd-order approximations of the mean may not be significant. For example, Bates and Townley (1988) found that the 2nd-order approximation offered little improvement in the mean value relative to the results of Monte Carlo simulation (Chap. 6) for an uncertainty analysis of the runoff routing program (RORB) rainfall-runoff model commonly used in Australia. Further, Melching (1995) states, with regard to rainfall-runoff modeling, that it has been his experience that the mean value of the model output is generally well estimated by the 1st-order approximation and the key to improving estimates offered by FOVE is to provide a better estimate of the variance of the output.

In practice, the first two moments are used in the uncertainty analysis for engineering problems. To estimate higher-order moments of W, the FOVE method can be implemented straightforwardly only when the stochastic basic variables are uncorrelated. The method does not require knowledge of the PDF of the stochastic basic variables, which simplifies the analysis. However, this advantage is also the weakness of the FOVE method because it is insensitive to the distributions of the stochastic basic variables in the uncertainty analysis.

The FOVE method is simple and straightforward. The computational effort associated with the method largely depends on the ways the sensitivity coefficients are calculated. For simple analytical functions, the computation of derivatives is a trivial task. However, for functions that are complex, or implicit, or both in the form of computer programs, or charts, or figures, the task of computing the derivatives could become cumbersome. Typically forward, backward, and central difference approaches have been used to numerically approximate the sensitivity coefficients with Δx values between 1 and 10 percent of the basic variable value depending on the sensitivity of the model to the parameters being varied (Melching 1995). In cases where numerical approximation of sensitivity coefficient values is cumbersome, probabilistic point estimation techniques can be used to circumvent the problems. Chowdhury and Xu (1994) demonstrated the use of the rational polynomial technique for more accurate evaluation of the sensitivity coefficients in reliability analysis at the expense of additional computation in terms of function evaluation.

5.2   Rosenblueth's Probabilistic Point Estimation Method

The *Rosenblueth probabilistic point estimation (PPE) method* is a computationally straightforward technique for the uncertainty analysis of engineering problems. Essentially, the method is based on the Taylor series expansion about the means of the stochastic basic variables in a model. It can be used to estimate statistical moments of any order of a model output involving several stochastic basic variables that are either correlated or uncorrelated. The method was originally developed for handling stochastic basic variables that are symmetric

(Rosenblueth 1975). It was later extended to treat nonsymmetric random variables (Rosenblueth 1981).

5.2.1   Univariate Rosenblueth PPE method

Consider a model $W = g(X)$ involving only a single stochastic basic variable X whose first three moments or PDF/PMF are known. Referring to Fig. 5.3, Rosenblueth's PE method approximates the original PDF or PMF of the random variable X by assuming that the entire probability mass of X is concentrated at two points, x_- and x_+. Using the two-point approximation, there are four unknowns to be determined, namely, the locations of x_- and x_+ and the corresponding probability masses p_- and p_+. Because $p_- + p_+ = 1$, three additional side conditions are needed to solve the four unknown quantities. The plausible three side conditions are that the first three moments of the original random variable X must be preserved by the two-point approximation. Without changing the nature of the original problem, it is easier to deal with the standardized variable $X' = (X - \mu_x)/\sigma_x$, which has a zero mean and unit variance. Hence, in terms of X', the following four simultaneous equations can be established to solve for x'_-, x'_+, p_-, and p_+:

$$p_+ + p_- = 1 \tag{5.30a}$$

$$p_+ \, x'_+ - p_- \, x'_- = \mu_{x'} = 0 \tag{5.30b}$$

$$p_+ \, x'^2_+ + p_- \, x'^2_- = \sigma^2_{x'} = 1 \tag{5.30c}$$

$$p_+ \, x'^3_+ - p_- \, x'^3_- = \gamma_x \tag{5.30d}$$

in which $x'_- = |\, x_- - \mu_x \,|/\sigma_x$, $x'_+ = |\, x_+ - \mu_x \,|/\sigma_x$ and γ_x is the skewness coefficient of the stochastic basic variable X. Solving Eqs. (5.30a) to (5.30d) simultaneously,

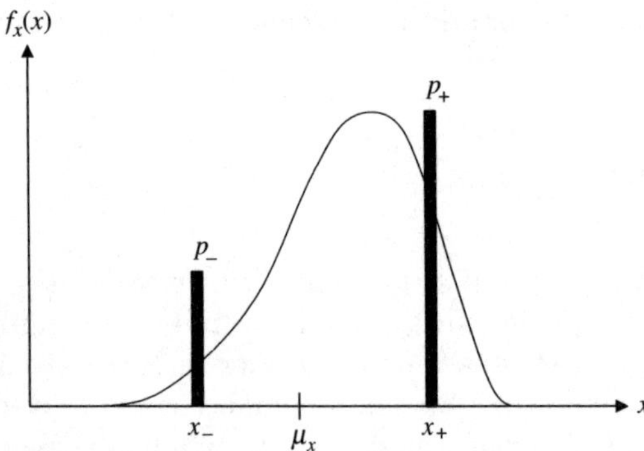

Figure 5.3   The Rosenblueth two-point representation of a PDF.

one obtains

$$x'_+ = \frac{\gamma_x}{2} + \sqrt{1 + \left(\frac{\gamma_x}{2}\right)^2} \tag{5.31a}$$

$$x'_- = x'_+ - \gamma_x \tag{5.31b}$$

$$p_+ = \frac{x'_-}{x'_+ + x'_-} \tag{5.31c}$$

$$p_- = 1 - p_+ \tag{5.31d}$$

When the distribution of the random variable X is symmetric, that is, $\gamma_x = 0$ then the solutions to Eqs. (5.30a) to (5.30d) are $x'_- = x'_+ = 1$ and $p_- = p_+ = 0.5$. This implies that for a symmetric random variable, the two points are located at one standard deviation to either side of the mean with an equal probability mass assigned at the two points.

From x'_- and x'_+, the two points in the original parameter space, x_- and x_+, can respectively be determined as

$$x_- = \mu_x - x'_- \sigma_x \tag{5.32a}$$

$$x_+ = \mu_x + x'_+ \sigma_x \tag{5.32b}$$

Based on x_- and x_+, the values of the model $W = g(X)$ at the two points can be computed, respectively, as $w_- = g(x_-)$ and $w_+ = g(x_+)$. Then, the rth-order moments about the origin of $W = g(X)$ can be estimated as

$$E[W^r] = \mu_{w,r} \approx p_+ w_+^r + p_- w_-^r \tag{5.33}$$

Schematically, the concept is shown in Fig. 5.4. Conversion from moments about the origin to central moments can be made by Eq. (2.21). Specifically, the mean and variance of $W = g(X)$ can, respectively, be expressed as

$$\mu_w = p_+ w_+ + p_- w_- \tag{5.34}$$

$$\sigma_w^2 = \mu'_2 - \mu_w^2 = [p_+ w_+^2 + p_- w_-^2] - \mu_w^2 \tag{5.35}$$

Unlike the FOVE method, the Rosenblueth PPE estimation method provides added capability allowing analysts to account for the asymmetry associated with the PDF of a stochastic basic variable. In addition, Example 5.4 shows that the FOVE method is a 1st-order approximation to the Rosenblueth PE method (Karmeshu and Lara Rosano 1987).

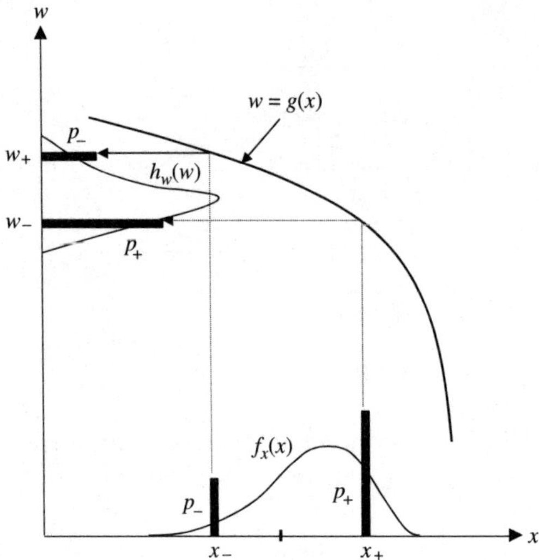

Figure 5.4 Information transfer in the Rosenblueth PE method.

Example 5.4 (Comparison between the FOVE and Rosenblueth PPE methods) Consider a univariate function $W = g(X)$ in which, X has a symmetric distribution with a mean μ_x and standard deviation σ_x. Since the PDF of X is symmetric, the skewness coefficient of X, $\gamma_x = 0$. Then from Eq. (5.31), $p_- = p_+ = 0.5$ and $x'_- = x'_+ = 1$. Hence,

$$x_+ = \mu_x + \sigma_x \qquad \text{and} \qquad x_- = \mu_x - \sigma_x$$

The mean of $W = g(X)$ can be calculated as

$$\mu_W = p_+ w_+ + p_- w_- = \frac{1}{2}\,[g(\mu_x + \sigma_x) + g(\mu_x - \sigma_x)]$$

The Taylor expansions of $g(\mu_x + \sigma_x)$ and $g(\mu_x - \sigma_x)$ with respect to μ_x, from Eq. (5.3), are

$$w_+ = g(\mu_x + \sigma_x) = \overline{w} + \sigma_x\, g'(\mu_x) + \frac{\sigma_x^2}{2!}\, g''(\mu_x) + \cdots$$

$$w_- = g(\mu_x - \sigma_x) = \overline{w} - \sigma_x\, g'(\mu_x) + \frac{\sigma_x^2}{2!}\, g''(\mu_x) + \cdots$$

in which $\overline{w} = g(\mu_x)$, $g'(\mu_x) = (\partial g/\partial X)_{\mu_x}$, and $g''(\mu_x) = (\partial^2 g/\partial^2 X)_{\mu_x}$. The expected value of $W = g(X)$ can be written as

$$\mu_w = \overline{w} + \frac{\sigma_x^2}{2!}\, g''(\mu_x) + \frac{\sigma_x^4}{4!}\, g^{(4)}(\mu_x) + \cdots$$

Similarly, the variance of $W = g(X)$ can be obtained as

$$\sigma_w^2 = \frac{1}{4}[g(\mu_x + \sigma_x) - g(\mu_x - \sigma_x)]^2$$

$$= \sigma_x^2 \, [g'(\mu_x)]^2 + \frac{\sigma_x^4}{3} \, g'(\mu_x) \, g^{(3)}(\mu_x) + \cdots$$

As can be seen, when the standard deviation of random variable X is small, higher-order terms of σ_x can be neglected. This reduces to Eqs. (5.9a) and (5.10), which is obtained from the FOVE method. Therefore, the Rosenblueth PPE method would yield results nearly the same as those from the FOVE method when the uncertainty of the stochastic basic variable is small. In other words, the Rosenblueth PPE method is less restricted by the magnitude of variance associated with the stochastic basic variables than the FOVE method. This conclusion can also be extended to problems involving multiple stochastic basic variables.

Example 5.5   Refer to Example 5.1, which considers Manning's formula for determining the flow capacity of a storm sewer

$$Q = 0.463 \, n^{-1} \, D^{2.67} \, S^{0.5}$$

For a section of a sewer with a diameter $D = 3.0$ ft and slope $S = 0.005$, quantify the uncertainty of the sewer flow capacity by Rosenblueth's PPE method, considering that the roughness coefficient is the only stochastic basic variable having the mean value 0.015 with a coefficient of variation 0.05. Furthermore, the roughness coefficient is assumed to be a symmetric random variable.

Solution   From Example 5.1, Manning's formula for a sewer of the specified pipe size and layout can be rewritten as

$$Q = 0.463 \, n^{-1} \, (3)^{2.67} \, (0.005)^{0.5} = 0.613 \, n^{-1}$$

The standard deviation of the roughness coefficient is

$$\sigma_n = (0.05)(0.015) = 0.00075$$

Since the roughness coefficient is a symmetric random variable, its skewness coefficient is equal to zero, that is, $\gamma_n = 0$. Hence, according to Eqs. (5.30a) to (5.30d), the standardized roughness coefficient $x'_- = x'_+ = 1$ and $p_- = p_+ = 0.5$ and the corresponding values of the roughness coefficient at the two points are

$$n_- = \mu_n - n'_- \, \sigma_n = 0.015 - (1)(0.00075) = 0.01425$$

$$n_+ = \mu_n + n'_+ \, \sigma_n = 0.015 + (1)(0.00075) = 0.01575$$

Substituting the values of n_- and n_+ in Manning's formula to compute the corresponding sewer capacities, one has

$$Q_- = 0.615 \, (n_-)^{-1} = 0.615(0.01425)^{-1} = 43.17 \text{ ft}^3/\text{s}$$

$$Q_+ = 0.615 \, (n_+)^{-1} = 0.615(0.01575)^{-1} = 39.06 \text{ ft}^3/\text{s}$$

Therefore, the rth-order moment about the origin of the sewer flow capacity can be estimated by

$$E(Q^r) = p_+ \, (Q_+)^r + p_- \, (Q_-)^r = 0.5 \, (Q_+)^r + 0.5 \, (Q_-)^r \qquad \text{for } r = 1, 2,\ldots.$$

Specifically, the mean of the sewer flow capacity can be estimated with $r = 1$ as

$$\mu_Q = 0.5(43.17) + 0.5(39.06) = 41.11 \text{ ft}^3/\text{s}$$

For the variance of the sewer flow capacity, the 2nd-order product-moment about the origin is first computed as

$$E(Q^2) = 0.5(43.17)^2 + 0.5(39.06)^2 = 1694.48 \ (\text{ft}^3/\text{s})^2$$

Then, the variance of the sewer flow capacity can be estimated by Eq.(2.36) as

$$\text{Var}(Q) = E(Q^2) - (\mu_Q)^2 = 1694.48 - (41.11)^2 = 4.23 \ (\text{ft}^3/\text{s})^2$$

Hence, the standard deviation of the sewer flow capacity is $\sqrt{4.23} = 2.06$ ft^3/s. Comparing with the results in Example 5.1, one observes that the results by the Rosenblueth PPE method, herein, are practically identical to those of the FOVE method. This is primarily due to a relatively small coefficient of variation associated with the random roughness coefficient. Also, the symmetry for the roughness coefficient is, in effect, the same as using the first two moments in the analysis as the FOVE method does. One can practice Prob. 5.15 to examine the effect of the skew coefficient on the results.

5.2.2   Bivariate Rosenblueth PPE method

In the bivariate case, a model or a function involves two stochastic basic variables, $W = g(X_1, X_2)$. By the Rosenblueth PPE method, each of the two random variables is represented by two points (x_{k-}, x_{k+}) for $k = 1, 2$. Therefore, there are four possible combinations of the points, namely, (x_{1-}, x_{2-}), (x_{1-}, x_{2+}), (x_{1+}, x_{2-}), and (x_{1+}, x_{2+}), in a two-dimensional random parameter sample space as shown in Fig. 5.5. For simplicity, the subscripts used for probabilities and locations are collapsed by dropping the first one. For example, the location x_{-+} denotes the point (x_{1-}, x_{2+}) in the parameter space and p_{-+} represents the corresponding probability mass at the location. Similar to Eq. (5.34), the rth-order product-moment about the origin of the model output W for a bivariate case can be expressed as

$$E[W^r] = p_{++} \, w_{++}^r + p_{+-} \, w_{+-}^r + p_{-+} \, w_{-+}^r + p_{--} \, w_{--}^r \tag{5.36}$$

in which $w_{++} = g(x_{1+}, x_{2+})$, $w_{+-} = g(x_{1+}, x_{2-})$ and so on, with $x_{k\pm} = \mu_k \pm x'_{k\pm} \, \sigma_k$ for $k = 1, 2$ and $x'_{k\pm}$ being computed according to Eqs.(5.31a) and (5.31b). A schematic diagram showing the information transfer by the Rosenblueth PPE method for bivariate problems is given in Fig. 5.6. To compute the probability

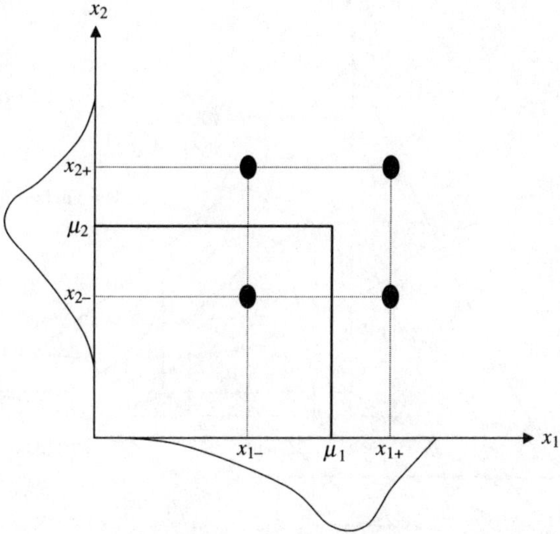

Figure 5.5   Locations of probability masses in two-dimensional sample space.

masses at the four locations, while taking into account the correlation between the two stochastic basic variables and their respective skewness coefficients, Rosenblueth (1981) made the following hypothesis:

$$p_{++} = p_{1+}\, p_{2+} + a \tag{5.37a}$$

$$p_{+-} = p_{1+}\, p_{2-} - a \tag{5.37b}$$

$$p_{-+} = p_{1-}\, p_{2+} - a \tag{5.37c}$$

$$p_{--} = p_{1-}\, p_{2-} + a \tag{5.37d}$$

in which the constant a is determined to preserve the correlation coefficient ρ between the two random variables. This constant a can be obtained by solving

$$E(X_1' X_2') = p_{++}\, x_{1+}' x_{2+}' - p_{+-}\, x_{1+}' x_{2-}' - p_{-+}\, x_{1-}' x_{2+}' + p_{--}\, x_{1-}' x_{2-}' = \rho$$

and the result is

$$a = \frac{\rho/4}{\sqrt{\left[1+\left(\frac{\gamma_1}{2}\right)^2\right]\left[1+\left(\frac{\gamma_2}{2}\right)^2\right]}} \tag{5.38}$$

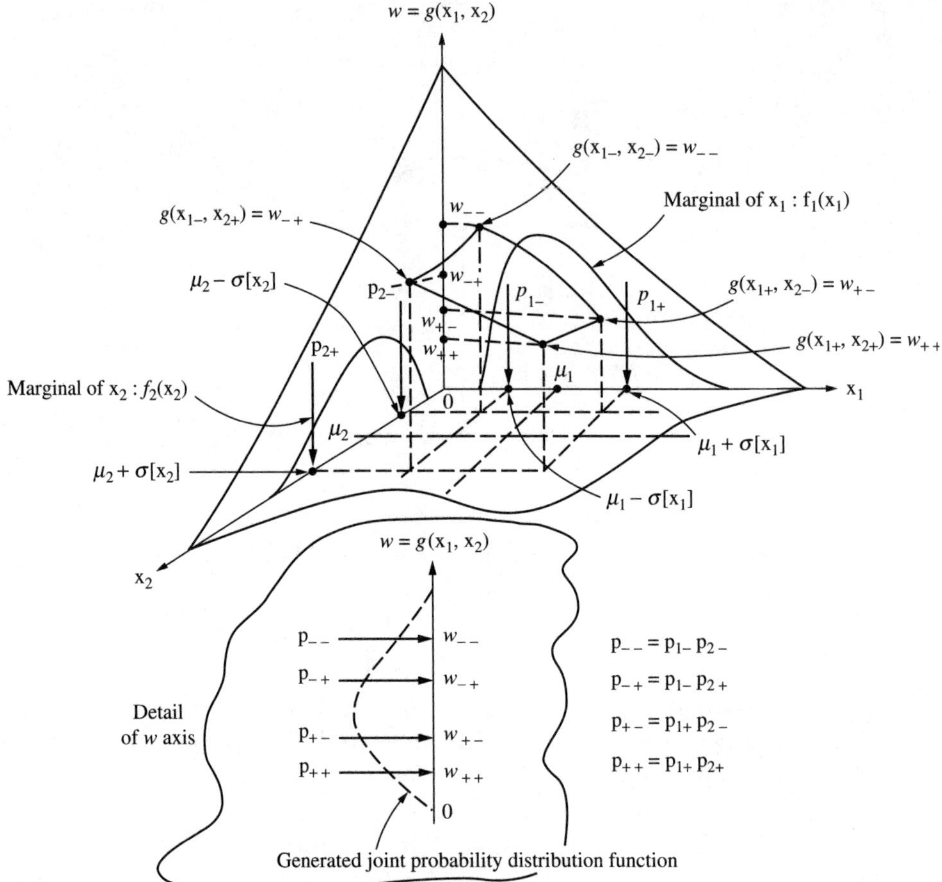

Figure 5.6 Information transfer by the Rosenblueth PPE method for bivariate problems (after Harr 1987).

When the two random variables are symmetric, Eqs. (5.37a) to (5.37d) reduce to

$$p_{++} = p_{--} = \frac{1+\rho}{4} \tag{5.39a}$$

$$p_{+-} = p_{-+} = \frac{1-\rho}{4} \tag{5.39b}$$

The effect of the correlation coefficient ρ on the probability mass concentration for symmetric bivariate cases is shown for a few cases in Fig. 5.7.

Example 5.6 Referring to Example 5.5 consider that both sewer diameter D and roughness coefficient are subject to uncertainty. It is known that the manufacturing error associated with the pipe diameter is about 5 percent of its nominal value. Determine the uncertainty of the sewer flow capacity using the Rosenblueth PPE

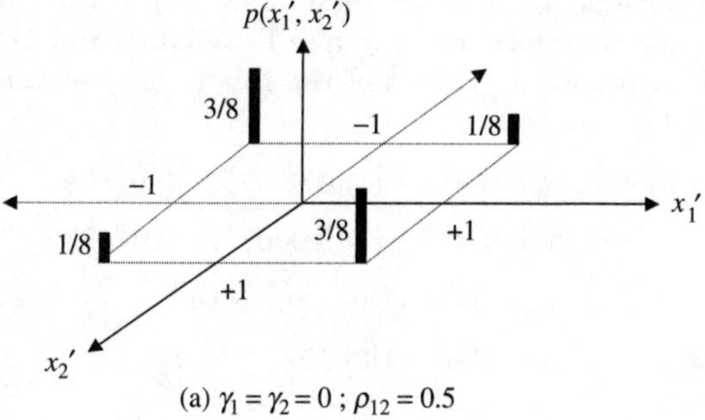

(a) $\gamma_1 = \gamma_2 = 0$; $\rho_{12} = 0.5$

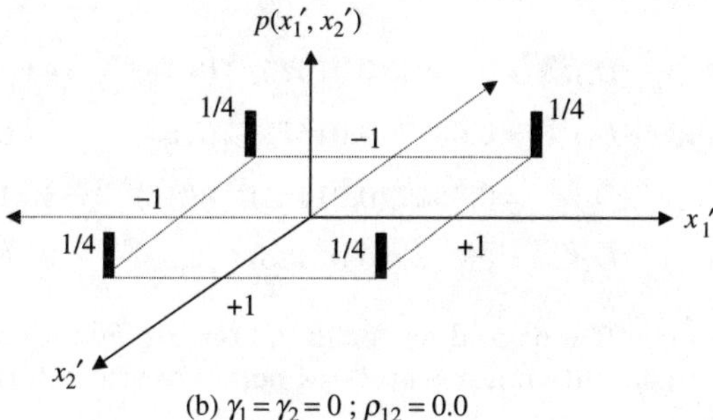

(b) $\gamma_1 = \gamma_2 = 0$; $\rho_{12} = 0.0$

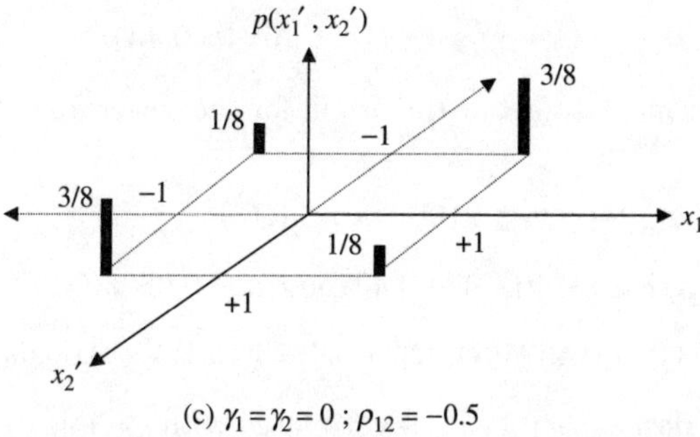

(c) $\gamma_1 = \gamma_2 = 0$; $\rho_{12} = -0.5$

Figure 5.7 Effect of correlation on probability mass distribution for symmetric bivariate random variables.

method for a section of sewer with a nominal diameter $D = 3.0$ ft and slope $S = 0.005$. The roughness coefficient has the mean value 0.015 with a coefficient of variation 0.05. Assume that both the sewer diameter and roughness coefficient are symmetric random variables with a correlation coefficient -0.75.

Solution   From Example 5.1, Manning's formula for the sewer of a specified slope can be rewritten as

$$Q = 0.463 \; n^{-1} \; D^{2.67} \; (0.005)^{0.5} = 0.0327 \; n^{-1} \; D^{2.67}$$

The standard deviations of the roughness coefficient and sewer diameter are, respectively, $\sigma_n = 0.00075$ and $\sigma_D = 0.15$. Since the roughness coefficient and sewer diameter are symmetric random variables, their skewness coefficients are equal to zero, that is, $\gamma_n = 0$, $\gamma_D = 0$. Therefore, according to Eqs.(5.30a) and (5.30b), $n'_- = n'_+ = D'_- = D'_+ = 1$ and the corresponding values of the roughness coefficient and sewer diameter are

$$n_+ = \mu_n + n'_+\, \sigma_n = 0.015 + (1)(0.00075) = 0.01575$$

$$n_- = \mu_n - n'_-\, \sigma_n = 0.015 - (1)(0.00075) = 0.01425$$

$$D_+ = \mu_D + D'_+\, \sigma_D = 3.0 + (1)(0.15) = 3.15 \text{ ft}$$

$$D_- = \mu_D - D'_-\, \sigma_D = 3.0 - (1)(0.15) = 2.85 \text{ ft}$$

Substituting the values of n_-, n_+, D_-, and D_+ into Manning's formula to compute the corresponding sewer capacities, one obtains

$$Q_{++} = 0.0327\,(n_+)^{-1}\,(D_+)^{2.67} = 0.0327\,(0.01575)^{-1}\,(3.15)^{2.67} = 44.44 \text{ ft}^3/\text{s}$$

$$Q_{+-} = 0.0327\,(n_+)^{-1}\,(D_-)^{2.67} = 0.0327\,(0.01575)^{-1}\,(2.85)^{2.67} = 34.02 \text{ ft}^3/\text{s}$$

$$Q_{-+} = 0.0327\,(n_-)^{-1}\,(D_+)^{2.67} = 0.0327\,(0.01425)^{-1}\,(3.15)^{2.67} = 49.12 \text{ ft}^3/\text{s}$$

$$Q_{--} = 0.0327\,(n_-)^{-1}\,(D_-)^{2.67} = 0.0327\,(0.01425)^{-1}\,(2.85)^{2.67} = 37.60 \text{ ft}^3/\text{s}$$

Because the roughness coefficient and sewer diameter are symmetric, correlated random variables, the probability masses at $2^2 = 4$ points can be determined, according to Eqs. (5.39a) and (5.39b), as

$$p_{++} = p_{--} = (1 + \rho_{n,D})/4 = (1 - 0.75)/4 = 0.0625$$

$$p_{+-} = p_{-+} = (1 - \rho_{n,D})/4 = (1 + 0.75)/4 = 0.4375$$

Therefore, the rth-order moment about the origin for the sewer flow capacity can be estimated by

$$E(Q^r) = p_{++}\,(Q_{++})^r + p_{+-}\,(Q_{+-})^r + p_{-+}\,(Q_{-+})^r + p_{--}\,(Q_{--})^r$$

$$= 0.0625(Q_{++})^r + 0.4375(Q_{+-})^r + 0.4375(Q_{-+})^r + 0.0625(Q_{--})^r$$

$$= 0.0625(44.44)^r + 0.4375(34.02)^r + 0.4375(49.12)^r + 0.0625(37.60)^r$$

The mean of the sewer flow capacity can be estimated with $r = 1$ as

$$\mu_Q = 0.0625(44.44) + 0.4375(34.02) + 0.4375(49.12) + 0.0625(37.60)$$

$$= 41.50 \text{ ft}^3/\text{s}$$

For the variance of the sewer flow capacity, the 2nd-order product-moment about the origin is first computed as

$$E(Q^2) = 0.0625(44.44)^2 + 0.4375(34.02)^2 + 0.4375(49.12)^2 + 0.0625(37.60)^2$$

$$= 1773.46 \text{ (ft}^3/\text{s)}^2$$

Then, the variance of the sewer flow capacity can be estimated as

$$\text{Var}(Q) = E(Q^2) - (\mu_Q)^2 = 1773.46 - (41.50)^2 = 51.36 \text{ (ft}^3/\text{s)}^2$$

Hence, the standard deviation of the sewer flow capacity is $\sqrt{51.36} = 7.17$ ft^3/s. Comparing with the results in Example 5.2, one observes that the results by the Rosenblueth PPE method, herein, yield slightly higher mean and variance of the sewer flow capacity than those of the FOVE method.

5.2.3   Multivariate Rosenblueth PPE method

In a general case where a model involves K correlated stochastic basic variables, the rth-order product-moment of the model output $W = g(X) = g(X_1, X_2,..., X_K)$ about the origin can be approximated as

$$E(W^m) \approx \sum p_{\delta_1,\delta_2,...,\delta_K} \times w^m_{\delta_1,\delta_2,...,\delta_K} \tag{5.40}$$

in which the subscript δ_k is a sign indicator and can only be $+$ or $-$ representing the stochastic basic variable X_k having the value of $x_{k+} = \mu_x + x'_{k+}\sigma_x$ or $x_{k-} = \mu_x - x'_{k-}\sigma_x$, respectively; the probability mass at each of the 2^K points, $p_{\delta_1,\delta_2,...,\delta_K}$, based on an extension of Eq. (5.37), can be approximated by

$$p_{\delta_1,\delta_2,...,\delta_K} = \prod_{k=1}^{K} p_{k,\delta_k} + \sum_{k=1}^{K-1}\left(\sum_{j=k+1}^{K} \delta_k\delta_j a_{kj} \right) \tag{5.41}$$

with

$$a_{ij} = \frac{\rho_{ij}/2^K}{\sqrt{\Pi_{k=1}^{K}\left[1 + \left(\frac{\gamma_k}{2}\right)^2\right]}} \tag{5.42}$$

where ρ_{ij} is the correlation coefficient between stochastic basic variables X_i and X_j. The number of terms in the summation of Eq. (5.40) is 2^K, which corresponds to the total number of possible combinations of $+$ and $-$ for all K stochastic basic variables.

It should be stressed that Eq. (5.41) for computing the probability masses is only an approximation. In fact, for a multivariate problem involving $K \geq 3$ skewed stochastic basic variables the number of unknowns in the Rosenblueth PPE method is $2K + 2^K$ where the term $2K$ represents the number of unknown coordinates for K stochastic basic variables and the term 2^K represents the number of unknown probability masses at each of the 2^K points in the K-dimensional parameter space. On the other hand, the number of side conditions allowing the determination of $2K + 2^K$ unknowns is $1 + 3K + K(K - 1)/2$, where the term 1 represents the condition that the sum of all 2^K probabilities equals unity, the term $3K$ arises from the preservation of the first three statistical moments (i.e., the mean, variance, and skewness coefficient) for each of the K stochastic basic variables, and the term $K(K - 1)/2$ represents the number of

TABLE 5.1   Number of Unknowns and Conditions by the Rosenblueth PE Method

Number of stochastic variables, K	Number of unknowns			Number of conditions, $1 + 3K + (K-1)/2$
	Unknown coordinates, $2K$	Unknown probability masses, 2^K	Total number of unknowns, $2K+2^K$	
2	4	4	8	8
3	6	8	14	13
4	8	16	24	19
5	10	32	42	26
10	20	1024	1044	76

correlation coefficients for all possible pairs of K stochastic basic variables. Table 5.1 shows the number of unknowns and conditions for different numbers of stochastic basic variables. As can be seen, when $K \geq 3$ the number of unknowns starts to exceed the number of available conditions. Consequently, the problem of determining the coordinates and the corresponding probability masses for $K \geq 3$ is indeterminate and the solution is not unique. Equations (5.41) and (5.42) represent only a simple approximated solution to the problem without extensive derivation. As shown in the various examples in this chapter, the accuracy of this simple approximation is quite compatible and desirable as compared with other methods. Panchalingam and Harr (1994) derived a much more complicated solution for this indeterminate problem. Nevertheless, their solution is still an approximation. To circumvent the nonuniqueness of the Rosenblueth method (for $K \geq 3$), Tsai and Franceschini (2002) modified the ways of preserving correlations by approximating the original function $W(\mathbf{X})$, according to Li (1992) (see Sec. 5.4), as

$$W(\mathbf{X}) \cong \overline{w} + \sum_{k=1}^{K} a_k(X_k - \mu_k) + \sum_{k=1}^{K} b_k(X_k - \mu_k)^2 + \sum_{k=1}^{K} c_k(X_k - \mu_k)^3$$

$$+ \sum_{k=1}^{K-1} \sum_{k'=k+1}^{K} e_{kk'}(X_k - \mu_k)(X_{k'} - \mu_{k'}) \tag{5.43}$$

In doing so the number of function evaluations by the methods of Tsai-Franceschini and Li is $(K^2 + 3K + 2)/2$, which will be smaller than the 2^K need for the Rosenblueth method for $K \geq 4$.

Example 5.7   Consider W as a function of three symmetric, correlated variables X_1, X_2, and X_3. The rth-order product-moment of W about the origin can be approximated as

$$E(W) \approx p_{+++}\, w^r_{+++} + p_{---}\, w^r_{---} + p_{++-}\, w^r_{++-} + p_{--+} + w^r_{--+}$$

$$+ p_{+-+}\, w^m_{+-+} + p_{-+-}\, w^m_{-+-} + p_{-++}\, w^m_{-++} + p_{+--}\, w^m_{+--} \tag{5.44}$$

where

$$p_{+++} = p_{---} = (1 + \rho_{12} + \rho_{13} + \rho_{23})/8$$
$$p_{++-} = p_{--+} = (1 + \rho_{12} - \rho_{13} - \rho_{23})/8$$
$$p_{+-+} = p_{-+-} = (1 - \rho_{12} + \rho_{13} - \rho_{23})/8 \qquad (5.45)$$
$$p_{-++} = p_{+--} = (1 - \rho_{12} - \rho_{13} + \rho_{23})/8$$

and

$$w_{+++} = g(\mu_1 + \sigma_1, \mu_2 + \sigma_2, \mu_3 + \sigma_3)$$
$$w_{+-+} = g(\mu_1 + \sigma_1, \mu_2 - \sigma_2, \mu_3 + \sigma_3)$$
$$\vdots$$

Example 5.8   Refer to Example 5.6 and consider that all model parameters, i.e., roughness coefficient n, sewer diameter D, and sewer slope S, are subject to uncertainty. The errors associated with the roughness coefficient and pipe diameter are 5 percent of their nominal values. Furthermore, the uncertainty associated with the sewer slope due to installation error is 5 percent of its intended value, $S = 0.005$. Determine the uncertainty of the sewer flow capacity using the Rosenblueth PPE method for a section of 3-ft sewer. The roughness coefficient has the nominal value 0.015 with a coefficient of variation 0.05. Assume that all three stochastic basic variables are symmetric random variables. The correlation coefficient between the roughness coefficient n and sewer diameter D is -0.75, whereas, the sewer slope S is uncorrelated with the other two stochastic basic variables.

Solution   From Example 5.1, Manning's formula for the sewer is

$$Q = 0.463 \; n^{-1} \; D^{2.67} \; S^{0.5}$$

The standard deviation of the roughness coefficient, sewer diameter, and pipe slope are

$$\sigma_n = 0.00075 \qquad \sigma_D = 0.15 \qquad \sigma_S = 0.00025$$

With $K = 3$ random variables, there are a total of $2^3 = 8$ possible points. Since all the three stochastic basic variables are symmetric random variables, their skewness coefficients are equal to zero, that is, $\gamma_n = 0$, $\gamma_D = 0$, and $\gamma_S = 0$. Therefore, according to Eqs. (5.30a) and (5.30b), $n'_- = n'_+ = D'_- = D'_+ = S'_- = S'_+ = 1$ and the corresponding values of the roughness coefficient, sewer diameter, and pipe slope are

$$n_+ = \mu_n + n'_+ \, \sigma_n = 0.015 + (1)(0.00075) = 0.01575$$

$$n_- = \mu_n + n'_- \, \sigma_n = 0.015 - (1)(0.00075) = 0.01425$$

$$D_+ = \mu_D + D'_+ \, \sigma_D = 3.0 + (1)(0.15) = 3.15 \text{ ft}$$

$$D_- = \mu_D - D'_- \, \sigma_D = 3.0 - (1)(0.15) = 2.85 \text{ ft}$$

$$S_+ = \mu_S + S'_+ \, \sigma_S = 0.005 + (1)(0.00025) = 0.00525$$

$$S_- = \mu_S + S'_- \, \sigma_S = 0.005 - (1)(0.00025) = 0.00475$$

Substituting the values of n_-, n_+, D_-, D_+, S_-, and S_+ in Manning's formula to compute the corresponding sewer capacities, one has, for instance,

$$Q_{+++} = 0.463 \, (n_+)^{-1} \, (D_+)^{2.67} \, (S_+)^{0.5}$$

$$= 0.463 \, (0.01575)^{-1} \, (3.15)^{2.67} \, (0.00525)^{0.5} = 45.59 \text{ ft}^3/\text{s}$$

Similarly, the values of sewer flow capacity for the other seven points are given in the following table:

Point	n	D	S	Q (ft^3/s)	p
1	+	+	+	45.59	0.03125
2	+	+	−	43.36	0.03125
3	+	−	+	34.90	0.21875
4	+	−	−	33.20	0.21875
5	−	+	+	50.39	0.21875
6	−	+	−	47.93	0.21875
7	−	−	+	38.57	0.03125
8	−	−	−	36.69	0.03125

Because the roughness coefficient and sewer diameter are symmetric, correlated random variables, the probability masses at $2^3 = 8$ points can be determined, according to Eqs. (5.45) in Example 5.7 as

$$p_{+++} = p_{---} = (1 + \rho_{nD} + \rho_{nS} + \rho_{DS})/8 = (1 - 0.75 + 0 + 0)/8 = 0.03125$$

$$p_{++-} = p_{--+} = (1 + \rho_{nD} - \rho_{nS} - \rho_{DS})/8 = (1 - 0.75 - 0 - 0)/8 = 0.03125$$

$$p_{+-+} = p_{-+-} = (1 - \rho_{nD} + \rho_{nS} - \rho_{DS})/8 = (1 + 0.75 + 0 - 0)/8 = 0.21875$$

$$p_{-++} = p_{+--} = (1 - \rho_{nD} - \rho_{nS} + \rho_{SD})/8 = (1 + 0.75 - 0 + 0)/8 = 0.21875$$

The values of probability masses are also tabulated in the last column of the preceding table. Therefore, the rth-order moment about the origin for the sewer flow capacity can be calculated by Eq. (5.44). The computations of the first two moments about the origin are shown in the following table, with columns 1 to 3 extracted from the above table and others self-explanatory:

Point (1)	Q (2)	p (3)	$Q \times p$ (4)	Q^2 (5)	$Q^2 \times p$ (6)
1	45.59	0.03125	1.42	2078.44	64.95
2	43.36	0.03125	1.36	1880.50	58.77
3	34.90	0.21875	7.63	1217.96	266.43
4	33.20	0.21875	7.26	1101.96	241.05
5	50.39	0.21875	11.02	2539.04	555.42
6	47.93	0.21875	10.48	2297.23	502.52
7	38.57	0.03125	1.21	1487.86	46.50
8	36.69	0.03125	1.15	1346.16	42.07
Sum	—	1.00	41.53	—	1777.70

The sums of columns 4 and 6 yield, respectively, $\mu_Q = E(Q) = 41.53$ ft^3/s and $E(Q^2) = 1777.70$ (ft^3/s)^2. Therefore, the variance of the sewer flow capacity can be estimated as

$$\text{Var}(Q) = E(Q^2) - (\mu_Q)^2 = 1777.70 - (41.53)^2 = 52.56 \ (\text{ft}^3/\text{s})^2$$

Hence, the standard deviation of the sewer flow capacity is $\sqrt{52.56} = 7.25$ ft^3/s. Comparing with the results in Example 5.3, one observes that the results by the Rosenblueth PE method, herein, yields slightly higher value of the mean and variance for the sewer flow capacity than those by the FOVE method.

For each term of the summation in Eq. (5.40), the model has to be evaluated once at the corresponding point in the parameter space. This indicates a potential difficulty of the Rosenblueth PPE method when applied to practical problems. When K is small, the method is practical for performing uncertainty analysis. However, as shown in Table 5.1, for a moderate or large K, the number of required function evaluations of $g(X)$ could be too numerous for practical implementation, even on a computer. To circumvent this shortcoming, the next section describes a method developed by Harr (1987) that reduces the 2^K function evaluations required by the Rosenblueth method down to $2K$.

There is another potential problem associated with Eqs. (5.41) and (5.42) that should be pointed out. In multivariate problems with $K \geq 3$, under some combinations of correlation coefficients and skewness coefficients among stochastic basic variables, the probabilities computed by Eq. (5.41) for some points could be negative. This could be attributed to the fact that the two equations are approximated solutions to an indeterminate problem. Numerical experiences gathered thus far indicate that these negative probabilities, when they occur, are relatively small and can be treated as zeros. However, adjustments of the probability masses are advisable before carrying out the computations for estimating the statistical moments. The modification made by Tsai and Francechini (2002) not only brings down the computation to a practically affordable level of $(K^2 + 3K + 2)/2$, but also circumvents the issue of determining the join probabilities as required by the Rosenblueth method.

5.3 Harr's Probabilistic Point Estimation Method

As described in the previous section, the Rosenblueth PPE method requires 2^K model evaluations when a model involves K stochastic basic variables. For a moderate or large K, the required computation could be very intensive. To circumvent this difficulty, Harr (1989) proposed an alternative probabilistic PPE method, which reduces the required model evaluations from 2^K to $2K$ and greatly enhances the applicability of the PPE method for the uncertainty analysis of practical problems. The method is a second-moment method, which is capable of taking into account the first two moments (i.e., the mean and variance) of the involved stochastic basic variables and their correlations. Skewness coefficients of the stochastic basic variables are ignored by the Harr method. Hence, the method is appropriate for treating stochastic basic normal variables. For problems that involve only a single stochastic basic variable, Harr's PPE method is identical to the Rosenblueth method with a zero skewness coefficient. The theoretical basis of the Harr PPE method is built on orthogonal transformations of the correlation (or variance-covariance) matrix. Therefore, this section starts with a description of the orthogonal transformation.

5.3.1 Orthogonal transformation

The *orthogonal transformation* is an important tool for treating problems with correlated stochastic basic variables. The main objective of the transformation is to map correlated stochastic basic variables from their original sample space to a new domain in which they become uncorrelated. Hence, the analysis to be conducted is greatly simplified.

Consider K multivariate stochastic basic variables $X = (X_1, X_2, \ldots, X_K)^t$ having a mean vector $\boldsymbol{\mu}_x = (\mu_1, \mu_2, \ldots, \mu_K)^t$ and covariance matrix C_x as

$$C_x = \begin{bmatrix} \sigma_{11} & \sigma_{12} & \sigma_{13} & \cdots & \sigma_{1K} \\ \sigma_{21} & \sigma_{22} & \sigma_{23} & \cdots & \sigma_{2K} \\ \vdots & \vdots & \vdots & \vdots & \vdots \\ \sigma_{K1} & \sigma_{K2} & \sigma_{K3} & \cdots & \sigma_{KK} \end{bmatrix}$$

in which $\sigma_{ij} = \text{Cov}[X_i, X_j]$, the covariance between stochastic basic variables X_i and X_j. The vector of correlated standardized stochastic basic variables $X' = D_x^{-1/2}(X - \mu_x)$, that is, $X' = (X'_1, X'_2, \ldots, X'_K)^t$ with $X'_k = (X_k - \mu_k)/\sigma_k$ for $k = 1, 2, \ldots, K$; and $D_x = \text{diag}(\sigma_1^2, \sigma_2^2, \ldots, \sigma_K^2)$ an $K \times K$ diagonal matrix of the variances of the stochastic basic variables, would have a mean vector of zero 0 and covariance matrix equal to the correlation matrix R_x

$$C_{x'} = R_x = \begin{bmatrix} 1 & \rho_{12} & \rho_{13} & \cdots & \rho_{1K} \\ \rho_{21} & 1 & \rho_{23} & \cdots & \rho_{2K} \\ \vdots & \vdots & \vdots & \cdots & \vdots \\ \rho_{K1} & \rho_{K2} & \rho_{K3} & \cdots & 1 \end{bmatrix}$$

Note that, from Sec. 2.4.5, the covariance matrix and correlation matrix are symmetric matrices, that is, $\sigma_{ij} = \sigma_{ji}$ and $\rho_{ij} = \rho_{ji}$ for $i \neq j$. Furthermore, both matrices should theoretically be positive-definite.

In the orthogonal transformation, a $K \times K$ square matrix T (called the *transformation matrix*) is used to transform the standardized correlated stochastic basic variables, X', into a set of uncorrelated standardized stochastic basic variables Y as

$$Y = T^{-1} X' \tag{5.46}$$

where Y is a vector with the mean vector 0 and covariance matrix I, a $K \times K$ identity matrix. Stochastic variables Y are uncorrelated because the off-diagonal elements of the covariance matrix are all zeros. If the original stochastic basic variables X are multivariate normal variables, then Y is a vector of uncorrelated, standardized normal variables, specifically designated as Z', because the right-hand side of Eq. (5.46) is a linear combination of the normal random vector.

It can be shown that, from Eq. (5.46) the transformation matrix T must satisfy

$$R_x = TT^t \qquad (5.47)$$

There are several methods that allow one to determine the transformation matrix in Eq. (5.47). Due to the fact that R_x is a symmetric and positive-definite matrix, it can be decomposed into

$$R_x = LL^t \qquad (5.48)$$

in which L is a $K \times K$ *lower triangular matrix* (Young and Gregory 1973; Golub and Van Loan 1989)

$$L = \begin{bmatrix} l_{11} & 0 & 0 & \cdots & 0 \\ l_{21} & l_{22} & 0 & \cdots & 0 \\ \vdots & \vdots & \vdots & \cdots & \vdots \\ l_{K1} & l_{K2} & l_{K3} & \cdots & l_{KK} \end{bmatrix}$$

which is unique. Comparing Eqs. (5.47) and (5.48), the transformation matrix T is the lower triangular matrix L. An efficient algorithm to obtain such a lower triangular matrix for a symmetric and positive-definite matrix is the *Cholesky decomposition* (or *Cholesky factorization*) method (see App. 5A).

The orthogonal transformation can alternatively be made using the *eigenvalue-eigenvector decomposition* or *spectral decomposition* by which R_x is decomposed as

$$R_x = C_{x'} = V \Lambda V^t \qquad (5.49)$$

where V is a $K \times K$ eigenvector matrix consisting of K eigenvectors as $V = (v_1, v_2, \ldots, v_K)$ with v_k being the kth eigenvector of the correlation matrix, R_x; and $\Lambda = $ diag $(\lambda_1, \lambda_2, \ldots, \lambda_K)$ being a diagonal eigenvalues matrix. Frequently, the eigenvectors v's are normalized such that the norm is equal to unity, that is, $v^t v = 1$. Furthermore, it should also be noted that the eigenvectors are orthogonal, that is, $v_i^t v_j = 0$ for $i \neq j$ and, therefore, eigenvector matrix V obtained from Eq. (5.49) is an orthogonal matrix satisfying $VV^t = V^t V = I$ where I is an identity matrix (Graybill 1983). The above orthogonal transform satisfies

$$V^t R_x V = \Lambda \qquad (5.50)$$

To achieve the objective of breaking the correlation among the standardized stochastic basic variables X', the following transformation based on the eigenvector matrix can be made:

$$U = V^t X' \qquad (5.51)$$

The resulting transformed stochastic variables U have the mean and covariance matrix as

$$E[U] = V^t E[X'] = 0 \qquad (5.52a)$$

and

$$C_u = V^t\, C_{x'}\, V = V^t\, R_x\, V = \Lambda \qquad (5.52b)$$

As can be seen, in the new vector of stochastic basic variables U obtained by Eq. (5.52) the variables are uncorrelated because their covariance matrix C_u is a diagonal matrix Λ. Hence, each new stochastic basic variable U_k has the standard deviation equal to $\sqrt{\lambda_k}$, for all $k = 1, 2,\ldots, K$.

The vector U can further be standardized as

$$Y = \Lambda^{-1/2}\, U \qquad (5.53)$$

Based on the definitions of the stochastic basic variable vectors $X \sim (\mu_x, C_x)$, $X' \sim (0, R_x)$, $U \sim (0, \Lambda)$, and $Y \sim (0, I)$ given previously, relationships between them can be summarized as the following:

$$Y = \Lambda^{-1/2}\, U = \Lambda^{-1/2}\, V^t\, X' \qquad (5.54)$$

Comparing Eqs. (5.46) and (5.54), it is clear that

$$T^{-1} = \Lambda^{-1/2}\, V^t$$

Applying an inverse operator on both sides of the equation, the transformation matrix T can alternatively, as opposed to Eq. (5.48), be obtained as

$$T = V\Lambda^{1/2} \qquad (5.55)$$

Using the transformation matrix T as given above, Eq. (5.46) can be expressed as

$$X' = TY = V\,\Lambda^{1/2}\, Y \qquad (5.56a)$$

and the random vector in the original parameter space is

$$X = \mu_x + D_x^{1/2}VL^{1/2}Y = \mu_x + D_x^{1/2}LY \qquad (5.56b)$$

Geometrically, the stages involved in the orthogonal transformation from the originally correlated parameter space to the standardized, uncorrelated parameter space are shown in Fig. 5.8 for a two-dimensional case.

From Eq. (5.46) the transformed variables are linear combinations of the standardized original stochastic basic variables. Therefore, if all the original stochastic basic variables X are normally distributed, then the transformed stochastic basic variables, by the reproductive property of the normal random variable described in Sec. 2.4.1, are also independent normal variables. More specifically

$$X \sim \mathrm{N}(\mu_x, C_x) \qquad X' \sim \mathrm{N}(0, R_x) \qquad U \sim \mathrm{N}(0, \Lambda) \qquad Y = Z' \sim \mathrm{N}(0, I)$$

The advantage of the orthogonal transformation is to convert the correlated stochastic basic variables into the uncorrelated ones so that the analysis can be made easier.

The orthogonal transformations described previously are applied to the standardized parameter space in which the lower triangular matrix and eigenvector matrix of the correlation matrix are computed. In fact, the orthogonal

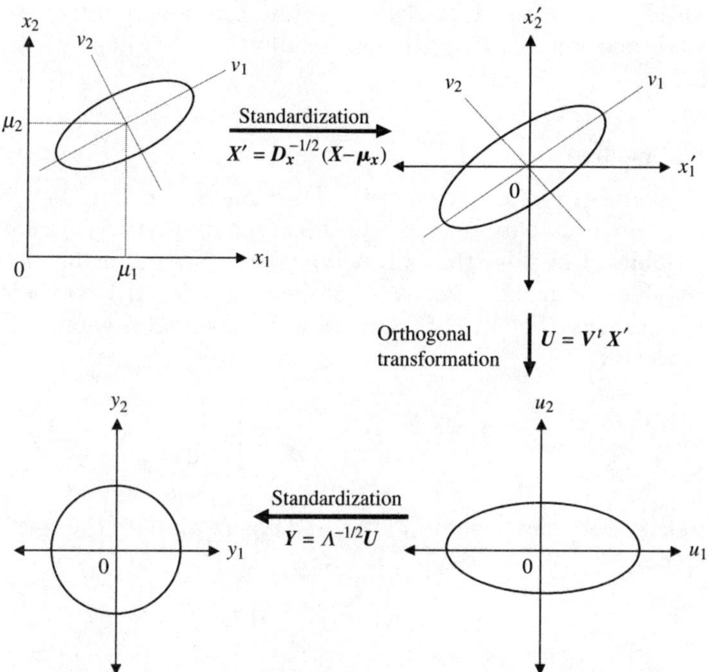

Figure 5.8 Geometric diagrams of various stages of transformations in spectral decomposition.

transformation can be directly applied to the variance-covariance matrix C_x. The lower triangular matrices of C_x, $\tilde{L}$, can be obtained from that of the correlation matrix L by

$$\tilde{L} = D_x^{1/2}\, L \tag{5.57}$$

Following a similar procedure as described for the spectral decomposition, the uncorrelated, standardized random vector Y can be obtained as

$$Y = \tilde{\Lambda}^{-1/2}\tilde{V}^t(X - \mu_x) = \tilde{\Lambda}^{-1/2}\tilde{U} \tag{5.58}$$

where $\tilde{V}$ and $\tilde{\Lambda}$ are the eigenvector matrix and diagonal eigenvalue matrix of the covariance matrix C_x satisfying

$$C_x = \tilde{V}\tilde{\Lambda}\tilde{V}^t$$

and $\tilde{U}$ is an uncorrelated vector of the random variables in the eigen-space having a zero mean 0 and covariance matrix $\tilde{\Lambda}$. Then, the original random vector X can be expressed in terms of Y and $\tilde{L}$

$$X = \mu_x + \tilde{V}\tilde{\Lambda}^{-1/2}Y = \mu_x + \tilde{L}Y \tag{5.59}$$

One should be aware that the eigenvectors and eigenvalues associated with the covariance matrix C_x will not be identical to those of the correlation matrix R_x.

5.3.2 Bivariate Harr PPE method

The orthogonal transformation of the Harr method is based on Eq.(5.51). The method is now illustrated using a problem involving two correlated stochastic basic variables. Consider that a function $W = g(X_1, X_2)$, involves two stochastic basic variables with a known vector of mean $\mu_x = (\mu_1, \mu_2)^t$, variances σ_1^2 and σ_2^2, and correlation coefficient ρ. By applying the spectral decomposition to the correlation matrix

$$R_x = \begin{bmatrix} 1 & \rho \\ \rho & 1 \end{bmatrix}$$

the corresponding eigenvector matrix and eigenvalues of the correlation matrix are

$$V = [v_1, v_2] = \begin{bmatrix} v_{11} & v_{12} \\ v_{21} & v_{22} \end{bmatrix} = \begin{bmatrix} 0.7071 & -0.7071 \\ 0.7071 & 0.7171 \end{bmatrix}$$

and

$$\Lambda = \operatorname{diag}(\lambda_1, \lambda_2) = \operatorname{diag}(1 + \rho, 1 - \rho) \tag{5.60}$$

Note that the eigenvectors of a bivariate correlation matrix have an angle of $45°$ with each parameter axis in the original space. Referring to Fig. 5.9(a) for a two-dimensional problem, Harr's PPE method selects the two points located at the intersections of each coordinate axis in the eigen-space with a circle of radius $\sqrt{2}$, i.e.,

Along u_1-axis:

$$\text{At point A}_+: \boldsymbol{u}_{A+} = (u_1, u_2)^t = (\sqrt{2}, 0)^t$$

$$\text{At point A}_-: \boldsymbol{u}_{A-} = (u_1, u_2)^t = (-\sqrt{2}, 0)^t$$

Along u_2-axis:

$$\text{At point B}_+: \boldsymbol{u}_{B+} = (u_1, u_2)^t = (0, \sqrt{2})^t$$

$$\text{At point B}_-: \boldsymbol{u}_{B-} = (u_1, u_2)^t = (0, -\sqrt{2})^t$$

The coordinates of A$_+$, A$_-$, B$_+$, and B$_-$ in the eigen-space, (u_1, u_2), can be transformed back to their corresponding coordinates in the standardized original space, (x_1', x_2'), using

$$x' = (V^t)^{-1} u = Vu \tag{5.61}$$

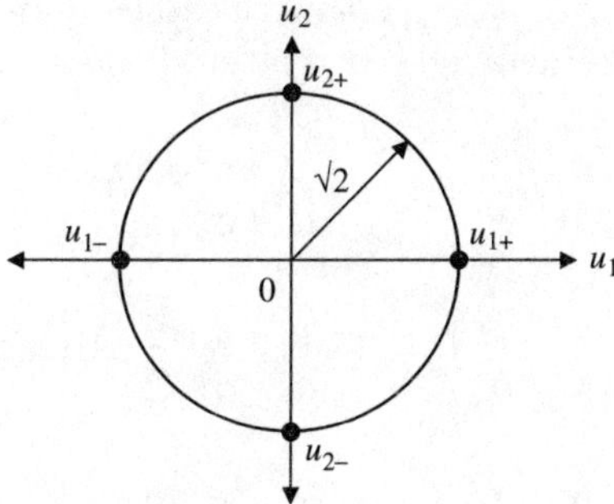

(a) Selected points in the nonstandardized eigen space

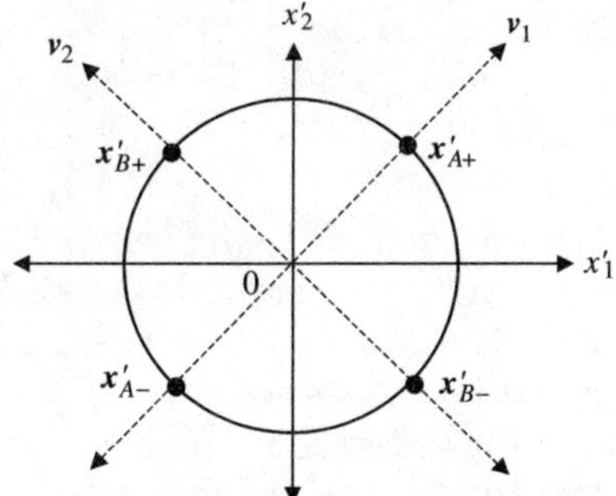

(b) Corresponding selected points in the original standardized space

Figure 5.9   Schematic diagram of the Harr PPE method for a bivariate problem.

resulting in

$$
\boldsymbol{x}'_{A+} = \boldsymbol{V}\boldsymbol{u}_{A+} =
\begin{bmatrix} v_{11} & v_{12} \\ v_{21} & v_{22} \end{bmatrix}
\begin{bmatrix} \sqrt{2} \\ 0 \end{bmatrix} =
\begin{bmatrix} \sqrt{2}\, v_{11} \\ \sqrt{2}\, v_{21} \end{bmatrix}
$$

$$
\boldsymbol{x}'_{A-} = \boldsymbol{V}\boldsymbol{u}_{A-} =
\begin{bmatrix} v_{11} & v_{12} \\ v_{21} & v_{22} \end{bmatrix}
\begin{bmatrix} -\sqrt{2} \\ 0 \end{bmatrix} =
\begin{bmatrix} -\sqrt{2}\, v_{11} \\ -\sqrt{2}\, v_{21} \end{bmatrix}
$$

$$
\boldsymbol{x}'_{B+} = \boldsymbol{V}\boldsymbol{u}_{B+} =
\begin{bmatrix} v_{11} & v_{12} \\ v_{21} & v_{22} \end{bmatrix}
\begin{bmatrix} 0 \\ \sqrt{2} \end{bmatrix} =
\begin{bmatrix} \sqrt{2}\, v_{12} \\ \sqrt{2}\, v_{22} \end{bmatrix}
$$

$$
\boldsymbol{x}'_{B-} = \boldsymbol{V}\boldsymbol{u}_{B-} =
\begin{bmatrix} v_{11} & v_{12} \\ v_{21} & v_{22} \end{bmatrix}
\begin{bmatrix} 0 \\ -\sqrt{2} \end{bmatrix} =
\begin{bmatrix} -\sqrt{2}\, v_{12} \\ -\sqrt{2}\, v_{22} \end{bmatrix}
$$

The four points $\boldsymbol{x}'_{A\pm}$ and $\boldsymbol{x}'_{B\pm}$ are the intersections of a circle with radius $\sqrt{2}$ and the two eigenvectors as shown in Fig. 5.9(*b*) for a two-dimensional problem. The values of the four points in the original (x_1, x_2)-space can be immediately obtained from the four points in the standardized parameter space as

$$\boldsymbol{x}_{A+} = \boldsymbol{\mu}_x + \begin{bmatrix} \sigma_1 & 0 \\ 0 & \sigma_2 \end{bmatrix} \boldsymbol{x}'_{A+} = \begin{bmatrix} \mu_1 + \sigma_1 \sqrt{2}\, v_{11} \\ \mu_2 + \sigma_2 \sqrt{2}\, v_{21} \end{bmatrix} \tag{5.62a}$$

$$\boldsymbol{x}_{A-} = \boldsymbol{\mu}_x + \begin{bmatrix} \sigma_1 & 0 \\ 0 & \sigma_2 \end{bmatrix} \boldsymbol{x}'_{A-} = \begin{bmatrix} \mu_1 - \sigma_1 \sqrt{2}\, v_{11} \\ \mu_2 - \sigma_2 \sqrt{2}\, v_{21} \end{bmatrix} \tag{5.62b}$$

$$\boldsymbol{x}_{B+} = \boldsymbol{\mu}_x + \begin{bmatrix} \sigma_1 & 0 \\ 0 & \sigma_2 \end{bmatrix} \boldsymbol{x}'_{B+} = \begin{bmatrix} \mu_1 + \sigma_1 \sqrt{2}\, v_{12} \\ \mu_2 + \sigma_2 \sqrt{2}\, v_{22} \end{bmatrix} \tag{5.62c}$$

$$\boldsymbol{x}_{B-} = \boldsymbol{\mu}_x + \begin{bmatrix} \sigma_1 & 0 \\ 0 & \sigma_2 \end{bmatrix} \boldsymbol{x}'_{B-} = \begin{bmatrix} \mu_1 - \sigma_1 \sqrt{2}\, v_{12} \\ \mu_2 - \sigma_2 \sqrt{2}\, v_{22} \end{bmatrix} \tag{5.62d}$$

Note that, in the process of transforming back to the original parameter space, the coordinates of the four points in the x_1- and x_2-axis are scaled differently by their respective standard deviations. Therefore, the four selected points in the original space may not lie on the eigenvectors unless σ_1 and σ_2 are equal. Figure 5.10 shows the possible arrangements of the selected four points in the original parameter space for a bivariate problem. In all cases, the four selected points are symmetric with respect to the mean point, (μ_1, μ_2).

Once the four points in the parameter space are determined, the statistical moments of model output W about the origin can be estimated by the following two steps. For the rth-order moment, the first step is to compute the arithmetic average of the model value raised to the power of r corresponding to the two eigenvectors, that is,

$$\overline{w^r_A} = g^r(\boldsymbol{x}_A) = \frac{g^r(\boldsymbol{x}_{A+}) + g^r(\boldsymbol{x}_{A-})}{2} = \frac{w^r_{A+} + w^r_{A-}}{2} \tag{5.63a}$$

$$\overline{w^r_B} = g^r(\boldsymbol{x}_B) = \frac{g^r(\boldsymbol{x}_{B+}) + g^r(\boldsymbol{x}_{B-})}{2} = \frac{w^r_{B+} + w^r_{B-}}{2} \tag{5.63b}$$

Note that w^r_A and w^r_B are determined, respectively, based on the two points lying on the axes of $\boldsymbol{v}_1$ and $\boldsymbol{v}_2$, along which the corresponding eigenvalues might be different. Realizing that the eigenvalues are the variances of the transformed stochastic basic variables $\boldsymbol{U}$, the rth-order moment of the model output W, in the second step, can be computed as the weighted average of $\overline{w^r_A}$ and $\overline{w^r_B}$ as

$$E[W^r] = \mu'_r(W) = \frac{\lambda_1 \overline{w^r_A} + \lambda_2 \overline{w^r_B}}{\lambda_1 + \lambda_2} = \frac{\lambda_1 \overline{w^r_A} + \lambda_2 \overline{w^r_B}}{2} \tag{5.64}$$

Once the moments about the origin for the model output W are computed, the central moments can be obtained by Eq. (2.21).

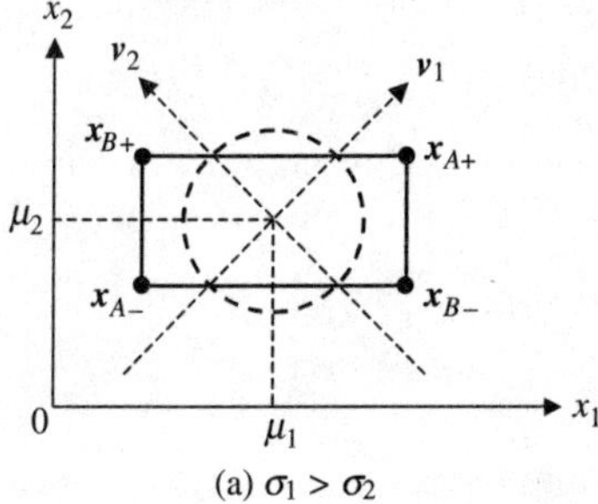

(a) $\sigma_1 > \sigma_2$

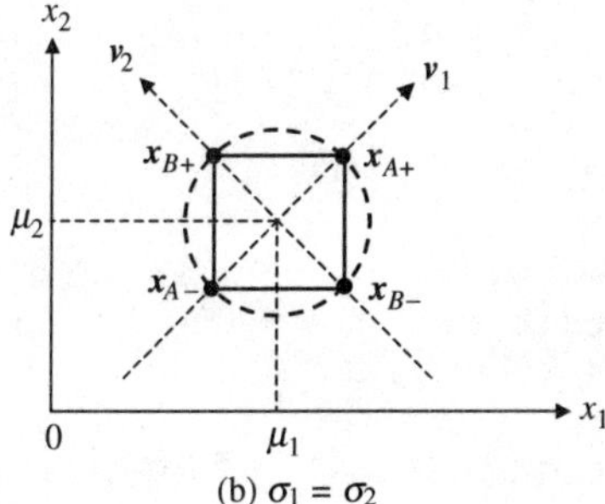

(b) $\sigma_1 = \sigma_2$

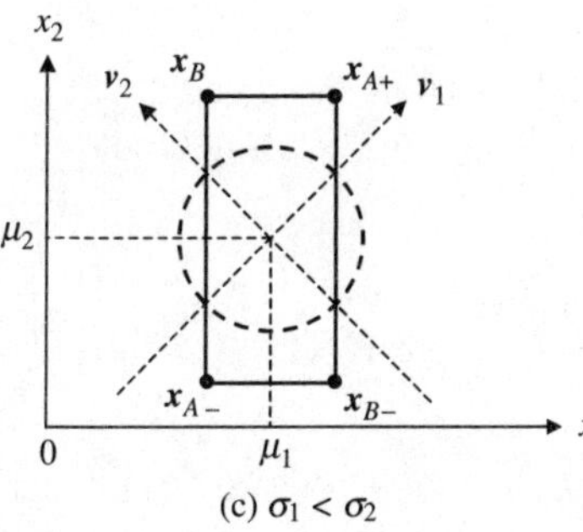

(c) $\sigma_1 < \sigma_2$

Figure 5.10   Selected points by the Harr PPE method in the original space for a bivariate problem.

Example 5.9   Refer to Example 5.6 and consider that both sewer diameter D and the roughness coefficient are subject to uncertainty. It is known that the manufacturing error associated with the pipe diameter is about 5 percent of its nominal diameter. Determine the uncertainty of the sewer flow capacity using the Harr PPE method for a section of a sewer with a nominal diameter $D = 3.0$ ft and slope $S = 0.005$. The roughness coefficient has the mean 0.015 with a coefficient of variation 0.05. Assume that sewer diameter and roughness coefficient have a correlation coefficient of -0.75.

Solution   From Example 5.6, Manning's formula for sewer of the specified layout can be written as

$$Q = 0.463\, n^{-1}\, D^{2.67}\, (0.005)^{0.5} = 0.0327\, n^{-1}\, D^{2.67}$$

The standard deviations of the roughness coefficient and sewer diameter are

$$\sigma_n = 0.00075 \qquad \sigma_D = 0.15$$

From the given correlation coefficient between the roughness coefficient and sewer diameter, the correlation matrix can be established as

$$\boldsymbol{R}(n, D) = \begin{bmatrix} 1.00 & -0.75 \\ -0.75 & 1.00 \end{bmatrix}$$

The corresponding eigenvector matrix and eigenvalue matrix are, respectively,

$$\boldsymbol{V} = [\boldsymbol{v}_1 \ \boldsymbol{v}_2] = \begin{bmatrix} v_{11} & v_{12} \\ v_{21} & v_{22} \end{bmatrix} = \begin{bmatrix} 0.7071 & 0.7071 \\ -0.7071 & 0.7071 \end{bmatrix}$$

$$\boldsymbol{L} = \mathrm{diag}(\lambda_1, \lambda_2) = \mathrm{diag}(1.75, 0.25)$$

According to Eq.(5.62), the coordinates of the four intersection points of the two principal axes defined by the eigenvectors and the circle with radius $\sqrt{2}$ can be determined as

$$\boldsymbol{x}_{A+} = \begin{bmatrix} n_{A+} \\ D_{A+} \end{bmatrix} = \begin{bmatrix} \mu_n + \sigma_n\sqrt{2}\ v_{11} \\ \mu_D + \sigma_D\sqrt{2}\ v_{21} \end{bmatrix} = \begin{bmatrix} 0.015 + 0.00075(\sqrt{2})(0.7071) \\ 3.0 + 0.15(\sqrt{2})(-0.7071) \end{bmatrix} = \begin{bmatrix} 0.01575 \\ 2.85 \end{bmatrix}$$

$$\boldsymbol{x}_{A-} = \begin{bmatrix} n_{A-} \\ D_{A-} \end{bmatrix} = \begin{bmatrix} \mu_n - \sigma_n\sqrt{2}\ v_{11} \\ \mu_D - \sigma_D\sqrt{2}\ v_{21} \end{bmatrix} = \begin{bmatrix} 0.015 - 0.00075(\sqrt{2})(0.7071) \\ 3.0 - 0.15(\sqrt{2})(-0.7071) \end{bmatrix} = \begin{bmatrix} 0.01425 \\ 3.15 \end{bmatrix}$$

$$\boldsymbol{x}_{B+} = \begin{bmatrix} n_{B+} \\ D_{B+} \end{bmatrix} = \begin{bmatrix} \mu_n + \sigma_n\sqrt{2}\ v_{12} \\ \mu_D + \sigma_D\sqrt{2}\ v_{22} \end{bmatrix} = \begin{bmatrix} 0.015 + 0.00075(\sqrt{2})(0.7071) \\ 3.0 + 0.15(\sqrt{2})(-0.7071) \end{bmatrix} = \begin{bmatrix} 0.01575 \\ 3.15 \end{bmatrix}$$

$$\boldsymbol{x}_{B-} = \begin{bmatrix} n_{B-} \\ D_{B-} \end{bmatrix} = \begin{bmatrix} \mu_n - \sigma_n\sqrt{2}\ v_{12} \\ \mu_D - \sigma_D\sqrt{2}\ v_{22} \end{bmatrix} = \begin{bmatrix} 0.015 + 0.00075(\sqrt{2})(0.7071) \\ 3.0 - 0.15(\sqrt{2})(-0.7071) \end{bmatrix} = \begin{bmatrix} 0.01425 \\ 2.85 \end{bmatrix}$$

Substituting the values of $\boldsymbol{x}_{A+}$, $\boldsymbol{x}_{A-}$, $\boldsymbol{x}_{B+}$, and $\boldsymbol{x}_{B-}$ in Manning's formula to compute the corresponding sewer capacities, one has

$$Q_{A+} = 0.0327(n_{A+})^{-1}(D_{A+})^{2.67} = 0.0327(0.01575)^{-1}(2.85)^{2.67} = 34.02 \text{ ft}^3/\text{s}$$

$$Q_{A-} = 0.0327(n_{A-})^{-1}(D_{A-})^{2.67} = 0.0327(0.01425)^{-1}(3.15)^{2.67} = 49.12 \text{ ft}^3/\text{s}$$

$$Q_{B+} = 0.0327(n_{B+})^{-1}(D_{B+})^{2.67} = 0.0327(0.01575)^{-1}(3.15)^{2.67} = 44.44 \text{ ft}^3/\text{s}$$

$$Q_{B-} = 0.0327\,(n_{B-})^{-1}(D_{B-})^{2.67} = 0.0327(0.01425)^{-1}(2.85)^{2.67} = 37.60 \text{ ft}^3/\text{s}$$

From the above four values of the sewer flow capacity, the mean value along each eigenvector can be computed as

$$\overline{Q}_A = (34.02 + 49.12)/2 = 41.57 \text{ ft}^3/\text{s} \qquad \overline{Q}_B = (44.44 + 37.60)/2 = 41.02 \text{ ft}^3/\text{s}.$$

The mean of the sewer flow capacity can be estimated, according to Eq. (5.64), as

$$\mu_Q = \frac{\lambda_1 \overline{Q_A} + \lambda_2 \overline{Q_B}}{\lambda_1 + \lambda_2} = \frac{1.75(41.57) + 0.25(41.02)}{2} = 41.50\,\text{ft}^3/\text{s}$$

Similarly, to compute the 2nd-order moment about the origin, the averages of the squared sewer flow capacity along the two eigenvector axes are made. That is, $\overline{Q_A^2} = (34.02^2 + 49.12^2)/2 = 1785.06\ (\text{ft}^3/\text{s})^2$ and $\overline{Q_B^2} = (44.44^2 + 37.60^2)/2 = 1694.34\ (\text{ft}^3/\text{s})^2$. For the variance of the sewer flow capacity, the 2nd-order moment about the origin is computed as

$$E(Q^2) = \frac{\lambda_1 \overline{Q_A^2} + \lambda_2 \overline{Q_B^2}}{\lambda_1 + \lambda_2} = \frac{1.75(1785.06) + 0.25(1694.34)}{2} = 1773.73\,(\text{ft}^3/\text{s})^2$$

Then, the variance of the sewer flow capacity can be computed as

$$\text{Var}(Q) = E(Q^2) - (\mu_Q)^2 = 1773.73 - (41.50)^2 = 51.48\ (\text{ft}^3/\text{s})^2$$

Hence, the standard deviation of the sewer flow capacity is $\sqrt{51.48} = 7.17\ \text{ft}^3/\text{s}$. Comparing with the results in Example 5.2, one observes that the mean and variance of sewer flow capacity obtained by Harr's method are almost the same as those of the Rosenblueth because of the selection of the identical four points for function evaluation (see Fig. 5.11). Compared with the FOVE method, the Harr and the Rosenblueth methods both yield slightly higher values of mean and variance for the sewer flow capacity.

5.3.3 Multivariate Harr PPE method

From the above descriptions for the bivariate case, Harr's method can be generalized for a multivariate model $W = g(X_1, X_2, \ldots, X_K)$ involving K stochastic

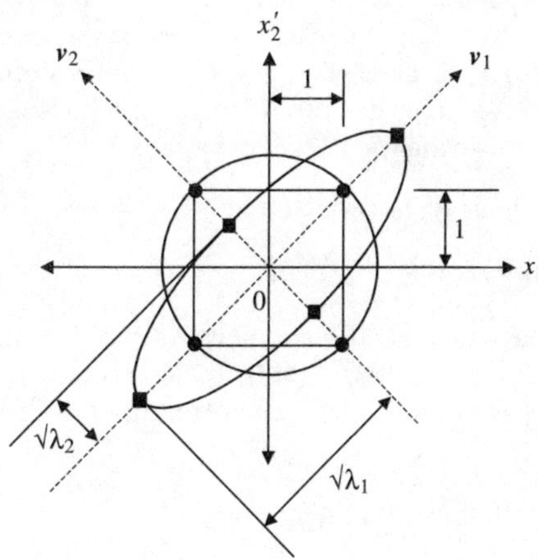

Figure 5.11 Illustration of selected points by different PPE methods in the standardized space.

• Points selected by the Resenblueth and Harr methods.
■ Points selected by the modified Harr method.

basic variables as

$$X_{k\pm} = \boldsymbol{\mu}_x \pm \sqrt{K}\,\boldsymbol{D}_x^{1/2}\boldsymbol{v}_k \qquad \text{for } k = 1, 2, \ldots, K \tag{5.65}$$

in which $\boldsymbol{x}_{k\pm}$ represents the vector of coordinates of the K stochastic basic variables in the parameter space corresponding to the kth eigenvector $\boldsymbol{v}_k$; $\boldsymbol{\mu}_x = (\mu_1, \mu_2, \ldots, \mu_K)^t$, a vector of means of K stochastic basic variables X; and $\boldsymbol{D}_x$ is a diagonal matrix of variances of K stochastic basic variables.

Based on the $2K$ points determined by Eq. (5.65), the function values at each of the $2K$ points can be computed. Then, the rth-order moment of the model output W can be calculated according to the following equations:

$$\overline{w_k^r} = \frac{w_{k+}^r + w_{k-}^r}{2} = \frac{g^r(\boldsymbol{x}_{k+}) + g^r(\boldsymbol{x}_{k-})}{2} \qquad \text{for } k = 1, 2, \ldots, K \qquad r = 1, 2, \ldots \tag{5.66}$$

$$E[W^r] = \mu_r'(W) = \frac{\sum_{k=1}^{K} \lambda_k \overline{w_k^r}}{K} \qquad \text{for } r = 1, 2, \ldots \tag{5.67}$$

If the decomposition was made to the covariance matrix, the corresponding eigenvectors $\tilde{\boldsymbol{v}}_k$, and eigenvalues $\tilde{\lambda}_k$ should be used in Eqs. (5.65) and (5.67), respectively. Recently, He and Sällfors (1994) showed that the orthogonal transformation made on the covariance matrix could improve the accuracy of Harr's method for uncertainty analysis. The following illustrates the original Harr PPE method by which the correlation matrix of the stochastic basic variables is decomposed. The alternative decomposition based on the covariance matrix is left as an exercise (Prob. 5.28).

Example 5.10   Refer to Example 5.8 and consider that all parameters, i.e., roughness coefficient n, sewer diameter D, and sewer slope S, are subject to 5 percent error associated with their nominal values. Determine the uncertainty of the sewer flow capacity using Harr's PPE method for a section of 3-ft sewer with nominal roughness coefficient 0.015 and slope 0.005. Assume that $\rho(n, D) = -0.75$ whereas sewer slope S is uncorrelated with the other two stochastic basic variables.

Solution   Manning's formula for the sewer is written as

$$Q = 0.463\, n^{-1} D^{2.67} S^{0.5}$$

The standard deviation of the roughness coefficient, sewer diameter, and sewer slope are, respectively, $\sigma_n = 0.00075$, $\sigma_D = 0.15$, and $\sigma_S = 0.00025$. From the given correlation relationship among the three stochastic basic variables, the correlation matrix can be established as

$$\boldsymbol{R}(n, D, S) = \begin{bmatrix} 1.00 & -0.75 & 0.00 \\ -0.75 & 1.00 & 0.00 \\ 0.00 & 0.00 & 1.00 \end{bmatrix}$$

The corresponding eigenvector matrix and eigenvalue matrix are, respectively,

$$V = [\boldsymbol{v}_1 \quad \boldsymbol{v}_2 \quad \boldsymbol{v}_3] = \begin{bmatrix} v_{11} & v_{12} & v_{13} \\ v_{21} & v_{22} & v_{23} \\ v_{31} & v_{32} & v_{33} \end{bmatrix} = \begin{bmatrix} 0.7071 & 0.7071 & 0.0000 \\ -0.7071 & 0.7071 & 0.0000 \\ 0.0000 & 0.0000 & 1.0000 \end{bmatrix}$$

and

$$\Lambda = \operatorname{diag}(\lambda_1, \lambda_2, \lambda_3) = \operatorname{diag}(1.75, 0.25, 1.00)$$

According to Eq. (5.65), the coordinates of $2 \times 3 = 6$ intersection points corresponding to the three eigenvectors and a sphere surface with radius $\sqrt{3}$ can be determined as

$$\boldsymbol{x}_{k\pm} = \begin{bmatrix} \mu_n \\ \mu_D \\ \mu_S \end{bmatrix} \pm \sqrt{3} \begin{bmatrix} \sigma_n & 0 & 0 \\ 0 & \sigma_D & 0 \\ 0 & 0 & \sigma_D \end{bmatrix} \boldsymbol{v}_k$$

$$= \begin{bmatrix} 0.015 \\ 3.0 \\ 0.005 \end{bmatrix} \pm \sqrt{3} \begin{bmatrix} 0.00075 & 0 & 0 \\ 0 & 0.06 & 0 \\ 0 & 0 & 0.00025 \end{bmatrix} \boldsymbol{v}_k \qquad \text{for } k = 1, 2, 3$$

The resulting coordinates at the six intersection points from the above equation are listed in column 2 of the table given below. Substituting the values of $\boldsymbol{x}$ in column 2 into Manning's formula, the corresponding sewer capacities are listed in column 3. Column 4 lists the value of Q^2 for computing the second moment about the origin later. After columns 3 and 4 are obtained, the averaged values of Q and Q^2 along each eigenvector are computed and listed in columns 5 and 6, respectively.

Point (1)	$\boldsymbol{x} = (n, D, S)$ (2)	Q (3)	Q^2 (4)	$\bar{Q}$ (5)	$\overline{Q^2}$ (6)
1+	(0.01592, 2.8163, 0.00500)	32.53	1058.20	41.77	1830.11
1−	(0.01408, 3.1837, 0.00500)	51.01	2602.02		
2+	(0.01592, 3.1837, 0.00500)	45.12	2035.81		
2−	(0.01408, 2.8163, 0.00500)	36.77	1352.03	40.95	1693.92
3+	(0.01500, 3.00, 0.00543)	42.59	1814.12		
3−	(0.01500, 3.00, 0.00457)	39.05	1524.95	40.82	1669.54

The mean of the sewer flow capacity can be calculated, according to Eq. (5.67), as

$$\mu_Q = \frac{\lambda_1 \bar{Q}_1 + \lambda_2 \bar{Q}_2 + \lambda_3 \bar{Q}_3}{\lambda_1 + \lambda_2 + \lambda_3}$$

$$= \frac{1.75(41.77) + 0.25(40.95) + 1.00(40.82)}{3} = 41.39 \ \text{ft}^3/\text{s}$$

The 2nd-order moment about the origin is calculated as

$$E(Q^2) = \frac{\lambda_1 \overline{Q_1^2} + \lambda_2 \overline{Q_2^2} + \lambda_3 \overline{Q_3^2}}{\lambda_1 + \lambda_2 + \lambda_3}$$

$$= \frac{1.75(1830.11) + 0.25(1693.92) + 1.00(1669.54)}{3} = 1765.24 \ (\text{ft}^3/\text{s})^2$$

The variance of the sewer flow capacity can then be calculated as

$$\text{Var}(Q) = E(Q^2) - (\mu_Q)^2 = 1765.24 - (41.39)^2 = 52.11 \ (\text{ft}^3/\text{s})^2$$

Hence, the standard deviation of the sewer flow capacity is $\sqrt{52.11} = 7.22 \ \text{ft}^3/\text{s}$. Comparing with the results in the previous examples, one observes that the Harr PPE method yields mean and variance for the sewer flow capacity that lie between those of the Rosenblueth and the FOVE methods.

5.3.4 Modified Harr PPE algorithm

From Eq. (5.67) the weighing factors based on the eigenvalues of the correlation matrix are used in estimating the statistical moments of the model output $W(X)$. Chang, Tung, and Yang (1995) proposed an algorithm to simplify the computation in the Harr PPE method. Their numerical experiments also showed that the modified algorithm improved the accuracy of Harr's method. The modified method selects the points for the model evaluation with an equal weight for a multivariate problem involving correlated normal stochastic basic variables.

From Eqs. (5.56b) and (5.59), the vector of multivariate normal stochastic basic variables X can be expressed in terms of the uncorrelated standard normal random vector Z' as

$$X = \mu_x + D_x^{1/2} \, V \, \Lambda^{1/2} \, Z' \tag{5.68}$$

In the modified Harr PPE method, a hypersphere with radius $\sqrt{K}$ centered at the origin in the K-dimensional standardized normal Z'-space is constructed. The points at which the model output is to be computed are located at the intersections of the hypersphere and the K eigenvectors of the correlation matrix R_x. Due to the normal distribution and the same scale on each component in the uncorrelated, standardized normal Z'-space, the $2K$ selected points in the original parameter space are located on a hypersurface with equal PDF values.

By Eq. (5.68), the selected points for model evaluation in the original parameter space can be obtained as

$$x_{k\pm} = \mu_x \pm D_x^{1/2} \, V\Lambda^{1/2} \, (\sqrt{K} \, e_k) = \mu_x \pm \sqrt{K} \, \sqrt{\lambda_k} \, D_x^{1/2} v_k \qquad \text{for } k = 1, \ldots, K \tag{5.69}$$

where $x_{k\pm}$ is a column vector containing the coordinates of the two intersection points corresponding to the kth eigenvector in the original parameter space and e_k is the unit base vector with the kth element equal to 1 and 0 elsewhere. At each selected point, the corresponding model output value can be computed.

Then, the rth-order moment about the origin of the model output W can be calculated as

$$E(W^r) = \frac{1}{K} \sum_{k=1}^{K} \overline{w_k^r} \qquad (5.70)$$

in which

$$\overline{w_k^r} = \frac{w_{k+}^r + w_{k-}^r}{2} \qquad (5.71)$$

As can be seen from Eq. (5.70), the weighting procedure in the computation of statistical moments of the model output by Harr's PE algorithm is not needed. When the spectral decomposition is applied to the covariance matrix, Eq. (5.69) can be modified as

$$\boldsymbol{x}_{k\pm} = \boldsymbol{\mu}_x \pm \sqrt{K}\sqrt{\tilde{\lambda}_k}\,\tilde{\boldsymbol{v}}_k \qquad \text{for } k = 1,\, 2,\ldots,\, K \qquad (5.72)$$

Figure 5.11 schematically shows the difference in point selection by the three probabilistic PPE methods for a bivariate case in the standardized correlated normal space $\boldsymbol{X}'$. Rosenblueth's points are located at the four corners of the square with side lengths of 2 centered at the origin. Harr's method selects the intersections of the eigenvectors and the circle with a radius of $\sqrt{2}$ centered at the origin of the standardized normal space. Therefore, for a bivariate correlated normal problem, Harr's method is identical to the Rosenblueth method. Note that the points selected by the modified Harr method are on an elliptic contour of equal density, which is a circle in the standardized eigen-space $\boldsymbol{Z}'$. When the stochastic basic variables are independent and normal, the modified algorithm is identical to the original Harr method. Chang, Tung, and Yang (1995) conducted an extensive numerical examination of the relative performance of the probabilistic point estimation methods and the results indicate that the modified Harr method outperforms the Rosenblueth and Harr techniques in estimating the product moments of the model output.

Example 5.11   Solve Example 5.10 using modified Harr PPE algorithm.

Solution   From Example 5.8, Manning's formula for the sewer is written as

$$Q = 0.463\, n^{-1}\, D^{2.67}\, S^{0.5}$$

The standard deviations of the roughness coefficient, sewer diameter, and sewer slope are, respectively, $\sigma_n = 0.00075$, $\sigma_D = 0.15$, and $\sigma_S = 0.00025$. From the given correlation relationship among the three stochastic basic variables, the correlation matrix can be established as

$$\boldsymbol{R}(n, D, S) = \begin{bmatrix} 1.00 & -0.75 & 0.00 \\ -0.75 & 1.00 & 0.00 \\ 0.00 & 0.00 & 1.00 \end{bmatrix}$$

The corresponding eigenvector matrix and eigenvalue matrix are, respectively,

$$V = [v_1 \quad v_2 \quad v_3] = \begin{bmatrix} v_{11} & v_{12} & v_{13} \\ v_{21} & v_{22} & v_{23} \\ v_{31} & v_{32} & v_{33} \end{bmatrix} = \begin{bmatrix} 0.7071 & 0.7071 & 0.0000 \\ -0.7071 & 0.7071 & 0.0000 \\ 0.0000 & 0.0000 & 1.0000 \end{bmatrix}$$

and

$$\Lambda = \mathrm{diag}(\lambda_1, \lambda_2, \lambda_3) = \mathrm{diag}(1.75, 0.25, 1.00)$$

According to Eq. (5.66), the coordinates of the six intersection points corresponding to the three eigenvectors and the sphere having radius $\sqrt{3}$ can be determined as

$$
x_{k\pm} = \begin{bmatrix} \mu_n \\ \mu_D \\ \mu_S \end{bmatrix} \pm \sqrt{3\lambda_k} \begin{bmatrix} \sigma_n & 0 & 0 \\ 0 & \sigma_D & 0 \\ 0 & 0 & \sigma_D \end{bmatrix} v_k
$$

$$
= \begin{bmatrix} 0.015 \\ 3.0 \\ 0.005 \end{bmatrix} \pm \sqrt{3\lambda_k} \begin{bmatrix} 0.00075 & 0 & 0 \\ 0 & 0.15 & 0 \\ 0 & 0 & 0.00025 \end{bmatrix} v_k \qquad \text{for } k = 1, 2, 3
$$

The resulting coordinates at the six intersection points from the preceding equation are listed in column 2 of the table for the required computations.

Point (1)	$x = (n, D, S)$ (2)	Q (3)	Q^2 (4)	$\bar{Q}$ (5)	$\overline{Q^2}$ (6)
1+	(0.0162, 2.757, 0.00500)	30.28	916.64	42.61	1967.61
1−	(0.0138, 3.243, 0.00500)	54.94	3018.59		
2+	(0.0155, 3.092, 0.00500)	43.13	1860.05		
2−	(0.0145, 2.908, 0.00500)	38.93	1536.16	40.97	1681.81
3+	(0.0150, 3.000, 0.00543)	42.75	1827.46		
3−	(0.0150, 3.000, 0.00457)	39.19	1515.90	41.03	1687.97

The mean of the sewer flow capacity can be calculated, according to Eq. (5.70), as

$$\mu_Q = \frac{\bar{Q}_1 + \bar{Q}_2 + \bar{Q}_3}{3} = \frac{42.61 + 41.03 + 40.97}{3} = 41.54 \text{ ft}^3\text{/s}$$

Similarly, the second moment about the origin is calculated as

$$E(Q^2) = \frac{\overline{Q_1^2} + \overline{Q_2^2} + \overline{Q_3^2}}{3} = \frac{1967.61 + 1687.97 + 1681.81}{3} = 1779.13 \, (\text{ft}^3\text{/s})^2$$

The variance of the sewer flow capacity then can be calculated as

$$\mathrm{Var}(Q) = E(Q^2) - (\mu_Q)^2 = 1779.13 - (41.54)^2 = 53.56 \, (\text{ft}^3\text{/s})^2$$

and the corresponding standard deviation of the sewer flow capacity is $\sqrt{53.56} = 7.32$ ft^3/s. Comparing with the results obtained earlier, the standard deviation computed by the modified Harr PPE method is slightly higher than all other methods and the mean is closer to that of the Rosenblueth method.

5.4   Li's Probabilistic Point Estimate Method

Li (1992) proposed a computationally practical PPE method that allows incorporation of the first four product-moments of individual stochastic basic variables and their correlations. Consider a univariate model $W = g(X)$ involving a single stochastic basic variable X whose PDF can be approximated by three discrete points at x_-, μ_x, and x_+ (Fig. 5.12). The five unknowns, namely, x_-, x_+, $p_- = P(X = x_-)$, $p_+ = P(X = x_+)$, and $p_o = P(X = \mu_x)$, are determined to preserve the first four moments of the stochastic basic variable. The system of five equations can be established, similar to Eqs. (5.30a) to (5.30d), in the standardized parameter space as

$$p_+ + p_o + p_- = 1 \tag{5.73a}$$

$$p_+ x'_+ - p_- x'_- = \mu_{x'} = 0 \tag{5.73b}$$

$$p_+ x'^2_+ + p_- x'^2_- = \sigma^2_{x'} = 1 \tag{5.73c}$$

$$p_+ x'^3_+ - p_- x'^3_- = \gamma_x \tag{5.73d}$$

$$p_+ x'^4_+ + p_- x'^4_- = \kappa_x \tag{5.73e}$$

in which $x'_- = |x_- - \mu_x|/\sigma_x$, $x'_+ = |x_+ - \mu_x|/\sigma_x$; γ_x and κ_x are the skewness coefficient and kurtosis of the stochastic basic variable X, respectively. The solutions to Eqs. (5.73a) to (5.73e) are

$$x'_+ = \frac{\gamma_x + \sqrt{4\kappa_x - 3\gamma_x^2}}{2} \qquad x'_- = \frac{-\gamma_x + \sqrt{4\kappa_x - 3\gamma_x^2}}{2} \tag{5.74a}$$

$$p_+ = \frac{1}{x'_+ \, (x'_+ + x'_-)} \qquad p_- = \frac{1}{x'_- \, (x'_+ + x'_-)} \qquad p_o = 1 - p_+ - p_- \tag{5.74b}$$

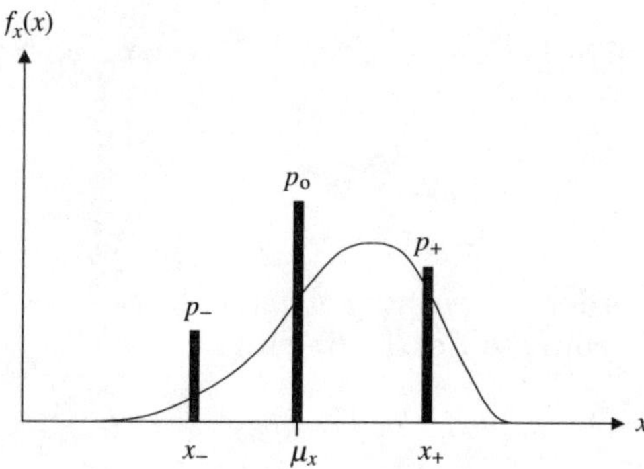

Figure 5.12   Li's three-point representation of a probability density function.

From Eq. (5.74a) the selected points for model evaluation in the original parameter space can be determined as

$$x_+ = \mu_x + x'_+ \, \sigma_x \qquad x_o = \mu_x \qquad x_- = \mu_x - x'_- \, \sigma_x \tag{5.75}$$

Substituting Eqs. (5.31a) to (5.31b) from the Rosenblueth solutions in Eqs. (5.74a) and (5.74b) gives the result of $p_o = 0$. This indicates that a two-point representation is sufficient for preserving the first three moments of a single stochastic basic variable. In fact, Eqs. (5.31a) and (5.31b) of the Rosenblueth solutions are special cases of Eqs. (5.74a) and (5.74b) when $\kappa_x = \gamma_x^2 + 1$, which are the boundary for all feasible probability distributions (Eq. (2.43)). The Rosenblueth solutions for a symmetric variable with $x'_- = x'_+ = 1$ corresponds to $\gamma_x = 0$ and $\kappa_x = 1$ in Li's solution of Eq. (5.74a).

Based on Eqs. (5.74b) and (5.75), the r-th order statistical moment of the model output $W = g(X)$ can be estimated as

$$E(W^r) = p_- w_-^r + p_o \, \overline{w}^r + p_+ w_+^r \tag{5.76}$$

where $w_- = g(x_-)$, $\overline{w} = g(\boldsymbol{\mu_x})$, and $w_+ = g(x_+)$.

Considering a multivariate model $W = g(\boldsymbol{X})$ has, or can be approximated by, the following form:

$$W = g(\boldsymbol{X}) \approx \overline{w} + \sum_{k=1}^{K} a_k(X_k - \mu_k) + \sum_{k=1}^{K} b_k(X_k - \mu_k)^2 + \sum_{k=1}^{K} c_k(X_k - \mu_k)^3$$

$$+ \sum_{k=1}^{K} d_k(X_k - \mu_k)^4 + \sum_{i=1}^{K-1} \sum_{j=i+1}^{K} e_{ij} \, (X_i - \mu_i)(X_j - \mu_j) \tag{5.77}$$

where $\overline{w} = g(\boldsymbol{\mu_x})$; a_k, b_k, c_k, and d_k are coefficients. Let $W_k = g_k(X_k) = g(\mu_1, \mu_2, \ldots, \mu_{k-1}, X_k, \mu_{k+1}, \ldots, \mu_K)$, which can be expressed as

$$W_k = g_k(X_k) = \overline{w} + a_k(X_k - \mu_k) + b_k(X_k - \mu_k)^2$$

$$+ c_k(X_k - \mu_k)^3 + d_k(X_k - \mu_k)^4 \tag{5.78}$$

The first five terms on the right-hand side of Eq. (5.77) can be expressed in terms of W_k as

$$(1 - K) \, \overline{w} + \sum_{k=1}^{K} W_k \tag{5.79}$$

Since Eq. (5.78) is the 4th-order univariate polynomial, the expectation of $W_k = g_k(X_k)$ can be obtained, according to Eq. (5.76) with $r = 1$, as

$$E(W_k) = p_{k-} w_{k-} + p_{ko} \overline{w} + p_{k+} w_{k+} \tag{5.80}$$

where $w_{k\pm} = g_k(\boldsymbol{x}_{k\pm})$ with $\boldsymbol{x}_{k\pm} = (\mu_1, \mu_2,\ldots,\mu_{k-1}, x_{k\pm}, \mu_{k+1},\ldots,\mu_K)^t$. Hence, the expectation of Eq. (5.79) is

$$
E\left[(1-K)\overline{w} + \sum_{k=1}^{K} W_k\right] = (1-K)\overline{w} + \sum_{k=1}^{K} E(W_k)
$$

$$
= (1-K)\overline{w} + \sum_{k=1}^{K} [p_{k-}w_{k-} + p_{ko}\overline{w} + p_{k+}w_{k+}] \qquad (5.81)
$$

Furthermore, referring to Eq. (5.16) it is easily seen that the coefficient e_{ij} in Eq. (5.77) is $\partial^2 W/\partial X_i \partial X_j$ and this partial derivative can be evaluated by the finite-difference formula as

$$
e_{ij} \approx \frac{g_{ij}(x_{i+}, x_{j+}) - g_i(x_{i+}) - g_j(x_{j+}) + g(\boldsymbol{\mu}_x)}{(x_{i+} - \mu_i)(x_{j+} - \mu_j)} = \frac{w_{ij} - w_{i+} - w_{j+} + \overline{w}}{(x'_{i+}\sigma_i)(x'_{j+}\sigma_j)} \qquad (5.82)
$$

where $w_{ij} = g_{ij}(x_{i+}, x_{j+}) = g(\mu_1, \mu_2,\ldots,\mu_{i-1}, x_{i+}, \mu_{i+1},\ldots,\mu_{j-1}, x_{j+}, \mu_{j+1},\ldots,\mu_K)$. Then, the expectation of the last terms on the right-hand side of Eq. (5.77) can be written as

$$
\sum_{i < j}\sum e_{ij}\sigma_i \sigma_j \rho_{ij} = \sum_{i < j}\sum e_{ij}[w_{ij} - w_{i+} - w_{j+} + \overline{w}]\eta_{ij} \qquad (5.83)
$$

where $\eta_{ij} = \rho_{ij}/(x'_{i+} x'_{j+})$. Combining Eqs. (5.81) and (5.83), the expectation of $W = g(\boldsymbol{X})$ given by Eq. (5.77), after simplification, yields

$$
E[W] = \left(1 - \frac{3K}{2} + \frac{\eta_{..}}{2} + \sum_{k=1}^{K} p_{ko}\right)\overline{w} + \sum_{k=1}^{K}[(p_{k+} - \eta_{k.} + 1)w_{k+} + p_- w_{k-}] + \sum_{i < j}\sum w_{ij}\eta_{ij} \qquad (5.84)
$$

where $\eta_{i.} = \Sigma_j \eta_{ij}$, $\eta_{..} = \Sigma_i \eta_{i.} = \Sigma_i \Sigma_j \eta_{ij}$ with subscript '·'representing the sum of η_{ij} over the associated subscript index, and $\eta_{ii} = 1$ by definition. As can be seen, the evaluation of $E[g(\boldsymbol{X})]$ by Eq. (5.84) requires $(K^2 + 3K + 2)/2$ evaluations of the model $W = g(\boldsymbol{X})$. In generalization, the rth-order moment of $W = g(\boldsymbol{X})$ about the origin can be estimated as

$$
E[W^r] = \left(1 - \frac{3K}{2} + \frac{\eta}{2} + \sum_{k=1}^{K} p_{ko}\right)\overline{w}^r + \sum_{k=1}^{K}\left[(p_{k+} - \eta_k + 1)w_{k+}^r + p_{k-}w_{k-}^r\right] + \sum_{i < j}\sum w_{ij}^r \eta_{ij}
$$

$$
(5.85)
$$

As can be seen that if only the first three moments are available, the Rosenblueth solutions by Eqs. (5.31a) to (5.31d) are used in Eq. (5.84) (Tsai and Francechini 2002). When only the first two moments are available, $x'_{k-} = x'_{k+} = 1$

and $p_{k-} = p_{k+} = 0.5$ by which Eq. (5.85) can be simplified as

$$E[W^r] = \left(1 - \frac{3K}{2} + \frac{\rho_{..}}{2}\right)\overline{w}^r + \frac{1}{2}\sum_{k=1}^{K}\left[(3 - 2\rho_k)w_{k+}^r + w_{k-}^r\right] + \sum\sum_{i<j}w_{ij}^r\,\eta_{ij} \quad (5.86)$$

where $\rho_{i.} = \Sigma_j\,\rho_{ij}$, $\rho_{..} = \Sigma_i\,\rho_{i.}$, and $\rho_{ii} = 1$ by definition.

Furthermore, when the stochastic basic variables are uncorrelated normal, that is, $\rho_{ij} = 0$ for all $i \neq j$, then, $\rho_i = 1$ for all $i = 1, 2,...,K$ and $\rho = \Sigma_i\,\rho_i = K$. Consequently, Eq. (5.86) can further be simplified as

$$E[W^r] = (1 - K)\overline{w}^r + \frac{1}{2}\sum_{k=1}^{K}\left[w_{k+}^r + w_{k-}^r\right] \quad (5.87)$$

where $w_{k\pm} = g(\mu_1, \mu_2,..., \mu_{k-1}, \mu_k \pm \sigma_k, \mu_{k+1},...,\mu_K)^t$. Equation (5.86) indicates that, under the condition that the stochastic basic variables are independent and normal, the Li PPE method requires only the $2K + 1$ model evaluation, which requires about the same amount of computations as the Harr method.

With Eq. (5.87) in mind, the idea can be extended to deal with problems involving multivariate normal variables. Through the orthogonal transformation as described in Sec. 5.2.1, the terms w_{k+} and w_{k-} in Eq. (5.87) can be computed as

$$w_{k\pm} = g(\boldsymbol{x}_{k\pm}) = g\left(\boldsymbol{\mu_x} + \boldsymbol{D}_x^{1/2}\,\boldsymbol{V}\boldsymbol{\Lambda}^{1/2}\boldsymbol{e}_{k\pm}\right) \qquad \text{for } k = 1, 2,...,K \quad (5.88)$$

where $\boldsymbol{e}_{k+}$ and $\boldsymbol{e}_{k-}$ are $K \times 1$ column unit vector with $+1$ and -1 for the kth element and zero elsewhere.

Referring to Eq. (5.77), when the polynomial order is four or less, Eq. (5.84) would yield the exact expected value of $W = g(\boldsymbol{X})$. However, for higher-order statistical moments, that is, $r \geq 2$, the computation of $E(W^r)$ by Eq. (5.85) is no longer exact. It should also be pointed out that Eq. (5.77) is an incomplete 4th-order Taylor series expansion. For a general model, the statistical moments of the model output computed by Eq. (5.85) are approximations only. An application of Eq. (5.85) to quantify the uncertainty in the backwater profile computation is given by Zoppou and Li (1993).

Example 5.12   Refer to Example 5.10 using Manning's formula to estimate storm sewer capacity. Using the same statistical properties of the model parameters listed below, estimate the first four moments of the sewer flow capacity by the Li point estimate method.

Variable	Distribution	Mean	COV
Roughness, n	Normal	0.015	0.05
Pipe diameter, D	Normal	3.0 ft	0.05
Pipe slope, S	Normal	0.005	0.05
Correlation	Corr $(n, D) = -0.75$; Corr$(n, S) = $ Corr$(D, S) = 0.0$		

Solution   Manning's formula for a sewer section is $Q = 0.463n^{-1}D^{2.67}S^{0.5}$. The standard deviations of the roughness coefficient, sewer diameter, and sewer slope are $\sigma_n = 0.00075$; $\sigma_D = 0.15$; $\sigma_S = 0.00025$, respectively. Furthermore, as all stochastic basic variables are normally distributed, their skewness coefficients are equal to zero and kurtosis values are equal to 3. That is, $\gamma_n = \gamma_D = \gamma_S = 0$ and $\kappa_n = \kappa_D = \kappa_S = 3$.

As all stochastic basic variables are normally distributed, the two points on either side of the mean will be symmetric to the mean. According to Eqs. (5.74a) and (5.74b), they are $x'_+ = x'_- = 1.732$ and the corresponding probability masses are $p_+ = p_- = 0.1667$ and $p_0 = 0.6667$. Hence, the locations of model variables and their corresponding probabilities in the original space are

Variable	Roughness, n	Diameter, D	Slope, S	Probability
x'_+	0.0163	3.260	0.0054	0.1667
μ_x	0.015	3.000	0.0050	0.6667
x'_-	0.0137	2.740	0.0046	0.1667

By the Li method, there are 10 points in the three-variable space where the values of storm sewer capacity have to be calculated as listed below:

Point	Roughness, n	Diameter, D	Slope, S	Flow cap., Q_c
1 (000)[*]	0.0150	3.000	0.0050	41.01
2 (+00)	0.0163	3.000	0.0050	37.74
3 (−00)	0.0137	3.000	0.0050	44.90
4 (0+0)	0.0150	3.260	0.0050	51.19
5 (0−0)	0.0150	2.740	0.0050	32.20
6 (00+)	0.0150	3.000	0.0054	42.75
7 (00−)	0.0150	3.000	0.0046	39.19
8 (++0)	0.0163	3.260	0.0050	47.11
9 (+0+)	0.0163	3.000	0.0054	39.34
10 (0++)	0.0150	3.260	0.0054	53.36

[*] (0+−) represents $n_1 = \mu_1$, $D_2 = d_{2+}$, and $S_3 = S_{3-}$.

Point 1 is used in Eq. (5.86) for computing $\overline{w}$ in the first term of Eq. (5.86), points 2 to 7 are for computing w_{k+} and w_{k-} in the 2nd terms, while points 8 to 10 are for computing w_{ij} in the 3rd terms. The resulting first four product-moments about the origin for the sewer flow capacity, according to Eq. (5.86), can be obtained as

$$E(Q_c) = 41.53 \qquad E(Q_c^2) = 1779.46 \qquad E(Q_c^3) = 78615.90 \qquad E(Q_c^4) = 3574{,}472.92$$

The standard deviation, skewness coefficient, and kurtosis of the sewer flow capacity then can be calculated as

$$\sigma_{Q_c} = 7.39 \text{ ft}^3/\text{s} \qquad \gamma_{Q_c} = 0.443 \qquad \kappa_{Q_c} = 1.599$$

One can compare the current results with those obtained in the previous examples by different methods.

5.5   Summary and Concluding Remarks

Each of the four uncertainty analysis techniques described in this chapter differs in the level of sophistication, computational complexity, and data requirements. In theory, all the four approximation methods discussed in this chapter do not require information on the marginal or joint PDF of the random variables involved in the problem. The FOVE and Harr point estimation methods require only the mean, standard deviation, and correlations of the stochastic basic variables involved in the models. As this is implied, the FOVE and Harr methods, theoretically, are appropriate only for dealing with normal random variables. The Rosenblueth method offers an added flexibility to consider the asymmetry of stochastic basic variables.

For the FOVE method, the correlation between stochastic basic variables can be incorporated to estimate the variance of the model output. However, such correlation cannot be used to estimate the mean value of the model output by the FOVE method without including the 2nd-order term in Eq. (5.11) or (5.16). Higher-order moments of a model output can be estimated by the method straightforwardly when the stochastic basic variables are statistically independent. However, the exercise could be cumbersome. Estimations of higher-order moments by the FOVE method involving correlated stochastic basic variables require knowing higher-order cross-product moments that can be obtained with additional computations when data for model parameters are available. In general, information on higher-order moments or cross-product moments is not available or cannot be reliably computed in most practical problems. A distinct advantage of the FOVE method over its competitors is that the method provides analysts insight about the sensitivity of each stochastic basic variable and their contributions to the overall uncertainty of the model output.

The practical implementation of the FOVE method hinges on the calculation of sensitivity coefficients that are the 1st-order partial derivative of the model output with respect to all the stochastic model parameters. If the model under consideration is complex or nonanalytical, computations of sensitivity coefficients would have to be done numerically. This could make the computations very time consuming and cumbersome. The FOVE method has been applied to the uncertainty and reliability analysis of hydrosystem engineering problems including open-channel flow (Huang 1986; Cesare 1991; Yeh and Tung 1993), groundwater flow (Dettinger and Wilson 1981; Nguyen and Chowdhury 1984), runoff model (Garen and Burges 1981; Melching 1992), National Weather rainfall frequency atlas (Tung 1987), levee systems (Tung and Mays 1981), storm drainage systems (Tang and Yen 1972; Yen and Tang 1976; Melching and Yen 1986), highway drainage structures such as culverts (Mays 1979; Yen, Chan, and Tung 1980; Lian and Yen 2003) and bridges (Tung and Mays 1982), benefit-cost analysis (Dandy 1986; Woods and Gulliver 1991; Tung 1992), water-quality modeling (Burges and Lettenmaier 1975; Chadderton, Millers, and McDonnell 1982; Tung and Hathhorn 1988a,b; 1989).

Computationally, probabilistic point estimation methods, generally, are simpler and more flexible than FOVE method especially when a model is either complex or nonanalytical in the forms of tables, figures, or computer programs. Another advantage of probabilistic point estimation methods is their simplicity in estimating the statistical moments of any order of the model output as one desires. This, as stated above, is not a luxury the FOVE method can always enjoy. For the Rosenblueth method the main factor, however, dictating its computational feasibility is the number of stochastic basic variables in a model. The method requires 2^K model evaluations, which could be computationally prohibitive when K is large. Furthermore, the determination of the coordinates of the points in the parameter space for model evaluation and their corresponding probability masses (or weights) is indeterminate when the number of stochastic basic variables $K \geq 3$. The probability assignment to points by the Rosenblueth method is nonunique. The simple equations presented in Sec. 5.2.3 could produce negative probabilities for some selected points when a problem involves asymmetric correlated random variables. The more complicated formulas developed by Panchalingam and Harr (1994) may correct this problem to a limited extent. A modification by Tsai and Francechini (2002) would circumvent this difficulty and reduce the computational burden. The Rosenblueth point estimation method has been applied to risk and uncertainty analyses of transporting dangerous gas (Van Aerde, Shortreed, and Saccomano 1987) and a fault tree (Van Aerde and Lind 1986). In hydrosystems engineering, applications have been made to bridge pier scouring (Chang 1994), groundwater flow (Nguyen and Chowdhury 1984; Emery 1990), fluvial fan modeling (Zhao and Mays 1996), culverts design (Lian and Yen 2003), hydrograph analysis (Yeh, Yang, and Tung 1997), and rainfall-runoff modeling (Melching 1992; Yu, Yang, and Chan 2001).

Along the same line of the Rosenblueth point estimation method, there is a three-point estimation method developed by Pearson and Tukey (1965). The method represents the PDF of a random variable by assuming that the probability masses are distributed symmetrically at 25-, 50-, and 75- percentile points. The method does not have the provision to account for the correlation among the stochastic basic variables and, thereby, assumes that they are statistically independent. Similar to the Rosenblueth PPE method, the method would require 3^K model evaluations for a model involving K stochastic basic variables. The computational practicality of the method is even less attractive than that of the Rosenblueth method. There are two applications of the Pearson-Tukey three-point method in the area of business (Keefer and Bodily 1983; Pfeifer, Bodily, and Frey 1991).

The Harr probabilistic point estimation method can be regarded as the remedy of the applicability of the Rosenblueth method to practical engineering problems when the number of stochastic basic variables is moderate or large. It reduces the number of model evaluations from 2^K, required by Rosenblueth's method, down to $2K$. The saving in computations for Harr's PPE method becomes more and more impressive as the number of stochastic basic variables K increases. Like the FOVE method, the Harr PPE method accounts for only the first two moments of the stochastic basic variables involved, including their correlations. Therefore, the method implicitly assumes that all stochastic basic variables involved are normal random variables. It may not be directly applicable to deal with problems involving several symmetric random variables; even their skewness coefficients are zero (Chang 1994) without applying a normal transformation. Chang (1994) proposed modifications to Harr's method to improve its capability in handling nonnormal stochastic basic variables while maintaining its computational advantage. Although the method provides information on the contribution of each of the transformed variables, defined by the eigenvalues of the correlation matrix, to the total uncertainty of the model output, it does not directly provide such information for the original stochastic basic variables. The inverse transformation for extracting such information is difficult. Applications of Harr's algorithm to the uncertainty analysis of hydrosystem engineering problems can be found for the gravel pit migration analysis (Yeh and Tung 1993), sediment transport modeling (Chang, Yang, and Tung 1993), bridge pier scouring (Chang 1994), ground water flow (Emery 1990), parameter estimation for a distributed hydrodynamic model (Zhao 1994), hydrograph analysis (Yeh, Yang, and Tung 1997), runoff modeling and hydrograph analysis (Yeh, Yang, and Tung 1997; Yu, Yang, and Chan 2001), and ground water flow (Guymon 1994).

Assuming that the model takes the form of or is approximated by a 4th-order polynomial of stochastic basic variables and their cross-products, Li's three-point method allows one to preserve the first four moments of each individual variable and the correlation among them. Under the identical condition of model functional form, the method of Tsai-Franceschini utilizes the Rosenblueth two-point representation to preserve the first three statistical moments and correlations to eliminate the nonunique solution of the Rosenblueth method. Both methods require $(K^2 + 3K + 2)/2$ model evaluations, which is more flexible and certainly computationally more efficient than the procedures of Pearson-Tukey and Rosenblueth, especially when the number of variables is large.

Similar to the analytical methods described in Chap. 4, once the statistical moments of the model output are estimated by the approximation methods described in this chapter, the probability distribution of the model output can be estimated by the two asymptotic expansions and the entropy distribution described in Sec. 4.5.

In uncertainty analysis, the methods described in this chapter provide approximations to the statistical moments of a model output in terms of the stochastic basic variables involved. The FOVE method has been known to yield rather accurate estimations of the first two moments if the nonlinearity of the model and/or the uncertainty of stochastic basic variables are not too large. Various authors have tried to define "not too large." For rainfall-runoff modeling Garen and Burges (1981) suggested that if the coefficient of variation (COV) of the key basic variables was less than or equal to 0.25 the FOVE method works well; whereas Cornell suggested a more general bound of 0.2 for the COV of the key basic variables. However, Melching (1995) found that the FOVE method yielded reasonable results for cases where COV values substantially exceeded these bounds. Thus, each researcher must define "not too large" for the problem under study. As the nonlinearity or parameter uncertainty increases, the accuracy of the FOVE method deteriorates rapidly. Nguyen and Chowdhury (1985) investigated the performance of the Rosenblueth PPE method under a multivariate normal condition in a geotechnical application. It was found to be more accurate than the FOVE method. Chang, Tung, and Yang (1995) conducted a systematic evaluation of the relative performance of the three probabilistic point estimation methods, namely, Rosenblueth, Harr, and modified Harr, by applying them to different models of varying degrees of nonlinearity. The model parameters are assumed to be correlated or independent normal random variables. In general, all three PPE algorithms are capable of yielding rather accurate estimations for the first two moments, especially when the model is close to linear. However, the Rosenblueth method produces better estimations for the skewness coefficient of the model output than the Harr procedure when a model is nonlinear, involving independent normal variables. For a model having correlated normal random variables, the modified Harr method outperforms the other two competitors in estimating the first three moments of the model output. In the evaluation, it is generally observed that the accuracy of the moment estimation decreases as the order of the moments increases, and as the degree of model nonlinearity increases.

Appendix 5A: Cholesky Decomposition

For any nonsingular square matrix A, it can be decomposed as

$$A = LU \tag{5A.1}$$

where L is a lower triangular matrix as shown in Sec. 5.3.1 and U is an upper triangular matrix. In general, the matrices L and U are not unique. However, Young and Gregory (1973) show that, if the diagonal elements of L or U are specified, the decomposition will be unique.

When the matrix A is real, symmetric, and positive-definite, then $U = L^t$, which means $A = LL^t$. This is called the *Cholesky decomposition*. Writing out $A = LL^t$ in components, one readily obtains the following relationships between the elements in matrices L and A as

$$l_{kk}^2 + \sum_{j=1}^{k-1} l_{kj}^2 = a_{kk} \qquad \text{for } k = 1,\ 2,\ldots,\ K \tag{5A.2}$$

$$l_{kk}l_{jj} + \sum_{i=1}^{j-1} l_{ki}l_{ji} = a_{kj} \qquad \text{for } k = j+1,\ldots,\ K \tag{5A.3}$$

in which l_{ij} and a_{ij} are elements in matrices L and A, respectively, and K is the size of the matrices. In terms of a_{ij}'s, l_{ij}'s can be expressed as

$$l_{ii} = \left(a_{ii} - \sum_{k=1}^{i-1} l_{ik}^2 \right)^{1/2} \tag{5A.4}$$

$$l_{ij} = \frac{1}{l_{ii}} \left(a_{ij} - \sum_{k=1}^{j-1} l_{ik}\ l_{jk} \right) \qquad \text{for } i = j+1,\ldots, K \tag{5A.5}$$

Computationally, the values of l_{ij}'s can be obtained by solving Eqs. (5A.4) and (5A.5) sequentially following the order $i = 1, 2, \ldots, K$. Numerical examples can be found in Wilkinson (1965). A simple computer program for the Cholesky decomposition is available from Press, Teukolsky, and Vetterling (1992). Note that the requirement of positive definite for matrix A is to ensure that the quantity in the square root of Eq. (5A.4) will always be positive throughout the computation. If A is not a positive definite matrix, the algorithm will fail.

For a real, symmetric, positive-definite matrix A, the Cholesky decomposition is sometimes expressed as

$$A = \tilde{L} \Lambda \tilde{L}^t \tag{5A.6}$$

in which L is a unit lower triangular matrix with all its diagonal elements having values of ones, and Λ is a diagonal eigenvalue matrix. Therefore, the eigenvalues associated with matrix A are the square roots of the diagonal elements in matrix L. If a matrix is positive-definite, all its eigenvalues will be positive, and vice versa.

In theory, the covariance and correlation matrices in any multivariate problems should be positive-definite. In practice, sample correlation and sample covariance are often used in the analysis. Due to the sampling errors, the resulting sample correlation matrix may not be positive-definite and, in such cases, the Cholesky decomposition may fail, whereas, the spectral decomposition described in Sec. 5.3.1 will be applicable.

Problems

5.1 Referring to Prob. 2.18, use the FOVE method to
  a. Estimate the mean and variance of the total drawdown at the observation point and to compare the result with that from part (b) of Prob. 2.18
  b. Estimate the covariance of the drawdowns from the two production wells
  c. Would the inclusion of 2nd-order terms in the Taylor expansion improve the estimation of the mean of the total drawdown?

5.2 Refer to Example 5.3 and estimate the mean sewer flow capacity by considering the 2nd-order expansion term in the Taylor series. Compare the resulting mean sewer flow capacity with those obtained in Examples 5.3, 5.8, and 5.10.

5.3 Consider the multivariate function $W = X_1 + 2X_1 + 3X_3$ in which X_1, X_2, and X_3 are normal random variables. The statistical properties of the three random variables are the following:

$$E(X_1) = 7.25 \qquad E(X_2) = 7.67 \qquad E(X_3) = 6.92$$

$$\text{Var}(X_1) = 7.45 \qquad \text{Var}(X_2) = 12.96 \qquad \text{Var}(X_3) = 14.59$$

$$\text{Corr}(X_1, X_2) = 0.89 \qquad \text{Corr}(X_1, X_3) = 0.75 \qquad \text{Corr}(X_2, X_3) = 0.89$$

Derive the exact values of the mean and variance of W.

5.4   Show that the FOVE method would yield the exact solution to Prob. 5.3.

5.5   Consider the multivariate function $W = X_1^2 + X_2^2 + X_3^2$ in which X_1, X_2, and X_3 are normal random variables. The three random variables X_1, X_2, and X_3 have the same statistical properties as given in Prob. 5.3. Apply the FOVE method to estimate the mean and variance of W. Also, determine the percentage contribution of each stochastic variable to the total variance of W.

5.6   Refer to Prob. 5.5 and estimate the mean of W by considering the 2nd-order expansion term in the Taylor series. Compare the results with those obtained in Prob. 5.5.

5.7   Consider the multivariate function $W = X_1 X_2 X_3$ in which X_1, X_2, and X_3 are lognormal random variables. The three random variables X_1, X_2, and X_3 have the same statistical properties as given in Prob. 5.3. Derive the exact values of the mean and variance of W.

5.8   Referring to Prob. 5.7, apply the FOVE method to estimate the mean and variance of W. Compare the estimated mean and variance with the exact values obtained in Prob. 5.7. Also, determine the percentage contribution of each stochastic basic variable to the total variance of W.

5.9   Refer to Prob. 5.7 and estimate the mean of W by considering the 2nd-order expansion term in the Taylor series. Compare the results with those obtained in Probs. 5.7 and 5.8.

5.10   Consider a model in which the model response is related to K independent random variables in a multiplicative form as

$$W = a_0 X_1^{a_1} X_2^{a_2} \cdots X_K^{a_K}$$

Show that, by using the FOVE method, the following is true:

$$\Omega_W^2 = \sum_{k=1}^{K} a_k^2\, \Omega_k^2$$

in which Ω is the coefficient of variation.

5.11   In the storm sewer design, the rational formula $Q_L = C\, i\, A$ is frequently used to determine the inflow (load) to the sewer system. On the other hand, the resistance of the sewer system is represented by the sewer flow capacity, which can be estimated by Manning's formula. One performance criterion in the reliability analysis is the safety margin (SM) that can be expressed as

$$SM = Q_C - Q_L$$

in which Q_C is the sewer flow capacity determined by Manning's formula as given in Example 5.1. Given the following data, estimate the mean and variance of the safety margin by the FOVE method.

Stochastic variable	Distribution	Mean	Standard deviation
C	Lognormal	0.825	0.0825
i (in/h)	Lognormal	4.000	1.0
A (acres)	Normal	10.000	0.5
Corr(C, i) = –0.50; Corr(C, A) = 0.0; Corr(i, A) = –0.6			
n	Lognormal	0.015	0.00075
D (ft)	Normal	3.0	0.06
S	Lognormal	0.005	0.00025
Corr(n, D) = –0.75; Corr(n, S) = 0.0; Corr(D, S) = 0.0			

5.12 In groundwater modeling, the drawdown of a confined aquifer can be estimated by the well-known Copper-Jacob equation

$$s = \xi \frac{Q_p}{4\pi T}\left[-0.5772 - \ln\left(\frac{r^2 S}{4Tt}\right)\right]$$

where ξ = model correction factor accounting for the error of approximation
  s = drawdown (in meters)
  S = storage coefficient
  T = transmissivity (in meter square per day)
  Q_p = pumping rate (in meter cube per day)
  t = elapsed time (in days)

Due to the nonhomogeneity of the geologic formation, the storage coefficient and transmissivity are, in fact, random variables. Furthermore, the model correction factor can be treated as a random variable. Given the following information about the stochastic variables in the Copper-Jacob equation, estimate the mean and standard deviation of the drawdown under the condition of $Q_p = 1000$ m^3/day, $r = 200$ m, and $t = 1$ day by the FOVE method.

Stochastic variable	Distribution	Mean	Coefficient of variation
ξ	Normal	1.0	0.10
T (m^3/day)	Lognormal	1000.0	0.15
S	Lognormal	0.0001	0.10
Corr(T, S) = –0.70; Corr(ξ, T) = 0.0; Corr(ξ, S) = 0.0			

5.13 The general rainfall intensity-duration-frequency (IDF) curve has the following form

$$i = \frac{a\,T^b}{c + t_d}$$

where i = rainfall intensity (in inch per hour)
  T = return period (in years)
  t_d = rainfall duration (in minutes)

a, b, and c are coefficients. The coefficients, in general, vary from location to location and are subject to error in rainfall data analysis. Therefore, they are random variables. Considering the following data for Urbana, Illinois, estimate the mean and variance of a 5-year, 30-minute rainfall intensity by the FOVE method.

Stochastic variable	Distribution	Mean	Coefficient of variation
a	Normal	120.0	0.10
b	Normal	0.175	0.20
c	Lognormal	27.0	0.15
Corr$(a, b) = -0.50$; Corr$(a, c) = 0.4$; Corr$(b, c) = 0.7$			

5.14 In hydraulic dredging, a frequently used formula for computing the friction loss in a pipe transporting slurry is that of Hazen-Williams, which is expressed as

$$h_L = 0.2082 \left(\frac{100}{C} \right)^{1.85} \frac{Q^{1.85}}{D^{4.866}}$$

where h_L = slurry friction loss per 100 ft of pipe expressed in feet of water
$\quad\quad D$ = inside diameter of pipe in inches
$\quad\quad Q$ = flow rate in gallons per minute (GPM)
$\quad\quad C$ = Hazen-Williams coefficient, which is a function of the median particle size (d_{50}) of the channel bed sediment and slurry specific gravity (SG) as shown in Fig. P5.1 (Turner 1984).

Considering that Q, D, and d_{50} are independent random variables with their statistical properties given in the table below, determine the mean and standard deviation of the head loss associated with a dredging pipe of 2000 ft by the FOVE method assuming that the specific gravity of slurry is 1.25.

Stochastic variable	Distribution	Mean	Coefficient of variation
Q (GPM)	Normal	5000	0.10
D (in)	Normal	18.0	0.05
d_{50} (mm)	Lognormal	0.4	0.25

5.15 Refer to Example 5.8 and assume that all the three stochastic variables in Manning's formula are correlated lognormal random variables. Determine the mean and variance of the sewer flow capacity by considering the skewness coefficients of the three stochastic variables using Rosenblueth's PPE method. Compare the exact solution and the results shown in Examples 5.3, 5.8, and 5.10, and Prob. 5.2.

5.16 Resolve Example 5.7 by Rosenblueth's method assuming that all stochastic model parameters are uncorrelated. Compare the results with those of Examples 5.8 and and discuss them.

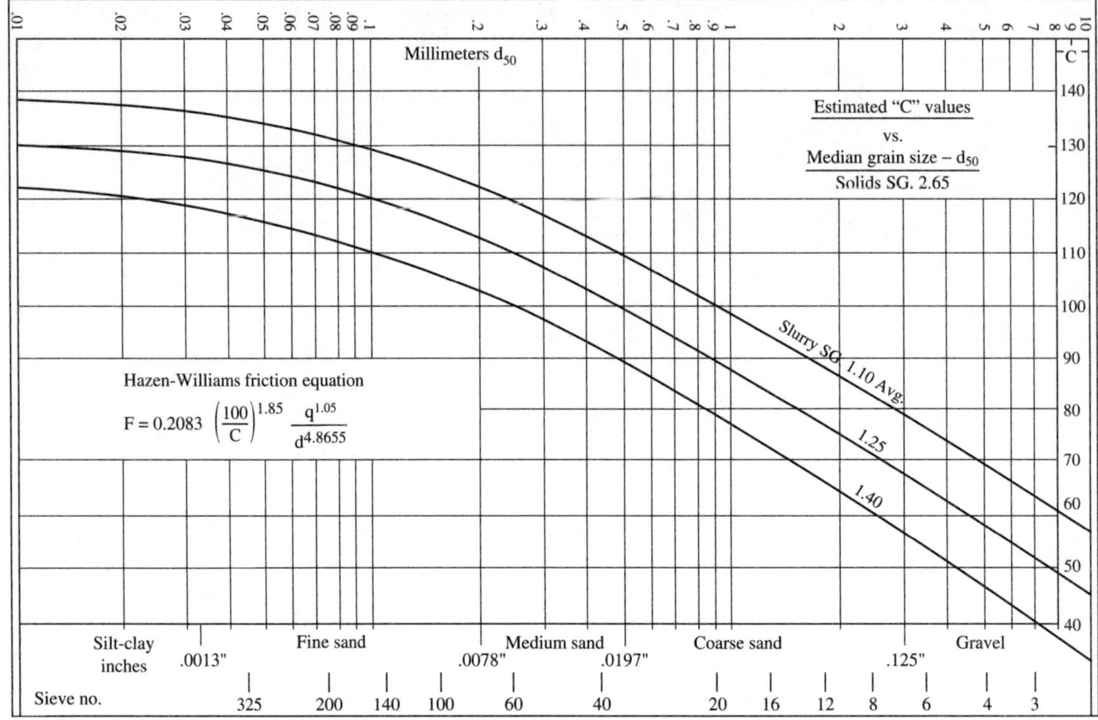

$$F = 0.2083 \left(\frac{100}{C}\right)^{1.85} \frac{q^{1.05}}{d^{4.8655}}$$

Figure P5.1   Hazen-Williams coefficient for slurry pipes (after Turner 1984).

5.17   Refer to Prob. 5.3. Estimate the mean and variance of W using the Rosenblueth PPE method.

5.18   Refer to Prob. 5.5. Estimate the mean and variance of W using the Rosenblueth PPE method. Compare the results with those from Probs. 5.5 and 5.6.

5.19   Referring to Prob. 5.7, estimate the mean and variance of W using the Rosenblueth PPE method. Compare the results with those from Probs. 5.7, 5.8, and 5.9.

5.20   Resolve Prob. 5.11 by the Rosenblueth PPE method and compare the answers with those obtained in Prob. 5.11.

5.21   Resolve Prob. 5.12 by the Rosenblueth PPE method and compare the answers with those obtained in Probs. 5.12.

5.22   Resolve Prob. 5.13 by the Rosenblueth PPE method and compare the answers with those obtained in Prob. 5.13.

5.23   Resolve Prob. 5.14 by the Rosenblueth PPE method and compare the answers with those obtained in Prob 5.14.

5.24   Derive Eq. (5.38).

5.25   Resolve Example 5.10 by the Harr PPE method assuming that all the stochastic variables are uncorrelated. Compare the results with those obtained in Prob. 5.10, Examples 5.8, 5.10, 5.11, and discuss them.

5.26   Resolve Example 5.9 by the Harr PPE method using the spectral decomposition of the covariance matrix. Compare the results with those obtained in Example 5.9.

5.27   Resolve Example 5.9 by the modified Harr PPE method using the spectral decomposition of the correlation matrix and covariance matrix. Compare the results with those obtained in Example 5.9 and Prob. 5.26.

Apply the following four procedures to solve Prob. 5.28 to 5.36.
a. Harr's PPE method with spectral decomposition of the correlation matrix

b. Harr's PPE method with spectral decomposition of the covariance matrix

c. Modified Harr PPE method with spectral decomposition of the correlation matrix

d. Modified Harr PPE method with spectral decomposition of the covariance matrix

5.28   Resolve Example 5.10 by procedures (b) and (d) listed previously and compare the results with those obtained in Example 5.10.

5.29   Refer to Prob. 5.3. Estimate the mean and variance of W. Compare the results with those obtained in Probs. 5.3 and 5.17.

5.30   Refer to Prob. 5.5. Estimate the mean and variance of W. Compare them to the results obtained from Probs. 5.5, 5.6, and 5.18.

5.31   Repeat Prob. 5.30 by considering the points on the two most dominant eigenvectors. Compare the results with those obtained in Prob. 5.30.

5.32   Referring to Prob. 5.7, estimate the mean and variance of W. Compare the results from Probs. 5.7, 5.8, 5.9, and 5.19.

5.33   Repeat Prob. 5.32 by considering the points on the two most dominant eigenvectors. Compare the results with those obtained in Prob. 5.32.

5.34   Resolve Prob. 5.11 and compare the answers with those obtained in Probs. 5.11 and 5.20.

5.35   Resolve Prob. 5.12 and compare the answers with those obtained in Probs. 5.12 and 5.21.

5.36   Resolve Prob. 5.13 and compare the answers with those obtained in Probs. 5.13 and 5.22.

5.37   Resolve Prob. 5.14 and compare the answers with those obtained in Probs. 5.14 and 5.23.

5.38 Resolve Example 5.26 by the Li PPE method and compare the results with those obtained previously.

5.39 Refer to Prob. 5.5. Estimate the mean and variance of W by the Li PPE method and compare the results with those obtained previously.

5.40 Refer to Prob. 5.6. Estimate the mean and variance of W by the Li PPE method and compare the results with those obtained previously.

5.41 Referring to Prob. 5.7, estimate the mean and variance of W by the Li PPE method and compare the results with those obtained previously.

5.42 Resolve Prob. 5.11 by the Li PPE method and compare the answers with those obtained previously.

5.43 Resolve Prob. 5.12 by the Li PPE method and compare with the answers obtained previously.

5.44 Resolve Prob. 5.13 by the Li PPE method and compare the answers with those obtained previously.

5.45 Resolve Prob. 5.14 by the Li PPE method and compare the answers with those obtained previously.

5.46 In growth forecasts, one of the commonly used models is the logistic growth model based on the following differential equation (Karmeshu and Lara-Rosano 1987)

$$\frac{dP}{dt} = RP\left(1 - \frac{P}{C}\right)$$

in which P is the population variable to be forecasted; R is the growth rate; and C is the carrying capacity of the system. The solution to the preceding differential equation is

$$P_t = \frac{P_o}{\dfrac{P_o}{C} + \left(1 - \dfrac{P_o}{C}\right)e^{-Rt}}$$

where P_o is the initial condition. Given the following information, construct the curves for temporal variation of $E(P_t)$ and coefficient of variation of P_t by the FOVE method and the three PPE methods described in this chapter.

$$
\begin{array}{lll}
P_o\colon & \mu_o = 0.1 & \sigma_o = 0.09 \\
R\colon & \mu_R = 0.15 & \sigma_R = 0.05 \\
C\colon & \mu_C = 20 & \sigma_C = 10 \\
\end{array}
$$

$$\text{Corr}(P_o, R) = 0.5 \qquad \text{Corr}(P_o, C) = 0 \qquad \text{Corr}(R, C) = -0.5$$

References

Bates, B. C., and L. R. Townley (1988). "Nonlinear, Discrete Flood Event Models, 1. Analysis of Prediction Uncertainty," *Journal of Hydrology*, **99**:91–101.

Berthouex, P. M. (1975)."Modeling Concepts Considering Process Performance, Variability, and Uncertainty," In *Mathematical Modeling for Water Pollution Control Processes*, 405–439, T. M. Keinath and M. P. Wanielista (eds.), Ann Arbor Science, Ann Arbor, MI.

Burges, S. J., and D. P. Lettenmaier (1975). "Probabilistic Methods in Stream Quality Management," *Water Resources Bulletin*, **11**:115–130.

Cesare, M. A. (1991). "First-Order Analysis of Open Channel Flow," *Journal of Hydraulic Engineering*, ASCE, **117**(2):242–247.

Chadderton, R. A., A. C. Miller, and A. J. McDonnell (1982). "Uncertainty Analysis of Dissolved Oxygen Model," *Journal of the Environmental Engineering Division*, ASCE, **108**(5):1003–1012.

Chang, C. H. (1994)."Incorporating Non-Normal Marginal Distributions in Uncertainty Analysis of Hydrosystems," *Ph.D. Dissertation*, National Chiao-Tung University, Hsinchu, Taiwan.

Chang, C. H., Y. K. Tung, and J. C. Yang (1995). "Evaluation of Probabilistic Point Estimate Methods," *Applied Mathematical Modelling*, **19**(2):95–105.

Chang, C. H., J. C. Yang, and Y. K. Tung (1993). "Sensitivity and Uncertainty Analyses of a Sediment Transport Model: A Global Approach," *Journal of Stochastic Hydrology and Hydraulics*, **7**(4):299–314.

Chowdhury, R. N., and D. W. Xu (1994). "Rational Polynomial Technique in Slope-Reliability Analysis." *Journal of Geotechnical Engineering*, ASCE, **119**(12):1910–1928.

Cornell, C. A. (1972). "First-Order Analysis of Model and Parameter Uncertainty," *Proceedings of International Symposium on Uncertainties in Hydrologic and Water Resources Systems*, **3**:1245–1272, Tucson, AZ.

Dandy, G. C. (1986). "An Approximate Method for the Analysis of Uncertainty in Benefit-Cost Ratios," *Water Resources Research*, **21**(3):267–271.

Dettinger, M. D., and J. L. Wilson (1981). "First Order Analysis of Uncertainty in Numerical Models of Groundwater Flow, Part 1. Mathematical Development," *Water Resources Research*, **17**(1):149–161.

Emery, J. (1990). "Comparison of Reliability Approximation Techniques." *M.S. Thesis*, Department of Statistics, University of Wyoming, Laramie, WY.

Garen, D. C., and S. M. Burges (1981). "Approximate Error Bounds for Simulated Hydrographs," *Journal of the Hydraulics Division*, ASCE, **107**(HY11):1519–1534.

Golub, G. H., and C. F. Van Loan (1989). *Matrix Computations*, 2d ed, The Johns Hopkins University Press, Baltimore, MD.

Graybill, F. A. (1983). *Matrices with Applications in Statistics*, 2d ed, Wadsworth Publishing Company, Belmont, CA.

Guymon, G. L. (1994). *Unsaturated Zone Hydrology*, Prentice-Hall, Englewood Cliffs, NJ.

Harr, M. E. (1987). *Reliability-Based Design in Civil Engineering*. McGraw-Hill, New York.

Harr, M. E. (1989). "Probabilistic Estimates for Multivariate Analyses," *Applied Mathematical Modelling*, **13**(5):313–318.

He, J., and G. Sällfors (1994). "An Optimal Point Estimate Method for Uncertainty Studies," *Applied Mathematical Modelling*, **18**(9):494–499.

Huang, K. -Z. (1986). "Reliability Analysis of Hydraulic Design of Open Channel," *Stochastic and Risk Analysis in Hydraulic Engineering*, B. C. Yen (ed.), 59–65, Water Resources Publications, Littleton, CO.

Karmeshu, B., and F. Lara-Rosano (1987). "Modelling Data Uncertainty in Growth Forecasts." *Applied Mathematical Modelling*, **11**: 62–68.

Keefer, D. L., and S. E. Bodily (1983). "Three-Point Approximations for Continuous Random Variables," *Management Science*, **29**(5):595–609.

Li, K. S. (1992). "Point Estimate Method for Calculating Statistical Moments," *Journal of Engineering Mechanics*, ASCE, **118**(7):1506–1511.

Lian, Y. Q., and B. C. Yen (2003). "Comparison of Risk Calculation Methods for a Culvert," *Journal of Hydraulic Engineering*, ASCE, **129**(2):140–152.

Mays, L. W. (1979). "Optimal Design of Culverts under Uncertainties," *Journal of the Hydraulics Division*, ASCE, **105**(HY5):443–460.

Melching, C. S., and B. C. Yen (1986). "Slope Influence on Storm Sewer Risk," In *Stochastic and Risk Analysis in Hydraulic Engineering*, B. C. Yen (ed.), 66–78, Water Resources Publications, Littleton, CO.

Melching, C. S. (1992). "A Comparison of Methods for Estimating Variance of Water Resources Model Predictions" In *Stochastic Hydraulics '92*, edited by J. T. Kuo and G. F. Lin, *Proceedings of Sixth IAHR International Symposium on Stochastic Hydraulics*, pp. 663–670.

Melching, C. S. (1995). "Reliability Estimation," Chap. 3, In: *Computer Models of Watershed Hydrology*, V. P. Singh (ed.), 69–118, Water Resources Publications, Littleton, CO.

Nguyen, V. U., and R. N. Chowdhury (1984). "Probabilistic Study of Spoil Pile Stability in Strip Coal Mines-Two Techniques Compared," *International Journal of Rock Mechanics and Mining. Sciences and Geomech. Abstr.*, **20**(6):303–312.

Nguyen, V. U., and R. N. Chowdhury (1985). "Simulation for Risk Analysis with Correlated Variables," *Geotechnique*, **35**(1):47–58.

Panchalingam, G., and M. E. Harr (1994). "Modelling of Many Correlated and Skewed Random Variables," *Applied Mathematical Modelling*, **18**(11):635–640.

Pearson, E. S., and J. W. Tukey (1965). "Approximate Means and Standard Deviations Based on Distances between Percentage Points of Frequency Curves," *Biometrika*, **52**: 533–546.

Pfeifer, P. E., S. E. Bodily, and S. C. Frey (1991). "Pearson-Tukey Three-Point Approximations versus Monte Carlo Simulation," *Decision Science*, **22**:74–90.

Press, W. H., S. A. Teukolsky, W. T. Vetterling, and B. P. Flannery (1992). *Numerical Recipes in Fortran*, 2d ed, Cambridge University Press, New York, p. 90.

Rosenblueth, E. (1975). "Point Estimates for Probability Moments," *Proceedings of National Academy of Science*, **72**(10):3812–3814.

Rosenblueth, E. (1981). "Two-Point Estimates in Probabilities," *Applied Mathematical Modelling*, **5**:329–335.

Tang, W. H., and B. C. Yen (1972). "Hydrologic and Hydraulic Design under Uncertainties," *Proceedings of International Symposium on Uncertainties in Hydrologic and Water Resources Systems*, 2:868–882, 3:1640–1641, Tucson, AZ.

Tyagi, A., and C. T. Haan (2001). "Uncertainty Analysis Using Correlated First-Order Approximation Method." *Water Resources Research*, **37**(6):1847–1858.

Tsai, C. W. S., and S. Franceschini (2002). "Uncertainty Analysis Using an Improved Point Estimate Method," AGU 2002 Fall Meeting, San Francisco, CA.

Tung, Y. K., and L. W. Mays (1981). "Risk Models for Levee Design," *Water Resources Research*, **17**(4):833–841.

Tung, Y. K., and L. W. Mays (1982). "Optimal Risk-Based Hydraulic Design of Bridges," *Journal of the Water Resources Planning and Management Division*, ASCE, **108**(WR2):191–202.

Tung, Y. K., and W. E. Hathhorn (1988a). "Probability distribution for critical DO location in streams." *Journal of Ecological Modelling*, **42**:45–60.

Tung, Y. K., and W. E. Hathhorn (1988b). "Assessment of Probability Distribution of Instream Dissolved Oxygen Deficit." *Journal of Environmental Engineering*, ASCE, **114**(6):1421–1435.

Tung, Y. K., and W. E. Hathhorn (1989). "Determination of Critical Locations in Stochastic Stream Environments," *Journal of Ecological Modelling*, 45:43–61.

Tung, Y. K. (1987). "Uncertainty Analysis of National Weather Service Rainfall Frequency Atlas," *Journal of Hydraulic Engineering*, ASCE, **113**(2):178–189.

Tung, Y. K. (1992). "Investigation of Probability Distribution of Benefit/Cost Ratio and Net Benefit," *Journal of Water Resources Planning and Management*, ASCE, **118**(2):133–150.

Turner, T. M. (1984). *Fundamentals of Hydraulic Dredging*, 1st ed. Cornell Maritime Press, Centreville, MD.

Van Aerde, M., and N. Lind (1986). "Uncertainty Analysis for Fault Tree Based on Mass-Point Representations of Probability Distributions," In *Proceedings of the Society for Risk Analysis Annual Conference*, Boston, MA.

Van Aerde, M., J. Shortreed, and F. Saccomano (1987). "Representing Parameter Variability and Uncertainty Within System Analyses and Evaluations," *Paper* presented at the 3rd Canadian Seminar on Systems Theory for the Civil Engineer, Ecole Polytechnique de Montreal.

Verdeman, S. B. (1994). *Statistics for Engineering Problem Solving*, PWS Publishing Company, Boston, MA.

Wilkinson, J. H. (1965). *The Algebraic Eigenvalue Problem*, Claredon Press, Oxford, England, p. 90.

Woods, J., and J. S. Gulliver (1991). "Economic and Financial Analysis," In *Hydropower Engineering Handbook*, J. S. Gulliver and R. E. A. Arndt (eds.), 9.1–9.37, McGraw-Hill, New York.

Yeh, K. C., and Y. K. Tung (1993). "Uncertainty and Sensitivity of a Pit Migration Model," *Journal of Hydraulic Engineering*, ASCE, **119**(2):262–281.

Yeh, K. C., J. C. Yang, and Y. K. Tung (1997). "Regionalization of Unit Hydrograph Parameters: 2. Uncertainty Analysis," *Journal of Stochastic Hydrology and Hydraulics*, **11**(2):173–191.

Yen, B. C., and W. H. Tang (1976). "Risk-Safety Factor Relation for Storm Sewer Design," *Journal of the Environmental Engineering Division*, ASCE, **102**(EE2):509–516.

Yen, B. C., S. T. Cheng, and W. H. Tang (1980). "Reliability of Hydraulic Design of Culverts," *Proceedings of International Conference on Water Resources Development*, IAHR Asian Pacific Division Second Congress, **2**:991–1001, Taipei, Taiwan.

Young, D. M., and R. T. Gregory (1973). *A Survey of Numerical Mathematics—Vol.2*, Dover Publications, New York.

Yu, P. S., T. C. Yang, and S. J. Chen (2001). "Comparison of Uncertainty Analysis Methods for a Distributed Rainfall-Runoff Model," *Journal of Hydrology*, **244**:43–59.

Zhao, B. (1994). "Stochastic Optimal Control and Parameter Estimation for an Estuary System," *Ph.D. Thesis*, Department of Civil Engineering, Arizona State University, Tempe, AZ.

Zhao, B., and L. W. Mays (1996). "Uncertainty and Risk Analyses for FEMA Alluvial-Fan Method," *Journal of Hydraulic Engineering*, ASCE, **122**(6):325–332.

Zoppou, C., and K. S. Li (1993). "New Point Estimate Method for Water Resources Modeling," *Journal of Hydraulic Engineering*, ASCE, **119**(11):1300–1307.

Monte Carlo Simulation

6.1 Introduction

As uncertainty and reliability related issues are becoming more critical in engineering design and analysis, proper assessment of the probabilistic behavior of an engineering system is essential. The true distribution for the system response subject to parameter uncertainty should be derived, if possible. However, due to the complexity of physical systems and mathematical functions, derivation of the exact solution for the probabilistic characteristics of the system response is difficult, if not impossible. In such cases, Monte Carlo simulation is a viable tool to provide numerical estimations of the stochastic features of the system response.

Simulation is a process of replicating the real world based on a set of assumptions and conceived models of reality (Ang and Tang 1984). As the purpose of a simulation model is to duplicate reality, it is an effective tool for evaluating the effects of different designs on a system's performance. *Monte Carlo simulation* is a numerical procedure to reproduce random variables that preserves the specified distributional properties. In Monte Carlo simulation the system response of interest is repeatedly measured under various system parameter sets generated from the known or assumed probabilistic laws. It offers a practical approach to the uncertainty analysis because the random behavior of the system response can be probabilistically duplicated.

Two major concerns in the practical applications of Monte Carlo simulation in uncertainty and reliability analyses are: (1) the requirement of a large amount of computations for generating random variates, and (2) the presence of correlation among stochastic basic parameters. However, as the computing power is increasing, the concern with the computation cost is diminishing, and Monte Carlo simulations are becoming more practical and viable for uncertainty analyses. In fact, Beck (1985) notes that "when the computing power is available, there can, in general, be no strong argument against the use of Monte Carlo simulation."

As noted previously, the accuracy of the model output statistics and probability distribution (e.g., probability that a specified safety level will be exceeded) obtained from Monte Carlo simulation is a function of the number of simulations performed. For models or problems with a large number of stochastic basic variables and for which low probabilities (<0.1) are of interest, tens of thousands of simulations may be required. Rules for determining the number of simulations required for convergence are not available, and thus, replication of Monte Carlo simulation runs for a given number of simulations is the only way to check convergence (Melching 1995). For example, Melching (1992) found that 1000 simulations were adequate to estimate the mean, standard deviation, and quantiles above 0.2 for an application of the HEC-1 (U.S. Army Corps of Engineers 1991) and *runoff routing program* RORB (Laurenson and Mein 1985) rainfall-runoff models, and that 10,000 simulations were needed to accurately estimate quantiles between 0.01 and 0.2. Brown and Barnwell (1987) report that for the QUAL2E multiple constituent (dissolved oxygen, nitrogen cycle, algae, and the like) steady-state, surface water-quality model 2000 simulations are required to obtain accurate estimates of the output standard deviation. With the computational speed of today's computers making even 10,000 runs is not prohibitive for simpler models. However, the increase in the computational speed has made the use of computational fluid dynamics codes in three dimensions for hydrosystems design work possible. When such codes are applied the variance reduction techniques described in Sec. 6.6 may be preferred to Monte Carlo simulation.

This chapter focuses on the basic principles and applications of Monte Carlo simulations in the uncertainty analysis of hydrosystem engineering problems. Section 6.2 describes some basic concepts of generating random numbers, followed by discussions on the classifications of random variates generation algorithms in Sec. 6.3. Algorithms for generating univariate random numbers are described in Sec. 6.4 for several commonly used distribution functions. In Sec. 6.5, attention is given to algorithms that generate multivariate random numbers. Due to the fact that Monte Carlo simulations, in essence, are sampling techniques, they provide only estimations that are inevitably subject to certain degrees of errors. To improve the accuracy of the Monte Carlo estimation, while reducing excessive computation time, several variance reduction techniques are discussed in Sec. 6.6. Finally, resampling techniques are described in Sec. 6.7, which allow for the assessment of the uncertainty of the quantity of interest based on the available random data, without having to make assumptions about the underlying probabilistic structures.

6.2   Generation of Random Numbers

The most commonly used techniques to generate a sequence of pseudorandom numbers are those that employ some form of recursive computation. In principle, such recursive formulas are based on calculating the residuals modulo of some integers of a linear transformation. The process of producing a random number sequence is completely deterministic. However, the generated sequence would appear to be uniformly distributed and independent.

Congruential methods for generating n random numbers are based on the fundamental congruence relationship, which can be expressed as (Lehmer 1951)

$$X_{i+1} = \{aX_i + c\} \ (\text{mod } m) \qquad i = 1, 2, ..., n \tag{6.1}$$

in which a is the multiplier, c is the increment, and m is an integer-valued modulus. The modulo notation (mod m) in Eq. (6.1) represents that

$$X_{i+1} = aX_i + c - mI_i \tag{6.2}$$

with $I_i = [(aX_i + c)/m]$ denoting the largest positive integer value in $(aX_i + c)/m$. In other words, X_{i+1} determined by Eq. (6.1) is the residual resulting from $(aX_i + c)/m$. Therefore, the values of the number sequence generated by Eq. (6.1) would satisfy $X_i < m$ for all $i = 1, 2, ..., n$. Random number generators that produce a number sequence according to Eq. (6.1) are called *mixed congruential generators*.

Applying Eq. (6.1) to generate a random number sequence requires the specification of a, c, and m, along with X_0, called the *seed*. Once the sequence of random number X's are generated, the random number from the unit interval $u_i \in [0,1]$ can be obtained as

$$U_i = \frac{X_i}{m} \qquad i = 1, 2, ..., n \tag{6.3}$$

It should be pointed out that the process of generating uniform random numbers is the building block in Monte Carlo simulation.

Due to the deterministic nature of number generation, it is clear that the number sequence produced by Eq. (6.1) is periodic, which will repeat itself in, at most, m steps. This implies that the sequence would contain, at most, m distinct numbers and will have a maximum period of length $m - 1$ beyond which the sequence will start repeating itself. For example, consider $X_{i+1} = 2X_i + 3$ (mod $m = 5$) with $X_0 = 3$, the number sequence generated would be 4, 1, 0, 3, 4, 1, 0,...

From the practical application viewpoint, it is desirable that the generated number sequence has a very long periodicity to ensure that sufficiently large amounts of distinct numbers are produced before the cycle occurs. Therefore, one would choose the value of modulus m to be as large as possible. However, the length of the periodicity in a sequence also depends on the values of multiplier a and increment c. Knuth (1981) derived three conditions under which a sequence from Eq. (6.1) has a *full period m*. Based on the three conditions of Knuth (1981), Rubinstein (1981) showed that, for a computer with a binary digit system, using $m = 2^\beta$, with β being the word length of the computer, along with an odd number for parameter c and $a = 2^r + 1$, $r \geq 2$ would produce a full period sequence. Literature (Hull and Dobell 1964; MacLaren and Marsagalia 1965; Olmstead 1946) indicates that good statistical results can be achieved by using $m = 2^{35}$, $a = 2^7+1$, and $c = 1$. Table 6.1 lists suggested values for the parameters in Eq. (6.1) for different computers.

TABLE 6.1   Suggested Values for Parameters in Congruential Methods (after Press et al. 1989)

Constants for portable random number generators							
Overflow at	m	a	c	Overflow at	m	a	c
2^{20}	6075	106	1283	2^{28}	117128	1277	24749
2^{21}	7875	211	1663		312500	741	66037
2^{22}	7875	421	1663		121500	2041	25673
2^{23}	11979	430	2531	2^{29}	120050	2311	25367
	6655	936	1399		214326	1807	45289
	6075	1366	1283		244944	1597	51749
2^{24}	53125	171	11213		233280	1861	49297
	11979	859	2531		175000	2661	36979
	29282	419	6173		121500	4081	25673
	14406	967	3041		145800	3661	30809
2^{25}	134456	141	28411	2^{30}	139968	3877	29573
	31104	625	6571		214326	3613	45289
	14000	1741	2957		714025	1366	150889
	12960	1741	2731	2^{31}	134456	8121	28411
	21870	1291	4621		243000	4561	51349
	139968	205	29573		259200	7141	54773
2^{26}	81000	421	17117	2^{32}	233280	9301	49297
	29282	1255	6173		714025	4096	150889
	134456	281	27411	2^{33}	1771875	2416	374441
2^{27}	86463	1093	18257	2^{34}	510300	17221	107839
	259200	421	54773		312500	36261	66037
	116640	1021	24631	2^{35}	217728	84589	45989
	121500	1021	25673				

A second commonly used generator is called the *multiplicative generator*

$$X_{i+1} = \{aX_i\} \pmod{m} \qquad i = 1, 2, \ldots, n \tag{6.4}$$

which is a special case of the mixed generator with $c = 0$. Knuth (1981) showed that a maximal period can be achieved for the multiplicative generator in a binary computer system when $m = 2^\beta$ and $a = 8r \pm 3$, with r being any positive integer.

Another type of generator is called the *additive congruential generator* having the recursive relationship as

$$X_{i+1} = \{X_i + X_{i-t}\} \pmod{m} \qquad t = 1, 2, \ldots, i-1 \tag{6.5}$$

As can be seen, the random numbers generated by the additive congruential generator depend on more than one of its preceding values. When $t = 1$, Eq. (6.5) would generate a sequence of Fibonacci numbers, which are not satisfactorily random. However, the statistical properties improve as k gets larger.

In summary, to ensure that a sequence of random numbers generated by the congruential methods would have satisfactory statistical properties, Knuth (1981) recommended the following principles to choose the parameters a, c, m, and X_0:

1. The seed X_0 can be chosen arbitrarily. If different random number sequences are to be generated, a practical way is to set X_0 equal to the date and time when the sequence is to be generated.

2. The modulus m must be large. It may be conveniently set to the word length of the computer because this would enhance computational efficiency. The computation of $\{aX + c\}(\text{mod } m)$ must be done exactly without round-off errors.

3. If modulus m is a power of 2 (for binary computers), select the multiplier a so that $a(\text{mod } 8) = 5$. If m is a power of 10 (for decimal computers), pick a such that $a(\text{mod } 200) = 21$. The selection of the multiplier a in this fashion, along with the choice of increment c described in condition 5, would ensure that the random number generator will produce all m distinct possible values in the sequence before repeating itself.

4. The multiplier a should be larger than $\sqrt{m}$, preferably larger than $m/100$, but smaller than $m - \sqrt{m}$. The best policy is to take some haphazard constant to be the multiplier satisfying both conditions 3 and 4.

5. The increment parameter c should be an odd number when the modulus m is a power of 2 and, when m is a power of 10, c should not be a multiple of 5.

6.3 Classifications of Random Variates Generation Algorithms

6.3.1 CDF-inverse method

Let a random variable X have the CDF $F_x(x)$. From Sec. 2.2.1, $F_x(x)$ is a nondecreasing function with respect to x, and $0 \leq F_x(x) \leq 1$. Therefore, $F_x^{-1}(u)$ may be defined for any value of u between 0 and 1 as $F_x^{-1}(u)$ is the smallest x satisfying $F_x(x) \geq u$.

For the majority of continuous probability distributions applied in hydrosystems engineering and analysis, $F_x(x)$ is a strictly increasing function of x. Hence, a unique relationship exists between $F_x(x)$ and u; that is, $u = F_x(x)$, as shown in Fig. 6.1. Furthermore, it can be shown that, if U is a standard uniform random variable defined over the unit interval [0, 1], denoted by $U \sim U(0, 1)$, the following relationship holds:

$$X = F_x^{-1}(U) \tag{6.6}$$

Note that X is a random variable because it is a function of the random variable U. From Eq. (6.6), the one-to-one correspondence between X and U, through the CDF, enables the generation of random numbers $X \sim F_x(x)$ from the standard uniform random numbers. The algorithm using the CDF-inverse method for generating continuous random numbers from a CDF, $F_x(x)$, can be stated as follows:

1. Generate n uniform random numbers $u_1, u_2,..., u_n$ from $U(0, 1)$

2. Solve for $x_i = F_x^{-1}(u_i)$, for $i = 1, 2,..., n$

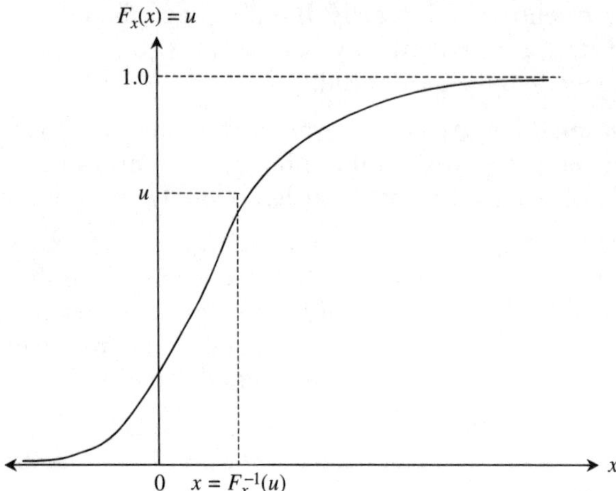

Figure 6.1   Schematic diagram of the inverse-CDF method for generating random variates.

Example 6.1   Consider that the Manning roughness coefficient (X) of a cast iron pipe is uncertain, having a uniform distribution $f_x(x) = 1/(b-a)$, $a \leq x \leq b$. Develop an algorithm using the CDF-inverse method to generate a random Manning roughness coefficient.

Solution   Using the CDF-inverse method, the expression of the CDF of the random variable is first sought. The CDF for this example can be derived as

$$F_x(x) = \int_a^x \frac{1}{b-a}\, dx = \frac{x-a}{b-a} \qquad \text{for } a \leq x \leq b$$

From the above expression for the CDF, the random variate of Manning's roughness coefficient x is obtained, in terms of $F_x(x)$, as

$$x = a + (b-a)\, F_x(x)$$

A simple algorithm for generating uniform random variates from $U(a, b)$ is

1. Generate n standard uniform random variates $u_1, u_2, \ldots, u_n$ from $U(0, 1)$
2. Calculate the corresponding uniform random variates $x_i = a + (b-a)u_i$, $i = 1, 2, \ldots, n$

To efficiently apply the CDF-inverse method for generating random numbers, an explicit expression between X and U is essential so that X can be obtained analytically from the generated U. The distributions whose inverse forms are analytically expressible include exponential, uniform, Weibull, and Gumbel. Table 6.2 lists some distributions that are used in hydrosystems whose CDF inverse are analytically expressible.

When the analytical forms of the CDF inverse are not available, applying the CDF-inverse method would require solving

$$u = \int_{-\infty}^x f_x(x')\, dx' \tag{6.7}$$

TABLE 6.2   List of Distributions Whose CDF Inverse Are Analytically Expressible

Distribution	$F_x(x) =$	$x = F_x^{-1}(u)$
Exponential	$1 - \exp(-\beta x),\ x > 0$	$-\beta \ln(1 - F)$
Uniform	$(x - a)/(b - a)$	$a + (b - a)F$
Gumbel	$\exp\{-\exp[-(x - \xi)/\beta)]\}$	$\xi - \beta \ln[-\ln(F)]$
Weibull	$1 - \exp\{-[(x - \xi)/\beta]^\alpha\}$	$\xi + \beta[-\ln(1 - F)]^{1/\alpha}$
Pareto	$1 - x^{-\alpha}$	$(1 - F)^{-(1/\alpha)}$
Wakeby	Not explicitly defined	$\xi + (\alpha/\beta)[1 - (1 - F)^\beta] - (\gamma/\delta)[1 - (1 - F)^{-\delta}]$
Kappa	$\{1 - h[1 - \alpha(x - \xi)/\beta]^{1/\alpha}\}^{1/h}$	$\xi + (\beta/\alpha)\{1 - [(1 - F^h)/h]^\alpha\}$
Burr	$1 - (1 + x^\alpha)^{-\beta}$	$[(1 - F)^{-1/\beta} - 1]^{1/\alpha}$
Cauchy	$0.5 + \tan^{-1}(x)/\pi$	$\tan[\pi(F - 0.5)]$
Rayleigh	$1 - \exp[-(x - \xi)^2/2\beta^2]$	$\xi + \{-2\beta^2 \ln(1 - F)\}^{1/2}$
Generalized lambda	Not explicitly defined	$\xi + \alpha F^\beta - \gamma(1 - F)^\delta$
Generalized extreme value	$\exp[-\exp(-y)]$ where $y = -\alpha^{-1} \ln\{1 - \alpha(x - \xi)/\beta\},\ \alpha \neq 0$ $\quad = (x - \xi)/\beta,\ \alpha = 0.$	$\xi + \beta\{1 - [-\ln(F)]^\alpha\}/\alpha,\ \alpha \neq 0$ $\xi - \beta \ln[-\ln(F)],\ \alpha = 0$
Generalized logistic	$1/[1 + \exp(-y)]$ where $y = -\alpha^{-1} \ln\{1 - \alpha(x - \xi)/\beta\},\ \alpha = 0$ $\quad = (x - \xi)/\beta,\ \alpha = 0.$	$\xi + \beta\{1 - [(1 - F)/F]^\alpha\}/\alpha,\ \alpha \neq 0$ $\xi - \beta \ln[(1 - F)/F],\ \alpha = 0$
Generalized pareto	$1 - \exp(-y)$ where $y = -\alpha^{-1} \ln\{1 - \alpha(x - \xi)/\beta\},\ \alpha \neq 0$ $\quad = (x - \xi)/\beta,\ \alpha = 0.$	$\xi + \beta[1 - (1 - F)^\alpha]/\alpha,\ \alpha \neq 0$ $\xi - \beta \ln[1 - F],\ \alpha = 0$

for x from the known u. For many commonly used distributions such as normal, lognormal, and gamma, solving Eq. (6.7) is inefficient and difficult. More efficient algorithms have been developed to generate random variates from those distributions; some of these are described in Sec. 6.4.

6.3.2 Acceptance-rejection methods

Consider the problem that random variates are to be generated from a specified probability density function (PDF), $f_x(x)$. The basic idea of the *acceptance-rejection (AR) method* is to replace the original $f_x(x)$ by an appropriate PDF, $h_x(x)$, from which random variates can be easily and efficiently produced. The generated random variate from $h_x(x)$, then, is subject to testing before it is accepted as one from the original $f_x(x)$. This approach for generating random numbers is being widely used.

In AR methods, the PDF $f_x(x)$ from which a random variate x to be generated is represented, in terms of $h_x(x)$, by

$$f_x(x) = \varepsilon h_x(x)\, g(x) \tag{6.8}$$

in which $\varepsilon \geq 1$ and $0 < g(x) \leq 1$. Figure 6.2 illustrates the AR method in that the constant $\varepsilon \geq 1$ is chosen such that $\psi(x) = \varepsilon h_x(x)$ over the sample space of the random variable X. The problem then is to find a function $\psi(x) = \varepsilon h_x(x)$ such that $\psi(x) \geq f_x(x)$ and a function $h_x(x) = \psi(x)/\varepsilon$, from which random variates are generated.

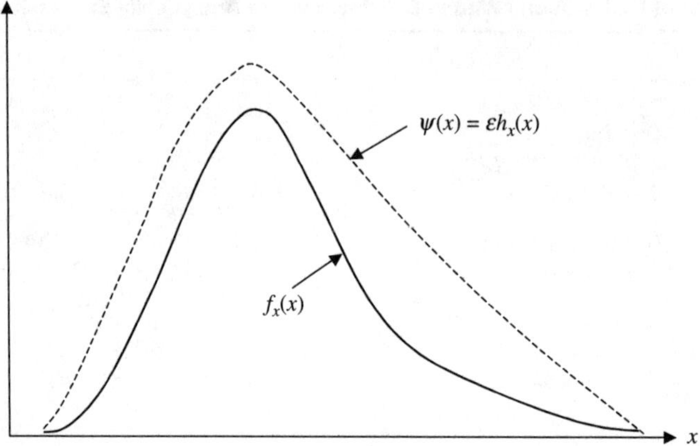

Figure 6.2   Illustration of von Neumann's acceptance-rejection procedure.

The constant ε that satisfies $\psi(x) \geq f_x(x)$ can be obtained from

$$\varepsilon = \max_x \left[\frac{f_x(x)}{h_x(x)} \right] \tag{6.9}$$

The algorithm of a generic AR method is as follows:

1. Generate a uniform random number u from $U(0, 1)$
2. Generate a random variate y from $h_x(x)$
3. If $u \leq g(y) = f_x(y)/\varepsilon h_x(y)$, accept y as the random variate from $f_x(x)$. Otherwise, reject both u and y and go to Step 1

The efficiency of an AR method is determined by $P\{U \leq g(Y)\}$, which represents the probability that each individually generated Y from $h_x(x)$ will be accepted by the test. The higher the probability, the faster the task of generating a random number can be accomplished. It can be shown that $P\{U \leq g(Y)\} = 1/\varepsilon$ (see Prob. 6.4). Intuitively, the maximum achievable efficiency for an AR method is when $\psi(x) = f_x(x)$. In this case $\varepsilon = 1$, $g(x) = 1$, and the corresponding probability of acceptance, $P\{U \leq g(Y)\} = 1$. Therefore, considerations must be given to two aspects when selecting $h_x(x)$ for AR methods: (1) the efficiency and exactness of generating a random number from $h_x(x)$, and (2) the closeness of $h_x(x)$ in imitating $f_x(x)$.

Example 6.2   Consider that Manning's roughness coefficient (X) of a cast iron pipe is uncertain and has a density function $f_x(x)$, $a \leq x \leq b$. Develop an AR algorithm using $\psi(x) = c$ and $h_x(x) = 1/(b - a)$ for $a \leq x \leq b$.

Solution   Since $\psi(x) = c$ and $h_x(x) = 1/(b - a)$, the efficiency constant ε and $g(x)$ are

$$\varepsilon = \frac{\psi(x)}{h_x(x)} = c(b - a)$$

$$g(x) = \frac{f_x(x)}{\psi(x)} = \frac{f_x(x)}{c} \qquad \text{for } a \le x \le b$$

The AR algorithm for this example, then, can be outlined as follows:

1. Generate u_1 from $U(0, 1)$.
2. Generate u_2 from $U(0, 1)$ from which $y = a + (b - a)u_2$.
3. Determine if

$$u_1 \le g(y) = \frac{f_x[a + (b - a)u_2]}{c}$$

holds. If yes, accept y; otherwise, reject (u_1, y) and return to Step 1.

In fact, this is the von Neumann (1951) algorithm for the AR method.

AR methods are important tools for random number generation because they can be very fast in comparison with the CDF-inverse method for distribution models whose analytical forms of CDF inverse are not available. The approach has been applied to some distributions, such as gamma, resulting in extremely simple and efficient algorithms (Dagpunar 1988).

6.3.3   Variable transformation method

The *variable transformation method* generates random variates of interest based on its known statistical relationship with other random variables whose variates can easily be produced. For example, one is interested in generating chi-square random variates with n degrees of freedom. The CDF-inverse method is not appropriate in this case since the chi-square CDF is not analytically expressible. However, knowing the fact that the sum of n squared independent standard normal random variables gives a chi-square random variable with n degrees of freedom, one could generate chi-square random variates by first producing n standard normal random variates, then squaring them, and finally adding them together. Therefore, the variable transformation method is sometimes effective for generating random variates from a complicated distribution, based on variates produced from simple distributions. In fact, many algorithms described in the next section are based on the idea of variable transformation.

6.4   Generation of Univariate Random Numbers for Some Distributions

This section briefly outlines efficient algorithms for generating random variates for some probability distributions commonly used in hydrosystems engineering and analysis.

6.4.1 Normal distribution

A normal random variable with a mean μ_x and standard deviation σ_x, denoted as $X \sim N(\mu_x, \sigma_x)$, has a PDF given in Eq. (2.50). The relationship between X and the standardized normal variable Z is

$$X = \mu_x + \sigma_x Z \tag{6.10}$$

in which Z is the standard normal random variable having a mean 0 and unit standard deviation, denoted as $Z \sim N(0, 1)$. Based on Eq. (6.10), normal random variates with a specified mean and standard deviation can be generated from standard normal variates. Herein, three simple algorithms for generating standard normal variates are described.

Box-Muller algorithm.   The algorithm (Box and Muller 1958) produces a pair of independent $N(0, 1)$ variates as

$$
\begin{aligned}
z_1 &= \sqrt{-2\ln(u_1)}\,\cos(2\pi u_2) \\
z_2 &= \sqrt{-2\ln(u_2)}\,\sin(2\pi u_2)
\end{aligned}
\tag{6.11}
$$

in which u_1 and u_2 are independent uniform variates from $U(0, 1)$. The algorithm involves the following two steps:

1. Generate two independent uniform random variates u_1 and u_2 from $U(0, 1)$
2. Compute z_1 and z_2 simultaneously, using u_1 and u_2 according to Eq. (6.11)

Marsagalia-Bray algorithm.   Marsagalia and Bray (1964) proposed an alternative algorithm that avoids using trigonometric evaluations. In their algorithm, two independent uniform random variates, u_1 and u_2, are produced to evaluate the following three expressions,

$$
\begin{aligned}
V_1 &= 2U_1 - 1 \\
V_2 &= 2U_2 - 1 \\
R &= V_1^2 + V_2^2
\end{aligned}
\tag{6.12}
$$

If $R > 1$, the pair (u_1, u_2) are rejected from further consideration and a new pair of (u_1, u_2) is generated. For the accepted pair, the corresponding standard normal variates are computed by

$$Z_1 = V_1\sqrt{\frac{-2\ln(R)}{R}} \qquad Z_2 = V_2\sqrt{\frac{-2\ln(R)}{R}} \tag{6.13}$$

The Marsagalia-Bray algorithm involves the following steps:

1. Generate two independent uniform random variates, u_1 and u_2, from $U(0, 1)$.

2. Compute V_1, V_2, and R according to Eq. (6.12).

3. Check if $R \leq 1$. If it is true, compute the two corresponding $N(0, 1)$ variates using Eq. (6.13). Otherwise, reject (u_1, u_2) and return to Step 1.

Algorithm based on the central-limit theorem.   This algorithm is based on the central-limit theorem, which states that the sum of independent random variables approaches a normal distribution as the number of random variables increases. Specifically, consider the sum of J independent standard uniform random variates from $U(0, 1)$. The following relationships are true:

$$E\left(\sum_{j=1}^{J} U_j\right) = \frac{J}{2} \tag{6.14}$$

$$\text{Var}\left(\sum_{j=1}^{J} U_j\right) = \frac{J}{12} \tag{6.15}$$

By the central-limit theorem, this sum of J independent U's would approach a normal distribution with the mean and variance given in Eqs. (6.14) and (6.15), respectively. Constrained by the unit variance of the standard normal variates, Eq. (6.15) yields $J = 12$. Then, a standard normal variate is generated by

$$Z = \left(\sum_{j=1}^{12} U_j\right) - 6 \tag{6.16}$$

The central-limit theorem-based algorithm can be implemented as

1. Generate 12 uniform random variates from $U(0, 1)$

2. Compute the corresponding standard normal variate by Eq. (6.16)

There are many other efficient algorithms developed for generating normal random variates using the variable transformation method and AR method. For these algorithms readers are referred to Rubinstein (1981).

6.4.2 Lognormal distribution

Consider a random variable X having a lognormal distribution with a mean μ_x and standard deviation σ_x, that is, $X \sim \text{LN}(\mu_x, \sigma_x)$. For a lognormal random variable X, its logarithmic transform, $Y = \ln(X)$, leads to a normal distribution for Y. The PDF of X is given in Eq. (2.57). In the log-transformed space, the mean and standard deviation of $\ln(X)$ can be computed, in terms of μ_x and σ_x, by Eqs. (2.59a) and (2.59b). Since $Y = \ln(X)$ is normally distributed, the generation of lognormal random variates from $X \sim \text{LN}(\mu_x, \sigma_x)$ can be obtained by the following steps:

1. Calculate the mean, $\mu_{\ln x}$, and standard deviation, $\sigma_{\ln x}$, of log-transformed variable $\ln(X)$ by Eqs. (2.59a) and (2.59b), respectively.

2. Generate the standard normal variate z from $N(0, 1)$.

3. Compute $y = \mu_{\ln x} + \sigma_{\ln x} z$.

4. Compute the lognormal random variate $x = e^y$.

6.4.3 Exponential distribution

Exponential distribution is frequently used in reliability computation in the framework of time-to-failure analysis. It is often used to describe the stochastic behavior of time-to-failure and time-to-repair of a system or component. A random variable X having an exponential distribution with parameter β, denoted by $X \sim \text{EXP}(\beta)$, is described by Eq. (2.71). By the CDF-inverse method,

$$u = F_x(x) = 1 - e^{-x/\beta} \tag{6.17}$$

so that

$$X = -\beta \ln (1 - U) \tag{6.18}$$

Since $1 - U$ is distributed in the same way as U, Eq. (6.18) is reduced to

$$X = -\beta \ln U \tag{6.19}$$

Equation (6.19) is also valid for random variables with the standard exponential distribution, $V \sim \text{EXP}(\beta = 1)$. The algorithm for generating exponential variates is

1. Generate uniform random variate u from $U(0, 1)$

2. Compute the standard exponential random variate $v = -\ln(u)$

3. Calculate $x = v\beta$

6.4.4 Gamma distribution

Gamma distribution is frequently used in the statistical analysis of hydrologic data. For example, Pearson type III and log-Pearson type III distributions used in the flood frequency analysis are members of the gamma distribution family. It is a very versatile distribution whose PDF can take many forms (Fig. 2.17). The PDF of a two-parameter gamma random variable, denoted by $X \sim \text{GAM}(\alpha, \beta)$, is given by Eq. (2.64). The standard gamma PDF involving one-parameter α can be derived, using variable transformation by letting $Y = X/\beta$. The PDF of the standard gamma random variable Y, denoted by $Y \sim \text{GAM}(\alpha)$, is shown in Eq. (2.70). The standard gamma distribution is used in all algorithms to generate gamma random variates Y's from which random variates from a two-parameter gamma distribution are obtained from $X = \beta Y$.

The simplest case in generating gamma random variates is when the shape parameter α is a positive integer (Erlang distribution). In such a case, the random variable $Y \sim \mathrm{GAM}(\alpha)$ is a sum of α independent and identical standard exponential random variables with parameter $\beta = 1$. The random variates from $Y \sim \mathrm{GAM}(\alpha)$, then, can be obtained as

$$Y = \sum_{i=1}^{\alpha} -\ln(U_i) \tag{6.20}$$

To avoid large numbers of logarithmic evaluations (when α is large), Eq. (6.20) can alternatively be expressed as

$$Y = -\ln\left[\prod_{i=1}^{\alpha} U_i\right] \tag{6.21}$$

Although simplicity is the idea, the above algorithm for generating gamma random variates has three disadvantages: (1) it is only applicable to the integer-valued shape parameter α, (2) the algorithm becomes extremely slow when α is large, and (3) for a large α, numerical underflow on a computer could occur.

Several algorithms have been developed for generating standard gamma random variates for a real-valued α. The algorithms can be classified into those which are applicable for the full range ($\alpha \geq 0$), $0 \leq \alpha \leq 1$, and $\alpha \geq 1$. Dagpunar (1988) shows that, through a numerical experiment, algorithms developed for a full range of α are not efficient in comparison with those especially tailored for sub-regions. The two efficient AR-based algorithms are presented in Dagpunar (1988).

6.4.5   Other univariate distributions and computer programs

Described earlier are algorithms for some probability distributions commonly used in hydrosystem engineering and analysis. One might encounter other types of probability distributions in an analysis that are not described herein. There are several books that have been written for generating univariate random numbers (Rubinstein 1981; Dagpunar 1988; Gould and Tobochnik 1988; Law and Kelton 1991). To facilitate the implementation of Monte Carlo simulation, computer subroutines in different languages are available (Press et al. 1989; 1992; 2002; IMSL 1980). In addition, many other spreadsheet-based computer software, such as Microsoft Excel, @Risk, and Crystal Ball, contain statistical functions allowing the generation of random variates of various distributions.

6.5   Generation of Vector of Multivariate Random Variables

In the previous sections, discussions are focused on generating univariate random variates. It is not uncommon for hydrosystem engineering problems to involve multiple random variables that are correlated and statistically dependent.

For example, many data show that the peak discharge and volume of a runoff hydrograph are positively correlated. To simulate systems involving correlated random variables, generated random variates must preserve the probabilistic characteristics of the variables and the correlation structure among them. Although multivariate random number generation is an extension of the univariate case, mathematical difficulty and complexity associated with multivariate problems increase rapidly as the dimension of the problem gets larger. Compared with generating univariate random variates, multivariate random variate generation is much more restricted to fewer joint distributions, such as multivariate normal, multivariate lognormal, and multivariate gamma (Ronning 1977; Johnson 1987; Parrish 1990). Nevertheless, the algorithms for generating univariate random variates serve as the foundation for many multivariate Monte Carlo algorithms.

6.5.1   CDF-inverse method

The method is an extension of the univariate case previously described in Sec. 6.3.1. Consider a vector of K random variables $X = (X_1, X_2,..., X_K)^t$ having a joint PDF of

$$f_x(x) = f_{1,2,...,K}(x_1, x_2,..., x_K) \tag{6.22}$$

The above joint PDF can be decomposed to

$$f_x(x) = f_1(x_1) \times f_2(x_2 \mid x_1) \times \cdots \times f_K(x_K \mid x_1, x_2,..., x_{K-1}) \tag{6.23}$$

in which $f_1(x_1)$ and $f_k(x_k \mid x_1, x_2,..., x_{k-1})$ are, respectively, the marginal PDF and the conditional PDF of random variables X_1 and X_k. In the case when all K random variables are statistically independent, Eq. (6.22) is simplified to

$$f_x(x) = \prod_{k=1}^{K} f_k(x_k) \tag{6.24}$$

One observes that, from Eq. (6.24), the joint PDF of several independent random variables is simply the product of the marginal PDF of the individual random variable. Therefore, generation of a vector of independent random variables can be accomplished by treating each individual random variable separately, as in the case of the univariate problem. However, treatment of random variables cannot be made separately in the case when they are correlated. Under such circumstances, as can be seen from Eq. (6.23), the joint PDF is the product of conditional distributions. Referring to Eq. (6.23), the generation of K random variates following the prescribed joint PDF can proceed as follows:

1. Generate random variates for X_1 from its marginal PDF, $f_1(x_1)$

2. Given $X_1 = x_1$ obtained from Step 1, generate X_2 from the conditional PDF, $f_2(x_2 \mid x_1)$

3. With $X_1 = x_1$ and $X_2 = x_2$ obtained from Steps 1 and 2, produce X_3 based on $f_3(x_3 | x_1, x_2)$

4. Repeat the procedure until all K random variables are generated

To generate multivariate random variates by the CDF-inverse method, it is required that the analytical relationship between the value of the variate and conditional distribution function is available. Following Eq. (6.23), the product relationship also holds in terms of CDFs as

$$F_x(x) = F_1(x_1) \times F_2(x_2 | x_1) \times \cdots \times F_K(x_K | x_1, x_2,..., x_{K-1}) \tag{6.25}$$

in which $F_1(x_1)$ and $F_k(x_k | x_1, x_2,..., x_{k-1})$ are the marginal CDF and conditional CDF of random variables X_1 and X_k, respectively. Based on Eq. (6.25), the algorithm using the CDF-inverse method to generate n sets of K multivariate random variates from a specified joint distribution is described as follows (Rosenblatt 1952):

1. Generate K standard uniform random variates $u_1, u_2, ..., u_K$ from $U(0, 1)$

2. Compute

$$x_1 = F_1^{-1}(u_1)$$

$$x_2 = F_2^{-1}(u_2 | x_1)$$

$$\vdots \tag{6.26}$$

$$x_K = F_K^{-1}(u_K | x_1, x_2, ..., x_{K-1})$$

3. Repeat Steps 1 and 2 for n sets of random vectors

There are $K!$ ways to implement the above algorithm in which different orders of random variates X_k, $k = 1, 2,..., K$, are taken to form the random vector X. In general, the order adopted could affect the efficiency of the algorithm.

Example 6.3 This example is extracted from Nguyen and Chowdhury (1985). Consider a box cut of an open-strip coal mine as shown in Fig. 6.3. The over-burden has a phreatic aquifer overlying the coal seam. In the next bench of operation, excavation is to be made 50 m ($d = 50$ m) behind the box-cut high wall. It is suggested that, for safety reasons of preventing slope instability, excavation should start at the time when the drawdown in the over-burden $d = 50$ m away from the excavation point has reached at least 50 percent of the total aquifer depth (h_o).

Nguyen and Raudkivi (1983) gave the transient drawdown equation for this problem

$$\frac{s}{h_o} = 1 - \mathrm{erf}\left[\frac{d}{2\sqrt{K_h h_o t / S}} \right] \tag{6.27}$$

where s = drawdown (in meters) at a distance d (in meters) from the toe of the embankment

h_o = original thickness of the water bearing aquifer

t = drawdown recess time (in days)

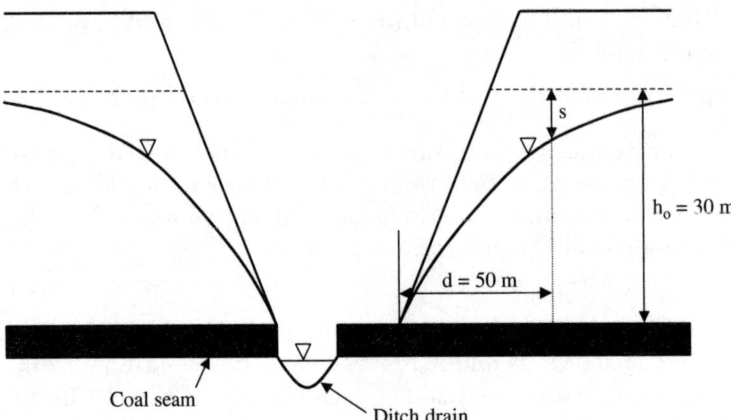

Figure 6.3 Box cut of an open-strip coal mine resulting in water drawdown (after Nguyen and Chowdhury 1985).

$$K_h = \text{aquifer permeability}$$
$$S = \text{aquifer storage coefficient}$$
$$\text{erf}(x) = \text{error function, referring to Eq. (2.61), as}$$

$$\text{erf}(x) = \frac{2}{\sqrt{\pi}} \int_0^x e^{-v^2} dv$$

with v being a dummy variable of integration.

From a field investigation through a pump test, data indicate that the aquifer permeability has, approximately, a normal distribution with a mean of 0.100 meter per day and coefficient of variation of 10 percent. The storage coefficient of the aquifer has a mean of 0.05 with a standard deviation of 0.005. Further, the correlation coefficient between the permeability and storage coefficient is about 0.5.

Since the aquifer properties are random variables, the time required for the drawdown to reach the safe level for excavation also is a random variable. Apply the CDF-inverse method (using $n = 400$ repetitions) to estimate the statistical properties of the time of recess, including its mean, standard deviation, and skewness coefficient.

Solution   The required drawdown recess time for a safe excavation can be obtained by solving Eq. (6.28), with $s/h_o = 0.5$ and $\text{erf}^{-1}(0.5) = 0.477$ (Abramowitz and Stegun 1972; or by Eq. (2.60) as

$$t = \left(\frac{d}{2 \times 0.477}\right)^2 \frac{S}{K_h h_o} \tag{6.28}$$

The problem is a bivariate normal distribution (Sec. 2.5.1) with two correlated random variables. The permeability K_h and storage coefficient S, referring to Eq. (2.101), have the joint PDF

$$f_{K_h, S}(k, s) = \frac{1}{2\pi \sigma_k \sigma_s \sqrt{1 - \rho^2}} e^{-Q}$$

with

$$Q = \frac{1}{2(1-\rho_{k,s}^2)}\left(\frac{(k-\mu_k)^2}{\sigma_k^2} - 2\rho_{k,s}\frac{(k-\mu_k)(s-\mu_s)}{\sigma_k\sigma_s} + \frac{(s-\mu_s)^2}{\sigma_s^2} \right)$$

where $\rho_{k,s}$ = correlation coefficient between K_h and S, which is 0.5
$\quad$ σ_k = standard deviation of permeability, $0.1 \times 0.1 = 0.01$ meter/day
$\quad$ σ_s = standard deviation of the storage coefficient, 0.005
$\quad$ μ_k = mean of permeability, 0.1 meter/day
$\quad$ μ_s = mean storage coefficient, 0.05

To generate bivariate random variates according to Eq. (6.26), the marginal PDF of permeability (K_h) and the conditional PDF of storage coefficient (S), or vice versa, are required. They can be derived, respectively, according to Eq. (2.102), as

$$f_K(k) = \frac{1}{\sqrt{2\pi}\sigma_k}\exp\left[-\frac{1}{2}\left(\frac{k-\mu_k}{\sigma_k}\right)^2 \right] \tag{6.29}$$

$$f_{s|k}(s|k) = \frac{1}{\sqrt{2\pi}\sigma_s\sqrt{1-\rho^2}}\exp\left[-\frac{1}{2}\left(\frac{(s-\mu_s)-\rho_{k,s}(\sigma_s/\sigma_k)(k-\mu_k)}{\sigma_s\sqrt{1-\rho_{k,s}}}\right)^2 \right] \tag{6.30}$$

From the conditional PDF given above, the conditional expectation and conditional standard deviation of storage coefficient S, given a specified value of permeability $K_h = k$, can be derived, respectively, according to Eqs. (2.103) and (2.104), as

$$\mu_{S|k} = E[S|K_h = k] = \mu_s + \rho_{k,s}\frac{\sigma_s}{\sigma_k}(k-\mu_k) \tag{6.31}$$

$$\sigma_{s|k} = \sigma_s\sqrt{1-\rho_{k,s}} \tag{6.32}$$

Therefore, the algorithm for generating bivariate normal random variates to estimate the statistical properties of the drawdown recess time can be outlined as follows:

1. Generate a pair of independent standard normal variates z_1' and z_2'
2. Compute the corresponding value of permeability $k = \mu_k + \sigma_k z_1'$
3. Based on the value of permeability obtained in Step 2, compute the conditional mean and conditional standard deviation of the storage coefficient according to Eqs. (6.31) and (6.32), respectively. Then, calculate the corresponding storage coefficient as $s = \mu_{s|k} + \sigma_{s|k} z_2'$
4. Use $K_h = k$ and $S = s$ generated in Steps 3 and 4 in Eq. (6.28) to compute the corresponding drawdown recess time t
5. Repeat Steps 1 to 4 $n = 400$ times to obtain 400 realizations of drawdown recess times $\{t_1, t_2, ..., t_{400}\}$
6. Compute the sample mean, standard deviation, and skewness coefficient of the drawdown recess time according to the last column of Table 2.1

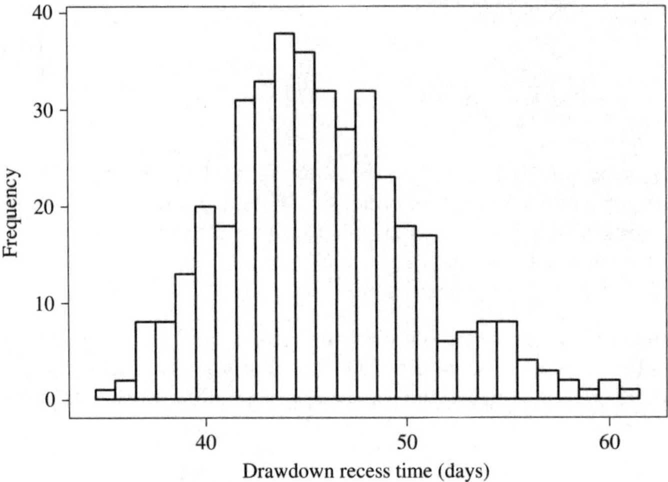

Figure 6.4   Histogram of simulated drawdown recess time for Example 6.3.

The histogram of the drawdown recess time resulting from 400 simulations is shown in Fig. 6.4. The statistical properties of the drawdown recess time are estimated as

Mean, μ_t $= 45.73$ days
Standard deviation, σ_t $= 4.72$ days
Skewness coefficient, $\gamma_t = 0.487$

6.5.2   Generating multivariate normal random variates

A random vector $X = (X_1, X_2,..., X_K)^t$ has a multivariate normal distribution with a mean vector μ_x and covariance matrix C_x, denoted as $X \sim N(\mu_x, C_x)$. The joint PDF of K normal random variables is given in Eq. (2.107). To generate multivariate normal random variates of higher dimensions with specified μ_x and C_x, the CDF-inverse algorithm described in Sec. 6.5.1 might not be efficient. In this section, two alternative algorithms for generating multivariate normal random variates are described. Both algorithms are based on orthogonal transformation, using the covariance matrix C_x or correlation matrix R_x described in Sec. 5.3.1. The result of the transformation is a vector of independent normal variables, which can be easily generated by the algorithms described in Sec. 6.4.1.

Square-root method.   The *square-root algorithm* decomposes the covariance matrix C_x or correlation matrix R_x into

$$R_x = LL^t \qquad C_x = \tilde{L}\tilde{L}^t$$

as shown in Sec. 3.1, in which L and $\tilde{L}$ are $K \times K$ lower triangular matrices associated with the correlation and covariance matrices, respectively. According to Eq. (5.57), $\tilde{L} = D_x^{1/2} L$ with D_x being the $K \times K$ diagonal matrix of variances of the K involved random variables.

In addition to being symmetric, if R_x or C_x is a positive-definite matrix, the Cholesky decomposition is an efficient method for finding the unique lower triangular matrices L or $\tilde{L}$ (Young and Gregory 1973; Golub and Van Loan 1989). Using the matrix L or $\tilde{L}$, the vector of multivariate normal random variables can be expressed as

$$X = \mu_x + \tilde{L}Z' = \mu_x + D^{1/2}LZ' \tag{6.33}$$

in which Z' is an $K \times 1$ column vector of independent standard normal variables. It was easily shown in Sec. 5.3.1 that the expectation vector and the covariance matrix of the right-hand side in Eq. (6.34) $E[\mu_x + \tilde{L}Z']$ is equal to μ_x and C_x, respectively. Based on Eq. (6.33), the square-root algorithm for generating multivariate normal random variates can be outlined as follows:

1. Compute the lower triangular matrix associated with the correlation or covariance matrix by the Cholesky decomposition method

2. Generate K independent standard normal random variates $z' = (z'_1, z'_2, \ldots, z'_K)^t$ from $N(0, 1)$

3. Compute the corresponding normal random variates by Eq. (6.33)

4. Repeat Steps 1 to 3 to generate the desired number of sets of normal random vectors

Example 6.4   Refer to Example 6.5.1. Apply the square-root algorithm to estimate the statistical properties of the drawdown recess time, including its mean, standard deviation, and skewness coefficient. Compare the results with Example 6.3.

Solution   By the square-root algorithm, the covariance matrix of permeability (K_h) and storage coefficient (S)

$$C(K_h, S) = \begin{bmatrix} 0.01^2 & 0.5(0.01)(0.005) \\ 0.5(0.01)(0.005) & 0.005^2 \end{bmatrix} = \begin{bmatrix} 0.0001 & 0.000025 \\ 0.000025 & 0.000025 \end{bmatrix}$$

is decomposed into the multiplication of the two lower triangular matrices, by the Cholesky decomposition, as

$$\tilde{L} = \begin{bmatrix} 0.01 & 0 \\ 0.0025 & 0.00443 \end{bmatrix}$$

The Monte Carlo simulation can be carried out by the following steps:

1. Generate a pair of standard normal variates z'_1 and z'_2
2. Compute the permeability (K_h) and storage coefficient (S) simultaneously as

$$\begin{bmatrix} k \\ s \end{bmatrix} = \begin{bmatrix} 0.1 \\ 0.05 \end{bmatrix} + \begin{bmatrix} 0.01 & 0 \\ 0.0025 & 0.00433 \end{bmatrix} \begin{bmatrix} z'_1 \\ z'_2 \end{bmatrix}$$

3. Use (k, s) generated in Step 2 in Eq. (6.28) to compute the corresponding drawdown recess time t
4. Repeat Steps 1 to 3 $n = 400$ times to obtain 400 realizations of drawdown recess times $\{t_1, t_2, ..., t_{400}\}$
5. Compute the mean, standard deviation, and skewness coefficient of the drawdown recess time

The results from carrying out the numerical simulation are:

Mean, μ_t = 45.94 days
Standard deviation, σ_t = 4.69 days
Skewness coefficient, γ_t = 0.301

The histogram of 400 simulated drawdown recess time is shown in Fig. 6.5. The mean and standard deviation are very close to those obtained in Example 6.3 whereas the skewness coefficient is 62 percent of that found in Example 6.3. This indicates that 400 simulations are sufficient to accurately estimate the mean and standard deviation, but more simulations are needed to accurately estimate the skewness coefficient.

Spectral decomposition method. The basic idea of spectral decomposition is described in Sec. 5.3.1. The method finds the eigenvalues and eigenvectors of the correlation or covariance matrix of the multivariate normal random variables. Through the spectral decomposition, the original vector of multivariate normal random variables X, then, is related to a vector of independent standard normal random variables $Z' \sim N(0, I)$ as

$$X = \mu_x + D_x^{1/2} \, V\Lambda^{1/2}Z' = \mu_x + \tilde{V}\tilde{\Lambda}^{1/2}Z' \tag{6.34}$$

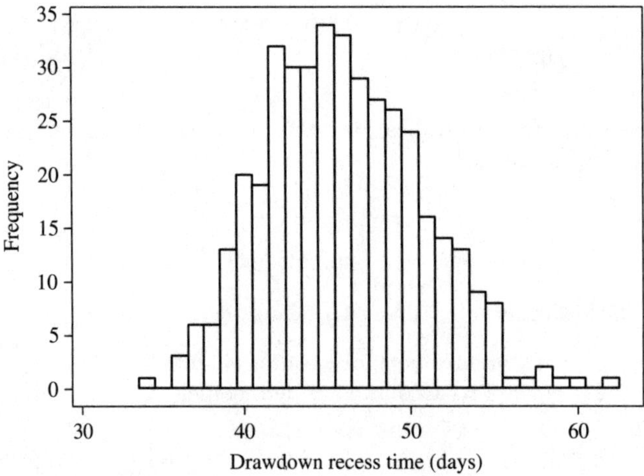

Figure 6.5 Histogram of simulated drawdown recess time for Example 6.4.

in which $\tilde{V}$ and $\tilde{\Lambda}$ are the eigenvector and diagonal eigenvalue matrices of C_x, respectively, whereas V and Λ are the eigenvector and diagonal eigenvalue matrices of R_x, respectively. Equation (6.33) clearly reveals the necessary computations for generating multivariate normal random vectors. The spectral decomposition algorithm for generating multivariate normal random variates involves the following steps:

1. Obtain the eigenvector matrix and diagonal eigenvalue matrix of the correlation matrix R_x or covariance matrix C_x
2. Generate K independent standard normal random variates $z' = (z'_1, z'_2, \ldots, z'_K)^t$
3. Compute the correlated normal random variates X by Eq. (6.33)

Many efficient algorithms have been developed to determine the eigenvalues and eigenvectors of a symmetric matrix. For the details of such techniques, readers are referred to Golub and Van Loan (1989) and Press et al. (1992).

6.5.3 Generating multivariate random variates with known marginal PDFs and correlations

In many practical hydrosystem engineering problems, random variables are often statistically and physically dependent. Furthermore, distribution types for the random variables involved can be a mixture of different distributions of which the corresponding joint PDF or CDF is difficult to establish. As a practical alternative, to properly replicate such systems, the Monte Carlo simulation should be able to preserve the correlation relationship among the stochastic variables and their marginal distributions.

In a multivariate setting, the joint PDF represents the complete information describing the probabilistic structures of the random variables involved. When the joint PDF or CDF is known, the marginal distribution and conditional distributions can be derived from which the generation of multivariate random variates can be made straightforwardly in the framework of Rosenblatt (1952). However, in most practical engineering problems involving multivariate random variables, the derivation of the joint CDF is generally difficult and the availability of such information is rare. The level of difficulty, in both theory and practice, increases with the number of random variables and, perhaps even more so by the type of corresponding distributions. Therefore, more often than not, one has to be content with preserving incomplete information represented by the marginal distribution of each individual random variable and their correlation structure. In doing so, the difficulty of requiring a complete joint PDF in the multivariate Monte Carlo simulation is circumvented.

To generate correlated random variables with a mixture of marginal distributions, a methodology adopting a bivariate distribution model was first suggested by Li and Hammond (1975). The practicality of the approach was advanced by Der Kiureghian and Liu (1985) who, based on the Nataf bivariate distribution model (Nataf 1962), developed a set of semiempirical formulas so that the necessary calculations to preserve the original correlation structure in the normal transformed space are reduced. Chang, Tung, and Yang (1994) used this set of formulas, which transforms the correlation coefficient of a pair of non-normal random variables to its equivalent correlation coefficient in the bivariate standard normal space, for multivariate simulation. Other practical alternatives, such as the polynomial normal transformation (Vale and Maurelli 1983; Chen and Tung 2003), can serve the same purpose. Through a proper normal transformation, the multivariate Monte Carlo simulation can be performed in a correlated standard normal space in which efficient algorithms, such as those described in Sec. 6.5.2, can be applied.

The Monte Carlo simulation that preserves marginal PDFs and correlation structure of the involved random variables consists of following two basic steps:

Step 1 Transformation to a standard normal space.   Through proper normal transformation the operational domain is transformed to a standard normal space in which the transformed random variables are treated as if they were multivariate standard normal with the correlation matrix R_z. As a result, multivariate normal random variates can be generated by the techniques described in Sec. 6.5.2.

Step 2 Inverse transformation.   Once the standardized multivariate normal random variates are generated, one can do the following inverse transformation:

$$X_k = F_k^{-1}[\Phi(Z_k)] \qquad \text{for } k = 1, 2, \ldots, k \qquad (6.35)$$

to compute the values of multivariate random variates in the original space.

6.6   Variance-Reduction Techniques

Since Monte Carlo simulation is a sampling procedure, results obtained from the procedure inevitably involve sampling errors that decrease as the sample size increases. Increasing the sample size to achieve a higher precision generally means an increase in the computer time for generating random variates and data processing. *Variance-reduction techniques* aim at obtaining high accuracy for the Monte Carlo simulation results without having to substantially increase the sample size. Hence, variance-reduction techniques enhance the statistical efficiency of Monte Carlo simulation. When applied properly, variance-reduction techniques sometimes can make the difference between an impossible expensive simulation study and a feasible, useful one.

Variance-reduction techniques attempt to reduce the error associated with the Monte Carlo simulation results by utilizing known information about the problem at hand. Naturally, such an objective cannot be attained if the analyst is completely ignorant about the problem. On the other extreme, the error is zero if the analyst has complete knowledge about the problem. Rubinstein (1981) states that "variance reduction cannot be obtained from nothing; it is merely a way of not wasting information." Therefore, for a problem that is not known at the initial stage of the study, pilot simulations can be performed for the purpose of gaining useful insight about the problem. The insight, then, can later be incorporated into the variance-reduction techniques for a more efficient simulation study. Therefore, most of the variance-reduction techniques require additional effort on the part of analysts.

6.6.1   Antithetic-variates technique

The *antithetic-variates technique* (Hammersley and Morton 1956) achieves the variance-reduction goal by attempting to generate random variates that would induce a negative correlation for the quantity of interest between separate simulation runs. Consider that $\hat{\Theta}_1$ and $\hat{\Theta}_2$ are two unbiased estimators of an unknown quantity θ to be estimated. The two estimators can be combined together to form another estimator as

$$\hat{\Theta}_a = \frac{1}{2}(\hat{\Theta}_1 + \hat{\Theta}_2) \tag{6.36}$$

The new estimator $\hat{\Theta}_a$ also is unbiased and has a variance as

$$\mathrm{Var}(\hat{\Theta}_a) = \frac{1}{4}\left[\mathrm{Var}(\hat{\Theta}_1) + \mathrm{Var}(\hat{\Theta}_2) + 2\mathrm{Cov}(\hat{\Theta}_1,\ \hat{\Theta}_2)\right] \tag{6.37}$$

If the two estimators $\hat{\Theta}_1$ and $\hat{\Theta}_2$ were computed by Monte Carlo simulation through generating two independent sets of random variates, they would be independent and the variance for $\hat{\Theta}_a$ would be

$$\mathrm{Var}(\hat{\Theta}_a) = \frac{1}{4}\left[\mathrm{Var}(\hat{\Theta}_1) + \mathrm{Var}(\hat{\Theta}_2)\right] \tag{6.38}$$

From Eq. (6.37) one realizes that the variance associated with $\hat{\Theta}_a$ could be reduced if the Monte Carlo simulation can generate random variates that result in a strong negative correlation between $\hat{\Theta}_1$ and $\hat{\Theta}_2$.

In a Monte Carlo simulation, the values of estimators $\hat{\Theta}_1$ and $\hat{\Theta}_2$ are functions of the generated random variates which, in turn, are related to the standard uniform random variates. Therefore, $\hat{\Theta}_1$ and $\hat{\Theta}_2$ are functions of the two standard uniform random variables U_1 and U_2. The objective to produce negative $\mathrm{Cov}[\hat{\Theta}_1(U_1), \hat{\Theta}_2(U_2)]$ can be achieved by producing U_1 and U_2, which are negatively correlated. However, it would not be desirable to complicate the computational procedure by generating two sets of uniform random variates subject to the constraint of being negatively correlated. One simple approach to generate negatively correlated uniform random variates with minimal computation is to let $U_1 = 1 - U_2$. It can be shown that $\mathrm{Cov}(U, 1 - U) = -1/12$ (see Prob. 6.18). Hence, a simple antithetic-variates algorithm is the following:

1. Generate u_i from $U(0, 1)$ and compute $1 - u_i$ for $i = 1, 2,..., n$.

2. Compute $\hat{\theta}_1(u_i)$, $\hat{\theta}_2(1 - u_i)$ and then $\hat{\theta}_a$ according to Eq. (6.18).

Example 6.5   Develop a Monte Carlo algorithm using the antithetic-variates technique to evaluate the integral G defined by

$$G = \int_a^b g(x)\,dx$$

in which $g(x)$ is a given function.

Solution   Applying the Monte Carlo method to estimate the value of G, the given integral can be rewritten as

$$G = \int_a^b \left[\frac{g(x)}{f_x(x)}\right] f_x(x)\,dx = E\left(\frac{g(X)}{f_x(X)}\right)$$

where $f_x(x)$ is the adopted distribution function based on which random variates are generated. As can be seen, the original integral becomes the calculation of the expectation of the ratio of $g(X)$ and $f_x(X)$. Hence, the two estimators for G using the antithetic-variates technique can be formulated as

$$\hat{G}_1 = \frac{1}{n}\sum_{i=1}^{n}\frac{g(X_{1i})}{f_x(X_{1i})} \tag{6.39a}$$

$$\hat{G}_2 = \frac{1}{n}\sum_{i=1}^{n}\frac{g(X_{2i})}{f_x(X_{2i})} \tag{6.39b}$$

in which $X_{1i} = F_x^{-1}(U_i)$ and $X_{2i} = F_x^{-1}(1 - U_i)$ with $F_x(\cdot)$ being the CDF of the random variable X. The algorithm for the Monte Carlo integral using the antithetic-variates technique is:

1. Generate n uniform random variates u_i from $U(0, 1)$ and compute the corresponding $1 - u_i$.
2. Compute $g(x_{1i})$, $f_x(x_{1i})$, $g(x_{2i})$, and $f_x(x_{2i})$ with $x_{1i} = F_x^{-1}(u_i)$ and $x_{2i} = F_x^{-1}(1-u_i)$.
3. Calculate the values of $\hat{G}_1$ and $\hat{G}_2$ by Eqs. (6.39a) and (6.39b), respectively. Then, estimate G by $\hat{G}_a = (\hat{G}_1 + \hat{G}_2)/2$.

In the case that X has a uniform distribution as $f_x(x) = 1/(b - a)$, $a \le x \le b$, the estimate of G by the antithetic-variates technique can be expressed as

$$\hat{g}_a = \frac{b-a}{2n}\sum_{i=1}^{n}[g(x_{1i}) + g(x_{2i})] \tag{6.40}$$

Rubinstein (1981) shows that the antithetic-variates estimator, in fact, is more efficient if $g(x)$ is a continuous monotonically increasing or decreasing function with continuous first derivatives.

Example 6.6   Refer to the strip coal mine in Example 6.3. Use the antithetic-variate Monte Carlo technique to estimate the first three product-moments of drawdown recess time corresponding to $s/h_o = 0.5$. Assume that permeability K_h is the only random variable having a lognormal distribution with the mean $\mu_k = 0.1$ m per day and coefficient of variation $\Omega_k = 10$ percent.

Solution   The drawdown recess time (in days), according to Eq. (6.29), under constant storage coefficient $S = 0.05$, $d = 50$ m, $h_o = 30$ m, and $s/h_o = 0.5$ can be determined as

$$T = \frac{4.579}{K_h}$$

To estimate the statistical moments of random drawdown recess time by the antithetic-variate technique, the following steps are implemented:

1. Generate n random variates from $U(0, 1)$, that is, u_i, $i = 1, 2,..., n$, from which the corresponding values of $u_i' = 1 - u_i$ are computed.
2. For each pair of (u_i, u_i'), the corresponding standard normal variates (z_i, z_i') are generated, that is, $z_i = \Theta^{-1}(u_i)$, $z_i' = \Theta^{-1}(u_i')$, $i = 1, 2,..., n$.
3. From (z_i, z_i'), determine the corresponding random variates of permeability as

$$k_i = \exp(\mu_{\ln k} + \sigma_{\ln k}\, z_i) \qquad k_i' = \exp(\mu_{\ln k} + \sigma_{\ln k} z_i')$$

   where $\mu_{\ln k}$ and $\sigma_{\ln k}$ can be determined by Eqs. (2.59a) and (2.59b) from μ_k and Ω_k.
4. Compute the drawdown recess time $t_i = 4.579/k_i$ and $t_i' = 4.579/k_i'$ for $i = 1, 2,..., n$.
5. For each generated drawdown recess time set $\{t_i\}$ and $\{t_i'\}$, $i = 1, 2,..., n$, one computes the corresponding values of mean, standard deviation, and skewness coefficient. The implementation of Steps 1 to 5 for $n = 400$ results in

	From $\{t_i\}$	From $\{t_i'\}$	Average
Mean, $\hat{\mu}_t$ (days)	46.22	46.30	46.26
Stdev, $\hat{\sigma}_t$ (days)	4.71	4.75	4.73
Skewness, $\hat{\gamma}_t$	0.258	0.380	0.319

The histogram of drawdown recess time corresponding to $\{t_i\}$ is shown in Fig. 6.6. One can easily show that the theoretical moments of lognormally distributed drawdown recess time (T) are $\mu_t = 45.79$ days, $\sigma_t = 4.58$ days, and $\gamma_t = 0.300$. When conducting this simulation, one would experience more variation of sample moments, especially skewness coefficient, for $\{t_i\}$ and $\{t_i'\}$ from one simulation to another. However, the averaged values of the statistical moments are much more stable.

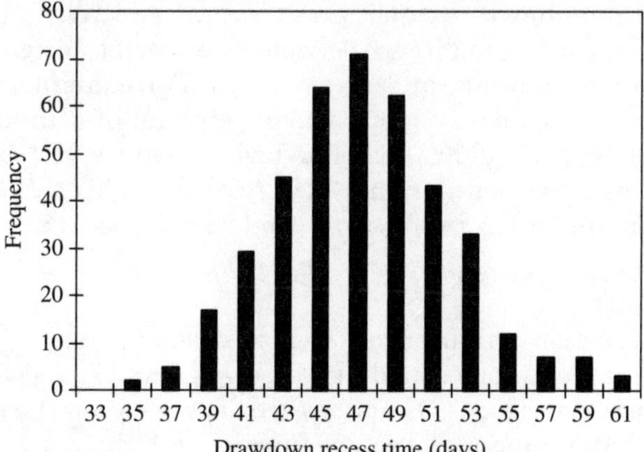

Figure 6.6 Histogram of simulated drawdown recess time for Example 6.6.

6.6.2 Correlated-sampling techniques

Correlated-sampling techniques are especially effective for variance reduction when the primary objective of the simulation study is to evaluate small changes in the system performance or to compare the difference in system performances between two specific designs (Rubinstein 1981; Ang and Tang 1984). Consider that one wishes to estimate

$$\Delta\Theta = \Theta_1 - \Theta_2 \tag{6.41}$$

in which

$$\Theta_1 = \int g_1(x) f_1(x)\, dx = E[g_1(X)]$$
$$\Theta_2 = \int g_2(y) f_2(y)\, dy = E[g_2(Y)] \tag{6.42}$$

with $f_1(x)$ and $f_2(y)$ being two different PDFs. By Monte Carlo simulation, $\Delta\Theta$ can be estimated as

$$\widehat{\Delta\Theta} = \hat{\Theta}_1 - \hat{\Theta}_2 = \frac{1}{n}\left[\sum_{i=1}^{n} g_1(X_i) - \sum_{i=1}^{n} g_2(Y_i)\right] = \frac{1}{n}\sum_{i=1}^{n}\widehat{\Delta\Theta}_i \tag{6.43}$$

in which X_i and Y_i are random samples generated from $f_1(x)$ and $f_2(y)$, respectively, and $\widehat{\Delta\Theta}_i = g_1(X_i) - g_2(Y_i)$.

The variance associated with $\widehat{\Delta\Theta}$ is

$$\mathrm{Var}(\widehat{\Delta\Theta}) = \mathrm{Var}(\hat{\Theta}_1) + \mathrm{Var}(\hat{\Theta}_2) - 2\mathrm{Cov}(\hat{\Theta}_1, \hat{\Theta}_2) \tag{6.44}$$

In the case that random variates X_i and Y_i are generated independently in the Monte Carlo algorithm, $\hat{\Theta}_1$ and $\hat{\Theta}_2$ would also be independent random variables. Hence, $\mathrm{Var}(\widehat{\Delta\Theta}) = \mathrm{Var}(\hat{\Theta}_1) + \mathrm{Var}(\hat{\Theta}_2)$.

Note that, from Eq. (6.44), $\mathrm{Var}(\widehat{\Delta\Theta})$ can be reduced if positively correlated random variables $\hat{\Theta}_1$ and $\hat{\Theta}_2$ can be produced to estimate $\Delta\Theta$. One easy way to obtain positively correlated samples is to use the same sequence of uniform random variates from $U(0, 1)$ in both simulations. That is, the random sequences $\{X_1, X_2,..., X_n\}$ and $\{Y_1, Y_2,..., Y_n\}$ are generated through $X_i = F_1^{-1}(U_i)$ and $Y_i = F_2^{-1}(U_i)$, respectively.

The correlated-sampling techniques are especially effective in reducing variance when the performance difference between two specific designs for a system involves the same or similar random variables. For example, consider two designs A and B for the same system involving a vector of K random variables $X = (X_1, X_2,..., X_K)$, which could be correlated with a joint PDF $f_x(x)$, or be independent of each other with a marginal PDF $f_k(x_k)$, $k = 1, 2,..., K$. The performance of the system under the two designs can be expressed as

$$\Theta_A = g(a, X) \qquad \Theta_B = g(b, X) \tag{6.45}$$

in which $g(\cdot)$ is a function defining the system performance; and a and b are vectors of design parameters corresponding to designs A and B, respectively. Since the two performance measures Θ_A and Θ_B are dependent on the same random variables through the same performance function $g(\cdot)$, their estimators will be positively correlated. In this case, independently generating two sets of K random variates, according to their probability laws for designs A and B, would still result in a positive correlation between $\hat{\Theta}_A$ and $\hat{\Theta}_B$. To further reduce $\text{Var}(\widehat{\Delta\Theta})$ an increase in the correlation between $\hat{\Theta}_A$ and $\hat{\Theta}_B$ can be achieved using a common set of standard uniform random variates for both designs A and B by assuming that system random variables are independent, as

$$\theta_{A,i} = g[\boldsymbol{a}, F_1^{-1}(u_{1i}), F_2^{-1}(u_{2i}),\ldots, F_K^{-1}(u_{Ki})] \qquad i = 1, 2,\ldots, n \qquad (6.46a)$$

$$\theta_{B,i} = g[\boldsymbol{b}, F_1^{-1}(u_{1i}), F_2^{-1}(u_{2i}),\ldots, F_K^{-1}(u_{Ki})] \qquad i = 1, 2,\ldots, n \qquad (6.46b)$$

in which $F_k^{-1}(u_{ki}) = x_{ki}$ is the inverse CDF for the kth random variable X_k operating on the kth standard uniform random variate for the ith simulation.

Example 6.7   Refer to the strip coal mine in Example 6.6. Suppose that engineers are also considering the possibility of starting excavation earlier. Evaluate the difference in the expected waiting time between the two options, that is, $s/h_o = 0.5$ and 0.6, by the correlated-sampling Monte Carlo simulation with $n = 400$. Assume that the only random variable is the permeability, K_h, having a lognormal distribution with the mean 0.1 m per day and coefficient of variation of 10 percent.

Solution   Referring to Eq. (6.43), the drawdown recess time (in days) under the constant storage coefficient $S = 0.05$, $d = 50$ m, $h_o = 30$ m can be determined as

$$\text{Option (a) for } s/h_o = 0.5: \; T_a = 4.579/K_h$$

$$\text{Option (b) for } s/h_o = 0.6: \; T_b = 2.941/K_h$$

To estimate the statistical moments of the difference in random drawdown recess times between the two options, the correlated-sampling Monte Carlo simulation can be implemented as follows:

1. Generate n random variates from $U(0, 1)$, that is, u_i, $i = 1, 2,\ldots, n$.
2. For each u_i generate the corresponding standard normal variate $z_i = \Phi^{-1}(u_i)$, $i = 1, 2,\ldots, n$.
3. From z_i determine the corresponding random permeability

$$k_i = \exp(\mu_{\ln k} + \sigma_{\ln k} z_i) \qquad i = 1, 2,\ldots, n$$

4. Compute drawdown recess time difference between the two options as

$$\Delta t_i = (4.579 - 2.941)/k_i \qquad \text{for } i = 1, 2,\ldots, n$$

5. Using each generated drawdown recess time differences, $\{\Delta t_i\}$, $i = 1, 2,\ldots, n$, compute the corresponding values of mean, standard deviation, and skewness coefficient. The results based on $n = 400$ are shown in column 1 of the following table.

	From $\{\Delta t_i\}$ (1)	From $\{\Delta t_i'\}$ (2)	Average (3)	Theoretical (4)
Mean, $\hat{\mu}_{\Delta T}$ (days)	16.69	16.41	16.55	16.38
Stdev, $\hat{\sigma}_{\Delta T}$ (days)	1.69	1.69	1.69	1.64
Skewness, $\hat{\gamma}_{\Delta T}$	0.112	0.532	0.322	0.300

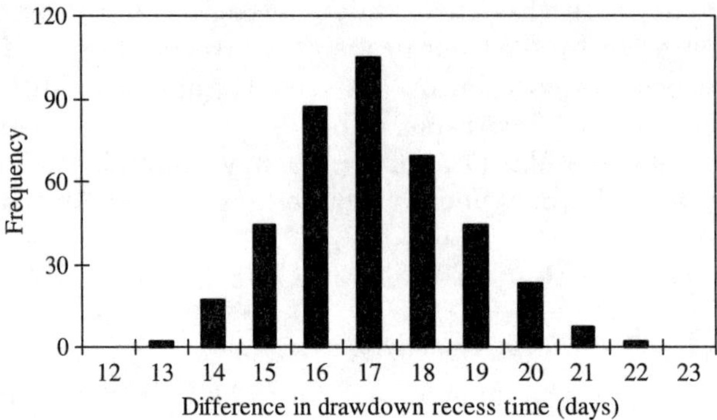

Figure 6.7   Histogram of the difference in drawdown recess time for Example 6.7.

The histogram corresponding to $\{t_i\}$ is shown in Fig. 6.7. For the purpose of comparison, column 4 of the preceding table lists the theoretical moments of the drawdown recess time difference (ΔT) that is also lognormally distributed. As can be seen, the first two moments of column 1 are quite close to those of the theoretical values whereas the skewness coefficient is not. To improve the estimation, the antithetic-variate technique can be implemented and the results listed in column 2 are obtained from the random series $\{1-u_i\}$. Neither column 1 nor 2 yields satisfactory estimation of the skewness coefficient. However, the average of the two, shown in column 3, clearly is superior.

6.6.3 Stratified sampling technique

The *stratified sampling* technique is a well-established area in statistical sampling (Cochran 1966). Variance reduction by the stratified sampling technique is achieved by taking more samples in important subregions. Consider a problem in which the expectation of a function $g(X)$ is sought where X is a random variable with a PDF $f_x(x)$, $x \in \Xi$. Referring to Fig. 6.8, the domain Ξ for the random variable X is divided into M disjoint subregions Ξ_m, $m = 1, 2,..., M$. That is,

$$\Xi = \bigcup_{m=1}^{M} \Xi_m \qquad \varnothing = \Xi_m \cap \Xi_{m'} \qquad m \neq m'$$

Let p_m be the probability that random variable X will fall within the subregion Ξ_m, that is, $\int_{x \in \Xi_m} f_x(x)\,dx = p_m$. Therefore, it is true that $\Sigma_m\, p_m = 1$. The expectation of $g(X)$ can be computed as

$$G = \int_{\Xi} g(x)f_x(x)\,dx = \sum_{m=1}^{M} \int_{\Xi_m} g(x)f_x(x)\,dx = \sum_{m=1}^{M} G_m \qquad (6.47)$$

where $G_m = \int_{\Xi m} g(x)\, f_x(x)\, dx$.

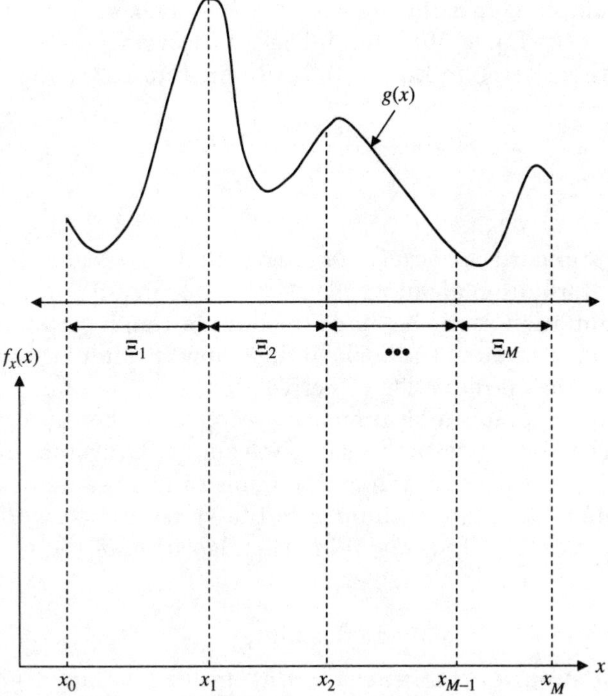

Figure 6.8 Schematic diagram of stratified sampling.

Note that the integral for G_m can be written as

$$G_m = p_m \int_{\Xi_m} g(x)\left[\frac{f_x(x)}{p_m}\right] dx = p_m E[g_m(X)] \qquad (6.48)$$

and it can be estimated by the Monte Carlo method as

$$\hat{G}_m = \frac{p_m}{n_m} \sum_{m=1}^{n_m} g(X_m) \qquad m = 1, 2, \dots, M \qquad (6.49)$$

where n_m is the number of sample points in the mth subregion; and $\Sigma_m n_m = n$, the total number of random variates to be generated. Therefore, the estimator for G in Eq. (6.47) can be obtained as

$$\hat{G} = \sum_{m=1}^{M} \hat{G}_m = \sum_{m=1}^{M} \frac{p_m}{n_m}\left[\sum_{i=1}^{n_m} g(X_{mi})\right] \qquad (6.50)$$

After the number of subregions M and the number of total samples n are determined, an interesting issue for the stratified sampling is how to allocate the total number of n sample points among the M subregions, such that the variance associated with $\hat{G}$ by Eq. (6.50) is minimized. A theorem shows that the optimal n_m^* that minimizes $\text{Var}(\hat{G})$ in Eq. (6.50) is (Rubinstein 1981)

$$n_m^* = n \left[\frac{p_m \sigma_m}{\displaystyle\sum_{m'=1}^{m} p_{m'} \sigma_{m'}} \right] \tag{6.51}$$

where σ_m is the standard deviation associated with the estimator $\hat{G}_m$ in Eq. (6.49).

In general, information about σ_m is not available in advance. It is suggested that a pilot simulation study be made to obtain a rough estimation about the value of σ_m, which serves as the basis in the follow-up simulation investigation to achieve the variance-reduction objective.

A simple plan for sample allocation is $n_m = np_m$ after the subregions are specified. It can be shown that, with this sampling plan, the variance associated with $\hat{G}$ by Eq. (6.50) is less than that from the simple random sample technique. One efficient stratified sampling technique is the *systematic sampling* (McGrath 1970) in which $p_m = 1/M$ and $n_m = n/M$. The algorithm of the systematic sampling can be described as follows:

1. Divide interval $[0, 1]$ into M equal subintervals

2. Within each subinterval, generate n/M uniform random numbers $u_{mi} \sim U[(m-1)/n, m/n]$, $m = 1, 2,..., M$ $i = 1, 2,..., n/m$

3. Compute $x_{mi} = F_x^{-1}(u_{mi})$

4. Calculate $\hat{G}$ according to Eq. (6.50)

Example 6.8   Repeat Example 6.5 using the systematic sampling technique to estimate the first three product-moments of drawdown recess time corresponding to $s/h_o = 0.5$.

Solution   The drawdown recess time (in days) under the condition of a constant storage coefficient $S = 0.05$, $d = 50$ m, $h_o = 30$ m, and $s/h_o = 0.5$ can be determined as $T = 4.579/K_h$. By the systematic sampling technique the statistical moments of random drawdown recess time can be estimated as follows:

1. Say, a total of n random variates is to be generated. By dividing interval $[0, 1]$ into M equal subintervals, n/M uniform random variates from $[(m-1)/n, m/n]$ are generated as

$$u_{mi} = \frac{m-1}{M} + \zeta_i$$

   in which $\zeta_i \sim U(0, 1/M)$ for $m = 1, 2,..., M$ $i = 1, 2,..., n/M$

2. For each u_{mi} determine the standard normal variate z_{mi}, that is, $z_{mi} = \Phi^{-1}(u_{mi})$

3. Compute the corresponding random permeability as

$$k_{mi} = \exp(\mu_{\ln k} + \sigma_{\ln k} z_{mi})$$

4. Compute drawdown recess time $t_{mi} = 4.579/k_{mi}$ for all m and i

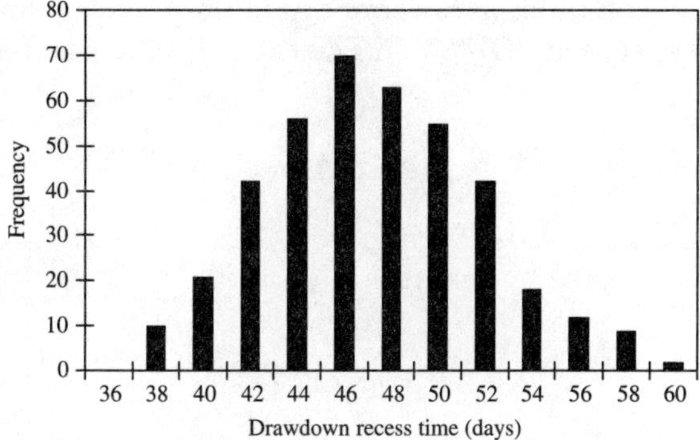

Figure 6.9 Histogram of drawdown recess time for Example 6.8.

5. Based on the total n generated drawdown recess time set $\{t_{mi}\}$ for $m = 1, 2,..., M$; $i = 1, 2,..., n/M$, one computes the values of mean, standard deviation, and skewness coefficient. Under $n = 400$ and $M = 10$ the resulting histogram for drawdown recess time is shown in Fig. 6.9 and its statistical moments are given as follows:

Statistics	Sampling	Theoretical
Mean, $\hat{\mu}_t$ (days)	46.23	45.79
Standard deviation, $\hat{\sigma}_t$ (days)	4.55	4.58
Skewness coefficient, $\hat{\gamma}_t$	0.310	0.300

6.6.4 Latin hypercube sampling technique

The *Latin hypercube sampling (LHS) technique* is a special method under the umbrella of stratified sampling which selects random samples of each random variable over its range in a stratified manner. Consider a multiple integral involving K random variables

$$G = \int_{x \in \Xi} g(\boldsymbol{x}) f_{\boldsymbol{x}}(\boldsymbol{x}) d\boldsymbol{x} = E[g(\boldsymbol{X})] \tag{6.52}$$

where $\boldsymbol{X} = (X_1, X_2,..., X_K)^t$ is an K-dimensional vector of random variables; and $f_{\boldsymbol{x}}(\boldsymbol{x})$ is their joint PDF.

The LHS technique divides the plausible range of each random variable into M ($M \geq K$ in practice) equal-probability intervals. Within each interval, a single random variate is generated resulting in M random variates for each random variable. The expected value of $g(\boldsymbol{X})$, then, is estimated as

$$\hat{G} = \frac{1}{M} \sum_{m=1}^{M} g(X_{1m}, X_{2m}, ..., X_{km}) \tag{6.53}$$

where X_{km} is the variate generated for the kth random variable X_k in the mth set.

More specifically, consider a random variable X_k over the interval of $[\underline{x}_k, \overline{x}_k]$ following a specified PDF $f_k(x_k)$. The range $[\underline{x}_k, \overline{x}_k]$ is partitioned into M intervals:

$$\underline{x}_k = x_{k0} < x_{k1} < x_{k2} < \cdots < x_{k,M-1} < x_{kM} = \overline{x}_k \tag{6.54}$$

in which $P(x_{km} \leq X_k \leq x_{k,m+1}) = 1/M$ for all $m = 0, 1, 2,..., M - 1$. The end points of the intervals are determined by solving

$$F_k(x_{km}) = \int_{\underline{x}_k}^{x_{km}} f_k(x_k)dx_k = \frac{m}{M} \tag{6.55}$$

where $F_k(\cdot)$ is the CDF of the random variable X_k. The LHS technique, once the end points for all intervals are determined, randomly selects a single value in each of the intervals to form the M samples set for X_k. The sample values can be obtained by the CDF-inverse or other appropriate method.

To generate M values of random variable X_k from each of the intervals, a sequence of probability values $\{p_{k1}, p_{k2},..., p_{k,M-1}, p_{kM}\}$ are generated as

$$p_{km} = \frac{m-1}{M} + \zeta_{km} \qquad m = 1, 2, ..., M \tag{6.56}$$

in which $\{\zeta_{k1}, \zeta_{k2},..., \zeta_{k,M-1}, \zeta_{kM}\}$ are independent uniform random numbers from $\zeta \sim U(0, 1/M)$. After $\{p_{k1}, p_{k2}, ..., p_{k,M-1}, p_{kM}\}$ are generated, the corresponding M random samples for X_k can be determined as

$$x_{km} = F_k^{-1}(p_{km}) \qquad m = 1, 2, ..., M \tag{6.57}$$

Note that p_{km} determined by Eq. (6.56) follows

$$p_{k1} < p_{k2} < \cdots < p_{km} < \cdots < p_{k,M-1} < p_{kM} \tag{6.58}$$

and, accordingly

$$x_{k1} \leq x_{k2} \leq \cdots \leq x_{km} \leq \cdots \leq x_{k,M-1} \leq x_{kM} \tag{6.59}$$

To make the generated $\{x_{k1}, x_{k2},..., x_{k,M-1}, x_{kM}\}$ a random sequence, random permutation can be applied to randomize the sequence. Alternatively, Latin Hypercube samples for K random variables with size M can be generated by (Pebesma and Heuvelink 1999)

$$x_{km} = F_k^{-1}\left(\frac{s_{km} - u_{km}}{M}\right) \tag{6.60}$$

where s_{km} is a random permutation of 1 to M; u_{km} is a uniformly distributed random variate in $[0, 1]$. Figure 6.10 shows the allocation of six samples by the

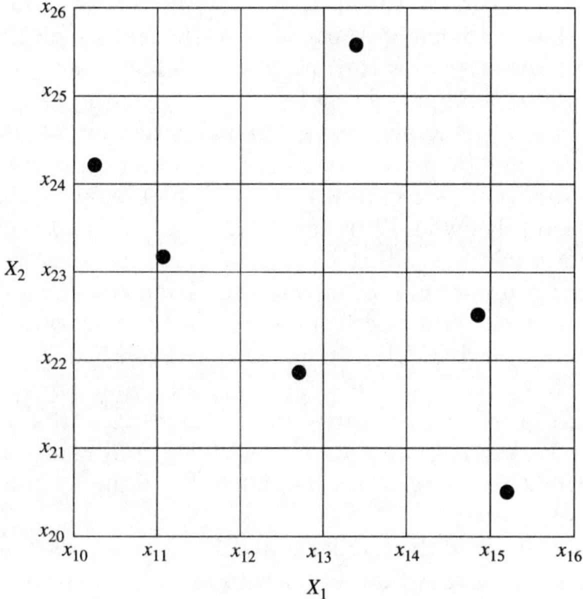

Figure 6.10   Schematic diagram of the LHS technique.

LHS technique for a problem involving two random variables. It is seen that, in each row or column of the 6×6 matrix, only one cell contains a generated sample. The LHS algorithm can implemented as follows:

1. Select the number of subinterval M for each random variable and divide the plausible range into M equal-probability intervals according to Eq. (6.55)
2. Generate M standard uniform random variates from $U(0, 1/M)$
3. Determine a sequence of probability values p_{km}, for $k = 1, 2,..., K$; $m = 1, 2,..., M$ using Eq. (6.56)
4. Generate random variates for each of the random variables using an appropriate method, such as Eq. (6.57)
5. Randomly permutate generated random sequences for all random variables
6. Estimate G by Eq. (6.53)

Using the LHS technique, the usual estimators of G and its distribution function are unbiased (McKay 1988). Moreover, when the function $g(X)$ is monotonic in each of the X_k, the variance of the estimators are no more than, and often less than, the variances when random variables are generated from simple random sampling. McKay (1988) suggested that the use of twice the number of involved random variables for sample size $(M \geq 2K)$ would be sufficient to yield accurate estimation of the statistics model output. Iman and Helton (1985) indicated that a choice of M equal to $4/3K$ usually gives satisfactory results. For a dynamic stream water-quality model over a 1-year simulation period, Manache (2001) compared results from LHS using $M = 4/3K$ and $M = 3K$ and found reasonable convergence in the identification of the most sensitive parameters but not in the calculation of the standard deviation of the model output. Thus, if it is computationally feasible, the generation of a larger number of samples would further enhance the accuracy of the estimation. Like all other variance-reduction Monte Carlo techniques, LHS generally would require fewer samples or model evaluations to achieve an accuracy level comparable to that obtained from a simple random sampling scheme. In hydrosystems engineering, LHS technique has been widely applied to sediment transport (Yeh and Tung 1993; Chang, Yang, and Tung 1993), water-quality modeling (Jaffe and Ferrara 1984; Melching and Bauwens 2001; Sohrabi et al. 2003; Manache and Melching 2004), and rainfall-runoff modeling (Melching 1995; Yu, Yang, and Chen 2001; Christiaens and Feyen 2002; Lu and Tung 2003).

Melching (1995) compared the results from LHS with $M = 50$ with those from Monte Carlo simulation with 10,000 simulations and also with those from FOVE and Rosenbleuth's method, for the case of using HEC-1 (U.S. Army Corps of Engineers 1991) to estimate flood peaks for a watershed in Illinois. All methods yielded similar estimates of the mean value of the predicted peak flow. The variation of standard-deviation estimates among the methods was much greater than that of the mean-value estimates. In the estimation of the standard deviation of the peak flow, LHS was found to provide the closest agreement to Monte Carlo simulation with an average error of 7.5 percent and 10 of 16 standard deviations within 10 percent of the value estimated with Monte Carlo simulation. This example indicates that LHS can yield accurate estimates of the mean and standard deviation of model output at a far smaller computational burden than Monte Carlo simulation.

Example 6.9 Repeat Example 6.8 using the LHS technique to estimate the first three product-moments of drawdown recess time corresponding to $s/h_o = 0.5$.

Solution According to the drawdown recess time (T, in days) equation $T = 4.579/K_h$, statistical moments of random drawdown recess time can be estimated by the LHS technique as:

1. In this example, $K = 1$. By setting $M = 400$, the interval [0, 1] is divided into 400 equal subintervals within each a single uniform random variate from $[(m-1)/M, m/M]$ are generated by

$$u_m = \frac{m-1}{M} + \zeta_m$$

   in which $\zeta_m \sim U(0, 1/M)$ for $m = 1, 2,..., M$
2. For each u_m determine the standard normal variate z_m, i.e., $z_m = \Phi^{-1}(u_m)$
3. Compute the corresponding random permeability as $k_m = \exp(\mu_{\ln k} + \sigma_{\ln k} z_m)$
4. Compute drawdown recess time $t_m = 4.579/k_m$ for all $m = 1, 2,..., M$

Based on the total M generated drawdown recess time set $\{t_m\}$ for $m = 1, 2, ..., M$, one computes the values of mean, standard deviation, and skewness coefficient and the results are

Statistics	LHS	Theoretical
Mean, $\hat{\mu}_t$ (days)	46.25	45.79
Standard deviation, $\hat{\sigma}_t$ (days)	4.61	4.58
Skewness coefficient, $\hat{\gamma}_t$	0.280	0.300

6.7   Resampling Techniques

Note that the Monte Carlo simulation described in the previous sections is conducted under the conditions that the probability distribution and the associated population parameters are known for the random variables involved in the system. The observed data are not directly utilized in the simulation. In many statistical estimation problems, the statistic of interest are often expressed as functions of random observations, that is,

$$\hat{\Theta} = \hat{\Theta}(X_1, X_2, ..., X_n) \tag{6.61}$$

The statistic $\hat{\Theta}$ could be estimators of unknown population parameters of interest. For example, consider that random observations X's are the annual maximum floods. The statistics $\hat{\Theta}$ could be the distribution of the floods; statistical properties such as mean, standard deviation, and skewness coefficient; the magnitude of the 100-year event; a probability of exceeding the capacity of a hydraulic structure; and so on.

Note that the statistic $\hat{\Theta}$ is a function of the random variables. It is also a random variable, having a PDF, mean, and standard deviation like any other random variable. After a set of n observations, $\{X_1 = x_1, X_2 = x_2, ..., X_n = x_n\}$, are available, the numerical value of the statistic $\hat{\Theta}$ can be computed. However, along with the estimation of $\hat{\Theta}$ values, a host of relevant issues can be raised with regard to the accuracy associated with the estimated $\hat{\Theta}$, its bias, confidence interval, and the like. These issues can be approached by the Monte Carlo simulation in that many sequences of random variates of size n are generated from each of which the value of the statistic of interest is computed $\hat{\Theta}$. Then, the statistical properties of $\hat{\Theta}$ can be summarized.

Unlike the Monte Carlo simulation approach, resampling techniques are developed, which reproduce random data exclusively on the basis of the observed ones. Two basic resampling techniques—the jackknife method and the bootstrap method—are described herein.

6.7.1   Jackknife method

The *jackknife method* was first proposed by Quenouille (1956) for estimating the bias and variance of any statistical estimator based on a set of observed data. Miller (1974) and Efron (1982) gave excellent reviews of this method. Refer to

Eq. (6.61), the estimated value of statistic $\hat{\Theta}$ is denoted as $\hat{\theta} = \hat{\Theta}(x_1, x_2, ..., x_n)$. By the jackknife method, the n observations are assumed to have an empirical distribution function $\hat{f}$

$$\hat{f}: P(X = x_i) = 1/n, \qquad \text{for } i = 1, 2, ..., n \tag{6.62}$$

In other words, $\hat{f}$ is the *non-parametric maximum likelihood estimator* of the unknown probability mass function $f_x(x)$ for each individual observation.

The Jackknife method sequentially deletes sample data point x_i and recomputes the value of $\hat{\Theta}$ assuming that the remaining $n - 1$ observations follow a empirical probability distribution $\hat{f}_{(i)}$

$$\hat{f}_{(i)}: P(X = x_j) = 1/(n-1) \qquad \text{for } j = 1, 2, ..., i - 1, i + 1, ..., n \tag{6.63}$$

with the subscript (i) indicating that the ith observation is removed. The corresponding value of statistic $\hat{\theta}_{(i)}$ then can be computed as $\hat{\theta}_{(i)} = \hat{\Theta}(x_1, x_2, ..., x_{i-1}, x_{i+1}, x_n)$. The jackknife estimate of the statistic $\hat{\Theta}_J$ is

$$\hat{\theta}_J = \frac{1}{n} \sum_{i=1}^{n} \hat{\theta}_{(i)} \tag{6.64}$$

The amount of bias associated with $\hat{\theta} = \hat{\Theta}(x_1, x_2, ..., x_n)$ can be estimated as

$$\Delta\hat{\theta} = (n-1)(\hat{\theta}_J - \hat{\theta}) \tag{6.65}$$

On the basis of $\hat{\theta}_{(i)}, i = 1, 2, ..., n$, the accuracy, in terms of the standard deviation, associated with $\hat{\Theta}$ can be computed as

$$s_{\hat{\theta}} = \sqrt{\frac{n-1}{n} \sum_{i=1}^{n} [\hat{\theta}_{(i)} - \hat{\theta}_J]^2} \tag{6.66}$$

In summary, the jackknife algorithm can be outlined as

1. Compute the value of statistic of the interest, $\hat{\theta} = \hat{\Theta}(x_1, x_2, ..., x_n)$, based on all n sample observations

2. For each observation $i = 1, 2, ..., n$, compute $\hat{\theta}_{(i)}$ from $n - 1$ observations with the ith observation removed from the data set

3. Compute the bias and accuracy associated with $\hat{\theta}$ according to Eqs. (6.65) and (6.66)

Example 6.10   Table 6.7.1 lists the annual maximum flood series (1929 to 1958) for Mill Creek at Los Molinos, California. Use the jackknife method to estimate the standard errors associated with the sample mean, standard deviation, and skewness coefficient of the log-transformed data.

Solution   In this example, the sample size is $n = 30$. The statistics of interest are the sample mean, standard deviation, and skewness coefficient of the flood magnitude in the log-space. Therefore, logarithmic transformation of the original data is made first to form a new data set $(y_1, y_2, ..., y_n)$ where $y_i = \ln(x_i)$. The jackknife method can be easily implemented on computers. The results of the computation are as follows:

$$s(\overline{Y}) = 0.05533 \qquad s(S_y) = 0.04217 \qquad s(G_y) = 0.45993$$

in which $s(\overline{Y})$, $s(S_y)$, and $s(G_y)$ are the standard error of the sample mean, sample standard deviation, and sample skewness coefficient of $Y = \ln(X)$, respectively.

6.7.2 Bootstrap technique

The *bootstrap technique* was first proposed by Efron (1979a,b) to deal with the variance estimation of sample statistics based on observations. The technique intends to be a more general and versatile procedure for sampling distribution problems without having to rely heavily on the normality condition on which classical statistical inferences are based. In fact, it is not uncommon to observe nonnormal data in hydrosystem engineering problems. Although the bootstrap technique is computationally intensive—a price to pay to break away from dependence of the normality theory—such concerns will be gradually diminished as the calculating power of the computers increases (Diaconis and Efron 1983).

Since the introduction of the bootstrap resampling technique, it has rapidly attracted the attention of statisticians and those who apply statistics in their research work. The bootstrap technique and its variations have been applied to various statistical problems such as bias estimation, regression analysis, time series analysis, and others. There are different variations of bootstrap resampling procedures for various situations. An excellent overall review and summary of bootstrap techniques, variations, and other resampling procedures are given by Efron (1982) and Efron and Tibshirani (1993). In hydrosystems engineering, bootstrap procedures have been applied to assess the uncertainty associated with the distributional parameters in flood frequency analysis (Tung and Mays 1981), optimal risk-based hydraulic design of bridges (Tung and Mays 1982), and unit hydrograph derivation (Zhao et al. 1997).

The basic algorithm of the bootstrap technique in estimating the standard deviation associated with any statistic of interest from a set of sample observations involves the following steps:

1. For a set of sample observations of size n, that is, $\boldsymbol{x} = \{x_1, x_2, ..., x_n\}$, assign a probability mass $1/n$ to each observation as Eq. (6.62)

2. Randomly draw n observations from the original sample set using $\hat{f}$ to form a *bootstrap sample*, $\boldsymbol{x}_\# = \{x_{1\#}, x_{2\#}, ..., x_{n\#}\}$. Note that the bootstrap sample $\boldsymbol{x}_\#$ is a subset of the original samples $\boldsymbol{x}$

3. Calculate the value of the sample statistic $\hat{\Theta}_\#$ of interest based on the bootstrap sample $\boldsymbol{x}_\#$

4. Independently repeat Steps 2 and 3 a number of times M, obtaining *bootstrap replications* of $\hat{\theta}_{\#} = \{\hat{\theta}_{\#1}, \hat{\theta}_{\#2}, ..., \hat{\theta}_{\#M}\}$ and calculate

$$\hat{\sigma}_{\hat{\theta}_{\#}} = \left[\frac{1}{M-1} \sum_{m=1}^{M} (\hat{\theta}_{\#m} - \hat{\theta}_{\#\bullet})^2 \right]^{0.5} \tag{6.67}$$

where $\hat{\theta}_{\#}$ is the average of the bootstrap replications of $\hat{\Theta}$, that is,

$$\hat{\theta}_{\#\bullet} = \sum_{m=1}^{M} \hat{\theta}_{\#m} \Big/ M \tag{6.68}$$

A flowchart for the basic bootstrap algorithm is shown in Fig. 6.11. The bootstrap algorithm, as shown previously, provides more information than just computing the standard deviation of a sample statistic. The histogram constructed on the basis of M bootstrap replications $\hat{\theta}_{\#} = \{\hat{\theta}_{\#1}, \hat{\theta}_{\#2}, ..., \hat{\theta}_{\#M}\}$ would give some ideas about the sampling distribution of the sample statistic $\hat{\Theta}$. Furthermore, based on the bootstrap replications $\hat{\theta}_{\#}$, one can construct confidence intervals for the sample statistic of interest. Similar to Monte Carlo simulation, the accuracy of estimation increases as the number of bootstrap samples gets larger. However, there exists a tradeoff between computational cost and the level of accuracy desired. Efron (1982) suggested that $M = 200$ is generally sufficient for estimating the standard errors of the sample statistics. However, to estimate the confidence interval with reasonable accuracy, one would need at least $M = 1000$.

The above algorithm is called *nonparametric, unbalanced bootstrapping*. Its parametric version can be made by replacing the nonparametric estimator $\hat{f}$ by a parametric distribution, in which the distribution parameters are estimated by the maximum likelihood method. More specifically, if one judges that, on the basis of the original data set, the random observations $\boldsymbol{x} = \{x_1, x_2, ..., x_n\}$ are from, say, a lognormal distribution, then the resampling of x's from $\boldsymbol{x}$ using the parametric mechanism would assume that $\hat{f}$ is a lognormal distribution.

Note that the theory of the unbalanced bootstrap algorithm described previously only ensures that the expected number to be resampled for each individual observation is equal to the number of bootstrap samples M generated. However, in actual implementations of the unbalanced simulation, the actual number of replications for each individual sample point might not be exactly equal to the number of bootstrap replications generated. For example, using the unbalanced bootstrap algorithm to generate 1000 bootstrap samples, the total number of times that observation x_1 is resampled might be 992, x_2 might be 1010, and so on. To improve the accuracy estimation associated with a statistical estimator of interest, Davison, Hinkley, and Schechtman (1986) proposed the *balanced bootstrap* simulation in which the number of appearance of each individual observation in the bootstrap data set must be exactly equal to the total number of bootstrap replications generated. This constrained bootstrap simulation has been found, in both theory and practical implementations, to be more efficient

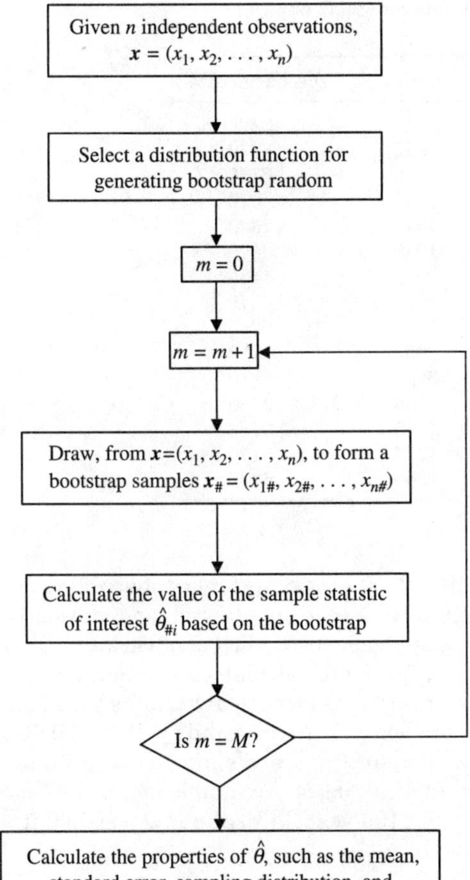

Figure 6.11   Flowchart of basic bootstrap resampling algorithm.

than the unbalanced algorithm in that the standard error associated with $\hat{\Theta}$ by the balanced algorithm is smaller. This implies that less bootstrap replications are needed by the balanced algorithm than the unbalanced approach to achieve the same accuracy level in estimation. Gleason (1988) discussed several computer algorithms for implementing the balanced bootstrap simulation.

Example 6.11   Refer to the annual maximum flood data listed in Table 6.3. Use the unbalanced bootstrap method to estimate the mean, standard errors, and 95 percent confidence interval associated with the following quantities:

1. Sample mean, standard deviation, and skewness coefficient of the log-transformed data
2. Magnitude of the 100-year flood assuming the annual maximum flood series follows a lognormal distribution
3. Annual probability that the flood magnitude exceeds 20,000 ft^3/s

**TABLE 6.3   Annual Maximum Floods for Mill Creek Near
Los Molinos, California**

Year	Discharge (ft³/s)	Year	Discharge (ft³/s)
1929	1500	1944	3220
1930	6000	1945	3230
1931	1500	1946	6180
1932	5440	1947	4070
1933	1080	1948	7320
1934	2630	1949	3870
1935	4010	1950	4430
1936	4380	1951	3870
1937	3310	1952	5280
1938	23000	1953	7710
1939	1260	9154	4910
1940	11400	1955	2480
1941	12200	1956	9180
1942	11000	1957	6140
1943	6970	1958	6880

Solution   In this example, $M = 2000$ bootstrap replications of size $n = 30$ from $\{y_i = \ln(x_i)\}$, $i = 1, 2,\dots, 30$, are generated by the unbalanced, nonparametric bootstrap procedure. In each replication, sample mean, coefficient of variation, and skewness coefficient of the log-transformed flows are calculated. Furthermore, the bootstrapped flows in each replication are treated as lognormal variates based on which the 100-year flow magnitude, Q_{100}, and exceedance probability, $P(Q > 20{,}000 \text{ ft}^3/\text{s})$, are computed. The results of the computations are shown in the following table. The histograms of bootstrapped replications of the sample mean, coefficient of variation, skewness coefficient, estimated 100-year flood, and $P(Q > 20{,}000 \text{ ft}^3/\text{s})$ are shown in Figs. 6.12(a) to (e), respectively.

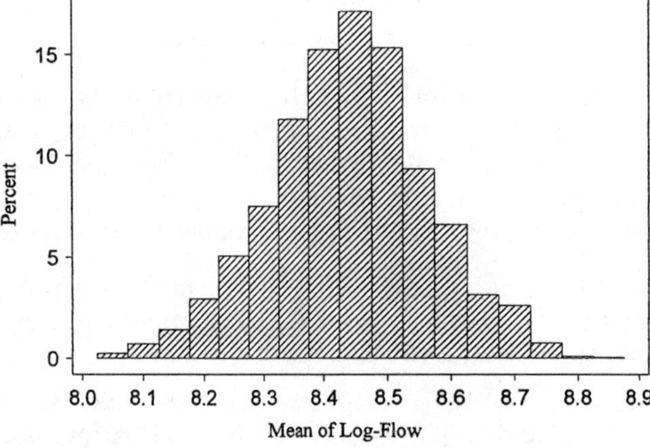

Figure 6.12(a)   Histogram of 2000 bootstrapped replications of the sample mean of log-flow for Example 6.10.

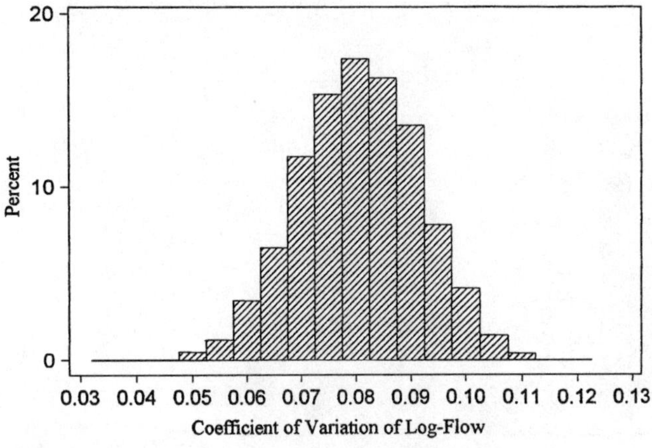

Figure 6.12(*b*) Histogram of 2000 bootstrapped replications of sample coefficient of variation of log-flow for Example 6.10.

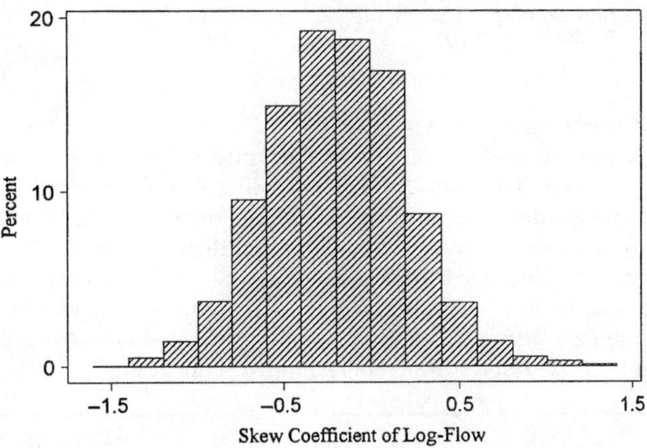

Figure 6.12(*c*) Histogram of 2000 bootstrapped replications of sample skewness coefficient of log-flow for Example 6.10.

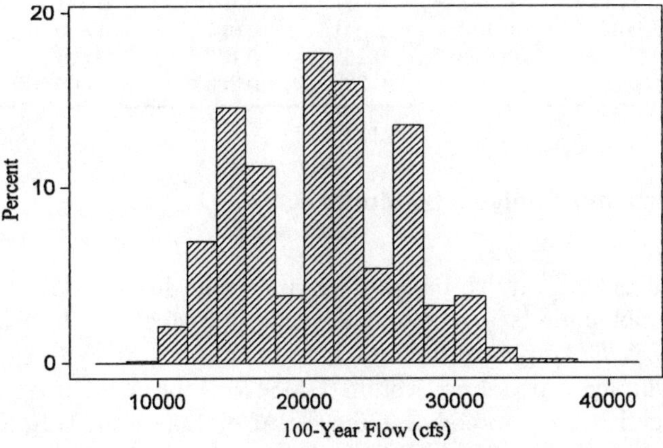

Figure 6.12(*d*) Histogram of 2000 bootstrapped replications of the 100-year flow for Example 6.10.

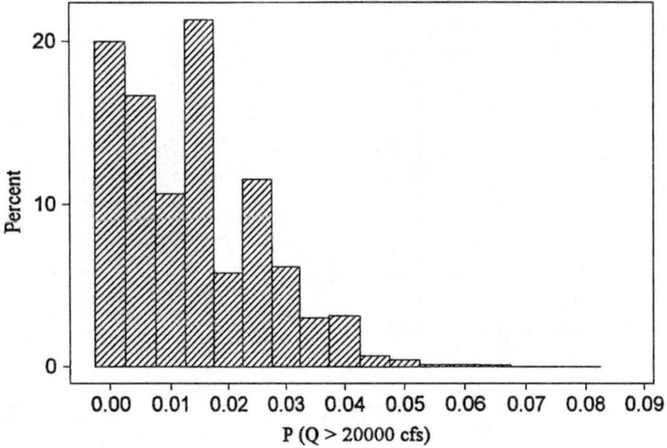

Figure 6.12(e)   Histogram of 2000 bootstrapped replications of $P(Q > 20,000 \text{ ft}^3/\text{s})$ for Example 6.10.

Examining the histograms of various sample statistics, it is interesting to observe that the sample mean, coefficient of variation, and skewness coefficient practically are normally distributed. The bootstrapped sampling distribution for Q_{100} is positively skewed (with a skewness coefficient of 0.175) and appears to be bimodal. Finally, the sampling distribution of the exceedance probability, $P(Q > 20,000 \text{ ft}^3/\text{s})$, is highly skewed to the right. Because the exceedance probability has to be bounded between 0 and 1, density functions, such as the Beta distribution may be applicable. The 95-percent confidence interval shown in the table is obtained by truncating 2.5 percent from both ends of the ranked 2000 bootstrapped replications.

Sample statistics	Mean	Coeff. of variation	Skew coeff.	95% Confidence interval Lower bound	Upper bound
Mean	8.437	0.015	−0.106	8.220	8.650
Coefficient of variation	0.081	0.011	−0.107	0.062	0.098
Skewness coefficient	−0.193	−2.041	0.097	−0.830	0.441
Q_{100} (ft^3/s)	20973	0.253	0.175	13246	30061
$P(Q > 20,000 \text{ ft}^3/\text{s})$	0.0143	0.829	0.900	0.000719	0.03722

6.8   Sensitivity and Uncertainty Analysis by Monte Carlo Simulation

In hydrosystems and environmental engineering designs and analyses, it is common to apply some type of model describing hydrologic, hydraulic, and biological processes. Examples of such models are the U.S. Army Corps of Engineers (USACoE) Hydrologic Modeling System (HMS) for simulating precipitation-runoff processes of dendritic watershed systems (USACoE 2001), the U.S. Environmental Protection Agency (EPA) QUAL2E (Brown and Barnwell 1987) for water-quality modeling in streams; MODFLOW for groundwater flow study, just to name a few. The typical features of these models are that they are conglomerations of complex mathematical operations that represent many interrelated physical processes described by various parameters, many of which are subject to uncertainty. As expected, outputs from the model application would be uncertain.

When facing uncertainties in engineering design problems, a commonly used approach is to conduct a sensitivity analysis to investigate the degree to which model responses are affected by possible variation of model parameters. *Sensitivity analysis* offers quantitative assessment of the influences of variation in parameters on the model output variability. Moreover, sensitivity analysis provides information about the relative importance of model parameters; which is essential for model parameter calibration and estimation. By conventional sensitivity analysis, model parameters are perturbed one at a time and the sensitivity of model outputs to such perturbation is observed. This type of sensitivity is described in the form of the sensitivity coefficient (see Sec. 5.1) as

$$s_k = \left(\frac{\partial W}{\partial x_k}\right)_{x_o} \approx \frac{\Delta W}{\Delta x_k} = \frac{W(x_o + \Delta x_k e_k) - W(x_o)}{\Delta x_k} \tag{6.69}$$

in which s_k is the sensitivity coefficient associated with the kth parameter evaluated at the selected point x_o; and x_k is an unit vector with the kth element equal to one and zero elsewhere. Other more accurate numerical approximations for the sensitivity coefficient can also be applied. A nondimensional form of the previously given sensitivity coefficient is

$$s_{k\%} = \left(\frac{\partial W / W(x)}{\partial x_k / x_k}\right)_{x_o} \approx \frac{\Delta W}{\Delta x_k}\left(\frac{x_{ko}}{W(x_o)}\right) = \frac{W(x_o + \Delta x_k e_k) - W(x_o)}{\Delta x_k}\left(\frac{x_{ko}}{W(x_o)}\right) \tag{6.70}$$

The term $s_{k\%}$ is called the *relative sensitivity coefficient* representing the percentage of change in model output due to 1-percent change in a model parameter. This type of sensitivity analysis provides information about the change in model output due to perturbation of one parameter in the neighborhood of its selected value in the parameter space. Therefore, it is also called *local sensitivity analysis*.

For a model whose sensitivity features varies from one region of the parameter space to another, the local sensitivity measures at a selected point do not shed much light in understanding the behavior of the model over the entire or selected domain of the parameter space. The validity of this argument is also true if one attempts to use local measures to address the global uncertainty features of a model output (see Fig. 5.2). Although local measures can provide, in principle, a more detailed description of model behavior, the use of local measures, in practice, is often restricted by the computational effort required for their evaluation. This is especially true when a model, such as those mentioned earlier is complex, the number of parameters is large, and the execution time for each model run is long.

Many researchers (Beck 1987; Gardner et al. 1981; Melching and Yoon 1996; Yeh and Tung 1993) have shown that the local sensitivity analysis is not appropriate for identifying the sources that significantly contribute to the model output uncertainty for more detailed study. This is because the traditional sensitivity analysis does not consider the likelihood of the parameter being different from its "best" value. As shown in Eq. (5.29) and Fig. 5.2, a highly sensitive parameter having a very well-known or defined value may have less influence on the model uncertainty than a much less sensitive parameter that is highly uncertain. Hence, uncertainty analysis is necessary to integrate the effects of sensitivity and uncertainty to determine the contribution of each individual model parameter to the overall uncertainty of model output (Melching 2001).

Analysis from a global perspective, on the other hand, focuses on the general model behavior over the defined parameter domain. *Global sensitivity analysis* is concerned with the pattern of change in model output due to changes in parameters over the concerned parameter range. In general, global sensitivity analysis can be accomplished with less computation. The lack of resolution could limit its usefulness, especially when the effect of a model parameter on an output is drastically different in various parts of the parameter space. However, if a global analysis is properly performed, the results could be much more valuable and useful than those from the local analysis. Examples of global sensitivity and uncertainty analyses using the Latin hypercube sampling technique (Sec. 6.6.4) can be found elsewhere (Chang et al. 1993; Yeh and Tung 1993; Melching and Bauwens 2001; Yu et al. 2001; Manache and Melching 2004).

Global sensitivity and uncertainty analyses by MCS-based schemes, in essence, generate M sets of K stochastic model parameters ($M > K$) according to their underlying statistical characteristics in a defined parameter range. With each generated parameter set, a model execution is made to produce the corresponding values of model outputs, that is,

$$(x_{1,m},\ x_{2,m},\dots,\ x_{K,m}) \rightarrow \text{Model} \rightarrow w_m \qquad m = 1,\ 2,\dots,\ M$$

in which $x_{k,m}$ is the kth model parameter generated in mth data set and w_m is the corresponding model output value. Based on the M sets of model parameters and outputs, one can define various sensitivity and uncertainty indicators by using correlation and regression analyses to show the relative importance of each of the K model parameters. As the computing power of computers increases, sensitivity and uncertainty analyses by MCS-based schemes, in conjunction with an efficient sampling procedure, are becoming practical, flexible, and robust (Iman and Helton 1985; Saltelli and Homa 1992; Saltelli, Andres, and Homa 1993). The accuracy of the MCS-based sensitivity analysis is dependent on the sample size. McKay (1988) suggested that using the LHS technique the sample size M should at least equal twice the number of perceived important input variables while Iman and Helton (1988) indicated that a choice of M equal to 4/3 times the perceived important model parameters would be sufficient to yield satisfactory results. Manache (2001) compared the results of LHS with M equal to 4/3 times the perceived important model parameters with those for LHS with M equal to 3 times the perceived important model parameters for a complex continuous simulation water quality model over a 1-year simulation period and found that the determination of the parameters significantly affecting the model output was nearly identical for the two cases.

Through a correlation analysis, the relative importance of each model parameter to the model output can be identified. Referring to Eq. (2.46), $\text{Corr}(X, Y) = \text{Cov}(X, Y)/\sigma_x\sigma_y$, the correlation coefficient is indicative of the strength of the linear relationship between two concerned random variables. Based on the M sets of Latin hypercube samples, the sample correlation coefficient (also known as the Pearson product-moment correlation coefficient), r_{w,x_k}, between individual random parameter X_k and model output W can be computed as

$$r_{w,x_k} = \frac{\displaystyle\sum_{m=1}^{M} x_{k,m}w_m - M\,\overline{x}_k\overline{w}}{\left(\displaystyle\sum_{m=1}^{M} x_{k,m}^2 - M\,\overline{x}_k^2\right)^{1/2}\left(\displaystyle\sum_{m=1}^{M} w_m^2 - M\,\overline{w}^2\right)^{1/2}} \qquad \text{for } k = 1, 2, ..., K \qquad (6.71)$$

in which the over-bar represents the sample mean of the corresponding variable. A higher value of the correlation coefficient indicates a stronger linear relationship between the model parameter and the output. To reveal the existence of a monotonically nonlinear relationship, sample data generated for individual model parameter and output can be ranked (either in ascending or descending order) separately and their ranks are used in Eq. (6.71) to calculate the correlation coefficient. The *Spearman's rank correlation coefficient* (Conover 1972) can be computed as

$$r_{R(w),R(x_k)} = \frac{\displaystyle\sum_{m=1}^{M}\left(R(x_{k,m}) - \frac{M+1}{2}\right)\left(R(w_m) - \frac{M+1}{2}\right)}{M(M^2 - 1)/12}$$

$$= 1 - \frac{6\displaystyle\sum_{m=1}^{M}[R(x_{k,m}) - R(w_m)]^2}{M(M^2 - 1)/12} \qquad \text{for } k = 1, 2, ..., K \qquad (6.72)$$

in which $r_{R(w),R(x_k)}$ is the rank correlation coefficient with $R(w_m)$ and $R(x_{k,m})$, respectively, representing the ranks of mth randomly generated values of model output W and parameter X_k. It is intuitively understood that for two variables having a strong monotonically increasing or decreasing relation, regardless of their degree of nonlinearity, the value of rank correlation coefficient will be high. Hence, a comparison of the relative magnitudes of the two types of correlation coefficients would give good indication about the strength and form of relation between the model output and a model parameter. In the case that the rank correlation coefficient is significantly higher than the Pearson product-moment correlation coefficient (from Eq. (6.71)), a nonlinear relationship is present between model output and the parameter and, then, the sensitivity coefficient would change with the value of the model parameter.

As most of the computerized models used in hydrosystems and environmental engineering studies are implicit and complex, multiple regression analysis (described in Chap. 3) can be applied to establish an equivalent explicit parameter-output relationship of the model to facilitate the global sensitivity and uncertainty analysis. Note that the credibility of global sensitivity and uncertainty analysis largely hinges on how well the equivalent model represents the actual model. Hence, for the credibility and practicality of the global analysis, it is essential to establish a reasonably accurate, yet simple, regression equation to represent the model parameter-output relationship based on the generated data. Indicators of goodness-of-fit, such as the coefficient of determination R^2 and its variations, standard error of estimate, and selection of explanatory variables by step-wise regression (see Secs. 3.4 and 3.9.1) help in achieving this objective.

For most practical problems, a 2nd-order regression equation is generally sufficient to represent the model parameter-output relationship:

$$W = a_0 + \sum_{k=1}^{K} b_k X_k + \sum_{k=1}^{K} c_k X_k^2 + \sum_{k=1}^{K-1} \sum_{j=k+1}^{K} d_{kj} X_k X_j + \varepsilon \tag{6.73}$$

where a, b, c and d are regression coefficients; and ε is the model error term representing the inaccuracy of the regression equation to emulate the actual model. The right-hand side of Eq. (6.73) consists of terms to account for linear, quadratic, and interactive relationships of model parameters in determining the value of model output. In practice, one would keep the model form as simple as possible without sacrificing too much accuracy in the approximation.

For the purpose of illustrating the global analysis, the following discussions assume that a linear representation of the regression equation, that is,

$$W = b_0 + \sum_{k=1}^{K} b_k X_k + \varepsilon \tag{6.74}$$

is sufficiently accurate in emulating the actual model. Melching (2001) and Manache and Melching (2004) found that for complex water-quality models over 1-year simulation periods key output features could be approximated by multiple linear regression models with coefficients of determination R^2 ranging from 0.91 to 0.985. Thus, such linear approximations may work well even for complex models. Clearly, one can immediately see that the regression coefficient b_k is the global sensitivity coefficient, $\partial W / \partial X_k$, associated with each model parameter X_k. It represents the "average" sensitivity of the model output with respect to a unit change in a parameter over the domain of the parameter space covered by the generated data. To facilitate the comparison of model output sensitivity to parameters of different units, model output and parameters in Eqs. (6.73) and (6.74) can be standardized as $X' = (X - \bar{x})/s$ with $\bar{x}$ and s, respectively, being the sample mean and standard deviation of a random variable. The regression equation, expressed in terms of standardized variables, can be written as

$$W' = \sum_{k=1}^{K} b'_k X'_k \tag{6.75}$$

in which b'_k is the *standardized regression coefficient* that is related to the regression coefficient b_k in Eq. (6.74) as

$$b'_k = \left(\frac{s_k}{s_w}\right) b_k \tag{6.76}$$

The standardized regression coefficient is a relative sensitivity measuring the effect of moving each model parameter away from its mean by one standard deviation on model output by a fraction of its standard deviation. The correlation coefficients indicate the strength of the association between model output and parameters whereas the regression coefficients represent the intensity of the relation.

The variance contribution of each stochastic model parameter to the total model output variability by Eq. (6.74) can be assessed from, according to Eq. (2.49),

$$s_w^2 = \sum_{j=1}^{K} \sum_{k=1}^{K} (b_j s_j)(b_k s_k) r_{x_j,x_k} + s_\varepsilon^2 \tag{6.77}$$

in which r_{x_j,x_k} is the correlation coefficient between model parameters X_j and X_k; and s_ε^2 is the mean squared error. From Eq. (6.77) it can be seen that in order for the approximated model to capture the totality of actual variability in model output, the approximation should be sufficiently accurate so that the value of mean squared error is reasonably small or $R^2 \approx 1$. When the model parameters are statistically independent, each term in the summation on the right-hand side of Eq. (6.77), that is, $b_k^2 s_k^2$ is the regression sum of squares (RSS) described in Sec. 3.8, each representing the contribution to overall model output uncertainty from the model parameter X_k.

When model parameters are not independent, correlation among the parameters could increase or decrease the total variability of model output, depending on the signs of the involved sensitivity coefficients and correlation coefficients. In such a circumstance, the assessment of the marginal contribution of each parameter to the total model output variability would not be clear. The break up of the RSS as shown in Eq. (3.27) is an alternative. However, one should know that each RSS would be affected by the presence of other model parameters and their sequences in the regression equation. Similar to the conditional RSS when model parameters are dependent, the *partial correlation coefficient (PCC)* can also be used to incorporate the influence of other correlated parameters (McKay 1988) as

$$PCC(W, X_k) = -c_{w,x_k} / (c_{w,w} c_{x_k,x_k})^{1/2} \tag{6.78}$$

in which $c_{x_k,x_k}, c_{x_k,w}$, and $c_{w,w}$ are elements of the inverse of the simple correlation matrix

$$\begin{bmatrix} R(X,X) & | & R(X,W) \\ \hline R^t(W,X) & | & 1 \end{bmatrix}^{-1} = \begin{bmatrix} c_{x_1,x_1} & c_{x_1,x_2} & \cdots & c_{x_1,x_K} & | & c_{x_1,w} \\ c_{x_2,x_1} & c_{x_2,x_2} & \cdots & c_{x_2,x_K} & | & c_{x_2,w} \\ \cdot & \cdot & \cdots & \cdot & | & \cdot \\ c_{x_K,x_1} & c_{x_K,x_1} & \cdots & c_{x_K,x_K} & | & c_{x_K,w} \\ \hline c_{x_1,w} & c_{x_2,w} & \cdots & c_{x_K,w} & | & c_{w,w} \end{bmatrix} \tag{6.79}$$

Hence, $PCC(W, X_k)$ quantifies the relative change in W with respect to change in model parameter X_k after removing the influence of the correlation between X_k and all other model parameters. When model parameters are statistically independent, the values of partial correlation coefficient will be identical to that of Pearson product-moment correlation coefficient, defined in Eqs. (2.46) or (6.71). Similar to the rank correlation coefficients, computation of the *partial rank correlation coefficient* can also be made from using ranked data.

Example 6.12   Sand and gravel mining from river beds are major sources of construction materials in many parts of the world. The migration of pits along the direction of flow might impose potential threats to the safety of downstream bridge piers and other in-stream hydraulic structures. Based on a series of laboratory experiments by Lee, Young, and Huang (1990) on pits of rectangular shape and uniform sand materials, a set of five empirical equations was established by regression analysis to estimate the maximum pit depth after the pit travels some distance downstream under specified flow conditions. A more detailed description of the model can be found in Yeh and Tung (1993).
Without listing all equations involved, the model, for simplicity, is expressed implicitly as

$$H_d = g(n,\ S_f,\ \gamma_s,\ d_s,\ \mathbf{b}_1,\ \mathbf{b}_2,\ \mathbf{b}_3,\ \mathbf{b}_4,\ \mathbf{b}_5,\ \varepsilon) \tag{6.80}$$

where H_d = maximum pit depth after traveling a specified distance
n = Manning roughness coefficient
S_f = frictional slope
γ_s = specific weight of sand
d_s = representative particle size
$\mathbf{b}_1, \mathbf{b}_2, \mathbf{b}_3, \mathbf{b}_4, \mathbf{b}_5$ = vectors containing the regression coefficients for each of the five regression equations in the pit-migration model
ε = vector containing errors associated with each of the five regression equations. In total, there are $K = 28$ stochastic model parameters subject to uncertainty

Uncertainty features of the channel hydraulic parameters (n and S_f) and channel bed materials (γ_s and d_s) in practice can be obtained from field survey and laboratory tests. As for regression coefficients and model error terms, their uncertainty features can be derived from a formal regression analysis described in Chap. 3. In Yeh and Tung (1993), the LHS technique was applied to generate 60 random samples of 28 stochastic model parameters based on which the corresponding maximum pit depth under the condition of a flow rate of 200 m³/s, channel width of 100 m, initial pit length of 40 m, pit depth of 3 m, and travel distance of 500 m.

Through the correlation study, various correlation coefficients between the maximum pit depth (H_d) and each of the 28 model parameters are listed in Table 6.4. Judging from the values of the four correlation coefficients, the model parameters defining channel and sediment characteristics (n, S_f, and d_s) and error terms associated with the 2nd and 4th regression equations are significantly more important than the remaining model parameters. Among the 19 regression coefficients, only one (that is, $b_{2,2}$) can be considered important. Comparing the values of partial correlation coefficient and partial rank correlation coefficient, practically all model parameters have about the same values. This indicates that the use of a nonlinear relation between the maximum pit depth and individual model parameters will not improve the association.

TABLE 6.4   Values of Various Correlation Coefficients of Maximum Pit Depth and 28 Model Parameters in the Pit-Migration Model (after Yeh and Tung 1993)

Model parameter	Correlation coefficient	Partial correlation coefficient	Rank correlation coefficient	Partial rank correlation coefficient
n	−0.531	−0.822 [4]	−0.512	−0.629 [5]
S_f	−0.270	−0.888 [3]	−0.305	−0.840 [3]
γ_s	0.162	0.432 [8]	0.168	0.248 [10]
d_s	0.468	0.947 [1]	0.477	0.900 [1]
$b_{1,1}$	−0.028	−0.165	−0.098	−0.303 [8]
$b_{1,2}$	0.064	−0.143	0.133	−0.235
$b_{1,3}$	−0.057	−0.142	−0.092	−0.084
$b_{1,4}$	0.017	−0.183	0.061	−0.197
$b_{1,5}$	−0.079	−0.128	−0.149	−0.073
$b_{1,6}$	0.072	−0.119	0.116	0.083
$b_{2,1}$	−0.044	0.337 [9]	−0.012	0.199
$b_{2,2}$	0.215	0.613 [6]	0.181	0.460 [6]
$b_{2,3}$	0.051	0.086	0.067	0.110
$b_{3,1}$	−0.210	−0.002	−0.194	−0.268 [9]
$b_{3,2}$	0.211	0.087	0.217	0.151
$b_{3,3}$	0.151	0.127	−0.155	0.420 [7]
$b_{3,4}$	0.184	0.079	0.194	−0.108
$b_{4,1}$	0.173	0.139	0.191	−0.001
$b_{4,2}$	−0.100	0.071	−0.120	−0.041
$b_{4,3}$	0.220	0.069	0.215	0.212
$b_{5,1}$	0.092	−0.235	0.012	0.197
$b_{5,2}$	−0.018	−0.245 [10]	−0.034	0.120
$b_{5,3}$	−0.093	−0.233	−0.010	0.252
ε_1	0.306	0.039	0.232	−0.112
ε_2	0.365	0.614 [5]	0.344	0.636 [4]
ε_3	−0.040	0.445 [7]	−0.051	0.035
ε_4	0.497	0.929 [2]	0.481	0.857 [2]
ε_5	0.088	0.073	0.057	−0.077

NOTE: Number in [] represents the rank of importance.

In addition to the use of various types of correlation coefficients, the relative importance of model parameters can be identified through regression analysis. Based on the form of Eq. (6.74), using centralized model output (i.e., $w_m^* = w_m - \bar{w}$) and standardized model parameters,

$$W^* = b_0^* + \sum_{k=1}^{K} b_k^* X_k' \tag{6.81}$$

the values of ordinary LS estimators of regression coefficients b_k^* along with their standard deviations and t-ratios, are listed in Table 6.5. The coefficient of determination R^2 associated with the regression model is 98.1 percent with a standard error of 0.0516 m. By Eq. (6.81) the regression coefficient b_k^* represents the change in model output due to one standard deviation change in model parameter X_k. The important model

TABLE 6.5   Results of Regressing Maximum Pit Depth on 28 Model Parameters in the Pit-Migration Model (after Yeh and Tung 1993)

Model parameter	Regression coefficient	Standard deviation	t-ratio	Rank
n	−0.074876	0.009274	−8.07	4
S_f	−0.096947	0.008682	−11.17	3
γ_s	0.027275	0.009302	2.93	7
d_s	0.143266	0.008654	16.56	1
$b_{1,1}$	−0.06020	0.09450	−0.64	−
$b_{1,2}$*	−	−	−	−
$b_{1,3}$	0.00349	0.09419	0.04	−
$b_{1,4}$	−0.05994	0.07086	−0.85	−
$b_{1,5}$	0.0496	0.1217	0.41	−
$b_{1,6}$	0.0519	0.1588	0.33	−
$b_{2,1}$	0.13106	0.07075	1.85	9
$b_{2,2}$	0.15130	0.03531	4.29	5
$b_{2,3}$	0.03259	0.04314	0.76	−
$b_{3,1}$	0.0234	0.1743	0.13	−
$b_{3,2}$	0.03292	0.04986	0.66	−
$b_{3,3}$	0.03011	0.05060	0.60	−
$b_{3,4}$	0.0722	0.1160	0.62	−
$b_{4,1}$	0.0930	0.1086	0.86	−
$b_{4,2}$	0.0537	0.1167	0.46	−
$b_{4,3}$	0.00804	0.03375	0.24	−
$b_{5,1}$	−0.1585	0.1166	−1.36	−
$b_{5,2}$	−0.02927	0.02001	−1.46	10
$b_{5,3}$	−0.1548	0.1157	−1.34	−
ε_1	0.0007	0.009201	0.08	−
ε_2	0.04490	0.01048	4.28	6
ε_3	0.026252	0.00967	2.72	8
ε_4	0.124358	0.00856	14.53	2
ε_5	0.007214	0.009781	0.74	−

*Deleted by the computer package due to its high correlation with other parameters.

parameters identified on the basis of PCC and PRCC (shown in Table 6.4) have exceptionally high values for the t-ratio. This indicates that they have very good predictive quality. In fact, from the viewpoint of model parsimony, parameters of less significance can be discarded from the regression model without jeopardizing the model's predictive quality. For example, parameters with t-ratios having absolute values less than 1.96 are not statistically significant at the 5-percent level and often are considered as not significantly affecting the uncertainty of model output.

Problems

6.1 Generate 100 random numbers from the Weibull distribution with parameters $\alpha = 2.0$, $\beta = 1.0$, and $\xi = 0$ by the CDF-inverse method. Check the consistency of the sample parameters based on the generated random numbers as compared with the population parameters used.

6.2 Generate 100 random numbers from the Gumbel (extreme value type I- max) distribution with parameters $\beta = 3.0$ and $\xi = 1.0$ by the CDF-inverse method. Check the consistency of the sample parameters based on the generated random numbers as compared with the population parameters used.

6.3 Generate 100 random numbers from a triangular distribution with lower bound $a = 2$, mode $m = 5$, and upper bound $b = 10$ by the CDF-inverse method. Check the consistency of the sample mean, mode, and standard deviation based on the generated random numbers as compared with the population values.

6.4 Prove that $P\{U \le g(Y)\} = 1/\varepsilon$ for the AR method.

6.5 Consider that the Hazen-Williams coefficient of a 5-year old, 24-in cast iron pipe is uncertain having a triangular distribution with lower bound $a = 115$, mode $m = 120$, and upper bound $b = 125$. Describe an algorithm to generate random numbers by the AR method with $\psi(x) = c$ and $\hat{f}_x(x) = 1/(b - a)$.

6.6 Refer to Prob. 6.5. Determine the efficient constant c and the corresponding acceptance probability for $c = 0.2$, 0.3, and 0.4.

6.7 Refer to Prob. 6.5. Develop computer programs to generate 100 random Hazen-Williams coefficients using $c = 0.2$, 0.3, and 0.4. Verify the theoretical acceptance probability for the different c values obtained in Prob. 6.6 by your numerical experiment. Discuss any discrepancy if exists in the results.

6.8 Develop an algorithm to generate random variable $Y = \max \{X_1, X_2,..., X_n\}$ where X_i are independent and identically distributed normal random variables with means μ_x and standard deviations σ_x.

6.9 Develop an algorithm to generate random variable $Y = \min \{X_1, X_2,..., X_n\}$ where X_i are independent and identically distributed lognormal random variables with means μ_x and standard deviations σ_x.

6.10 Based on the algorithm developed in Prob. 6.8, estimate the mean, standard deviation, and the magnitude of the 100-year event for a 10-year maximum rainfall ($n = 10$) in which the parent distribution for the annual rainfall is normal with a mean 3 in/h and standard deviation 0.5 in/h.

6.11 Based on the algorithm developed in Prob. 6.9 estimate the mean, standard deviation, and the magnitude of 100-year event for a 10-year minimum water supply ($n = 10$) in which the parent distribution for annual water supply is lognormal with mean 30,000 acre-feet (AF) and standard deviation 10,000 AF.

6.12 Repeat Example 6.4 by the spectral decomposition method.

6.13 Refer to the strip mine excavation problem described in Example 6.3. Suppose that a decision is made to start the excavation on the 50th day ($t = 50$ days). Using the CDF-inverse method, determine the first three product-moments that the excavation operation poses no safety threat on the excavation point.

6.14 Resolve Prob. 6.13 using the square-root algorithm.

6.15 Resolve Prob. 6.13 using the spectral decomposition algorithm.

6.16 Resolve Prob. 6.13 assuming that the conductivity and storage coefficient are correlated lognormal random variables. Compare the simulated result with the exact solution.

6.17 Develop a Monte Carlo simulation algorithm to solve Prob. 5.13 and compare the simulated results with those obtained in Probs. 5.13, 5.32 and 5.57.

6.18 Show that $\mathrm{Cov}(U, 1 - U) = -1/12$ in which $U \sim U(0, 1)$.

6.19 Resolve Prob. 6.13 by incorporating the antithetic-variates method.

6.20 Referring to Prob. 6.13, use the correlated-sampling method to determine the difference in drawdown at the excavation point when $t = 30$ days and $t = 50$ days.

6.21 Resolve Prob. 6.13 using the Latin hypercube sampling method.

6.22 Refer to the annual maximum flood data in Table 6.3. Use the jackknife method to estimate the magnitude of a 100-year flood and its associated error assuming that the flood data follow a lognormal distribution.

6.23 Refer to the annual maximum flood data in Table 6.3. Use the jackknife method to estimate $P[\text{flood peak} \geq 15{,}000 \ \text{ft}^3/\text{s}]$ and its associated error assuming that the flood data follow a lognormal distribution.

6.24 Refer to the annual maximum flood data in Table 6.3. Since the flood magnitude of a specified return period (Q_T) is determined on the basis of sample statistics, such as sample mean $(\overline{Q})$ and sample standard deviation (s_Q), the estimated Q_T also is subject to uncertainty. Assuming that the flood data follow a lognormal distribution, use the nonparametric, unbalanced bootstrap algorithm (with 1000 replications) to estimate the magnitude of the 100-year flood $(\Theta = Q_{T=100})$ and its associated error. Compare with the results obtained from Prob. 6.22. Furthermore, based on the 1000 bootstrap samples generated, assess the probability distribution and 90 percent confidence interval for $Q_{T=100}$.

6.25 Refer to the annual maximum flood data in Table 6.3. Assuming the flood data follow a lognormal distribution, use the nonparametric, unbalanced bootstrap algorithm to estimate $\Theta = P[\text{flood peak} \geq 15{,}000 \ \text{ft}^3/\text{s}]$ and its associated error and compare these results with those obtained from Prob. 6.23. Furthermore, based on the 1000 bootstrap samples generated, assess the probability distribution and 90 percent confidence interval for $\Theta = P[\text{flood peak} \geq 15{,}000 \ \text{ft}^3/\text{s}]$.

References

Abramowitz, M., and I. A. Stegun. (1972). *Handbook of Mathematical Functions*, Dover Publications, New York.

Ang, A. H. -S., and W. H. Tang. (1984). *Probability Concepts in Engineering Planning and Design, Vol. II: Decision, Risk, and Reliability*, John Wiley and Sons, New York, p. 562.

Beck, M. B. (1985). "Water Quality Management: A Review of the Development and Application of Mathematical Models," *Lecture Notes in Engineering 11*, C.A. Brebbia and S.A. Orszag (eds.), Springer-Verlag, New York.

Beck, M. B. (1987). "Water Quality Modeling: A Review of the Analysis of Uncertainty," *Water Resources Research*, 23(5):1393–1441.

Brown, L. C., and Barnwell, T. O., Jr. (1987). "The Enhanced Stream Water Quality Models QUAL2E and QUAL2E-UNCAS: Documentation and User Manual," *Report EPA/600/3-87/007*, U.S. Environmental Protection Agency, Athens, GA.

Box, G. E. P., and M. E. Muller. (1958). "A Note on Generation of Random Normal Deviates," *Annals of Mathematical Statistics*, **29**:610–611.

Chang, C. H., J. C. Yang, and Y. K. Tung. (1993). "Sensitivity and Uncertainty Analyses of a Sediment Transport Model: A Global Approach," *Journal of Stochastic Hydrology and Hydraulics*, **7**(4):299–314.

Chang, C. H., Y. K. Tung, and J. C. Yang. (1994). "Monte Carlo Simulation for Correlated Variables with Marginal Distributions," *Journal of Hydraulic Engineering*, ASCE, **120**(2):313–331.

Chen, X. Y., and Y. K. Tung. (2003). "Investigation of Polynomial Normal Transformation," *Journal of Structural Safety*, **25**:423–445.

Conover, W. J. (1972). *Practical Nonparametric Statistics*, John Wiley and Sons, New York.

Christiaens, K., and J. Feyen. (2002). "Use of Sensitivity and Uncertainty Measures in Distributed Hydrological Modeling with an Application to the MIKE SHE model," *Water Resources Research*, **38**(9):1169, doi:10.1029/2001 WR000478.

Cochran, W. (1966). *Sampling Techniques*, 2d ed., John Wiley and Sons, New York.

Dagpunar, J. (1988). *Principles of Random Variates Generation*, Oxford University Press, New York, p. 228.

Davison, A. C., D. V. Hinkley, and E. Schechtman. (1986). "Efficient Bootstrap Simulation," *Biometrika*, **73**(3): 555–566.

Der Kiureghian, A., and P. L. Liu. (1985). "Structural Reliability under Incomplete Probability Information." *Journal of Engineering Mechanics*, ASCE. **112**(1): 85–104.

Diaconis, P., and B. Efron. (1983). "Computer-Intensive Methods in Statistics," *Scientific American*, 116–131, May.

Efron, B. (1979a). "Bootstrap Methods: Another Look at the Jackknife." *The Annals of Statistics*, **3**:1189–1242.

Efron, B. (1979b). "Computers and Theory of Statistics: Thinking the Unthinkable," *SIAM Reviews*, **21**:460–480.

Efron, B. (1982). "The Jackknife, the Bootstrap, and Other Resampling Plans," In *CBMS-NSF Regional Conference Series in Applied Mathematics*, No. 38. SIAM, Philadelphia, PA.

Efron, B., and R. J. Tibshirani (1993). *An Introduction to the Bootstrap*, Chapmann and Hall, New York.

Gardner, R. H., O'Neill, R. V., Mankin, J. B., and Carney, J. H. (1981). "A Comparison of Sensitivity and Error Analysis Based on a Stream Ecosystem Model," *Ecological Modeling*, 12:173–190.

Gleason, J. R. (1988). "Algorithms for Balanced Bootstrap Simulations," *The American Statistician*, **42**(4): 263–266.

Golub, G. H., and C. F. Van Loan. (1989). *Matrix Computations*, 2d edition, The John Hopkins University Press, Baltimore, MD, p. 642.

Gould, H., and J. Tobochnik. (1988). *An Introduction to Computer Simulation Methods: Applications to Physical Systems, Part 2*, Addison-Wesley, Reading, MA, p. 372.

Hammersley, J. M., and K. W. Morton. (1956). "A New Monte Carlo Technique Antithetic-Variates," *Proceedings of Cambridge Physics Society*, **52**:449–474.

Hull, T. E., and A. R. Dobell. (1964). "Mixed Congruential Random Number Generators for Binary Machines," *Journal of the Association of Computing Machinery*, **11**:31–40.

Iman, R. L., and J. C. Helton. (1985). "A Investigation of Uncertainty and Sensitivity Analysis Techniques for Computer Models," *Risk Analysis*, 8(1):71–90.

International Mathematical and Statistical Library (IMSL) IMSL, Houston, TX, 1980.

Jaffe, P. R., and R. A. Ferrara. (1984). "Modeling Sediment and Water Column Interactions for Hydrophobic Pollutants, Parameter Discrimination and Model Response to Input Uncertainty," *Water Research*, 18(9):1169–1174.

Johnson, M. E. (1987). *Multivariate Statistical Simulation*, John Wiley and Sons, New York.

Knuth, D. E. (1981). *The Art of Computer Programming: Seminumerical Algorithms*, Vol. 2, 2nd ed., Addison-Wesley, Reading, MA.

Laurenson, E. M., and R. G. Mein. (1985). RORB-Version 3 Runoff Routing Program User Manual, Department of Civil Engineering, Monash University, Clayton, VIC, Australia.

Law, A. M., and W. D. Kelton. (1991). *Simulation Modeling and Analysis*, McGraw-Hill, New York, p. 759.

Lee, H.Y., D. L. Young. and L. H. Huang. (1990). "A Study of the Effects of Sand and Gravel Mining Operation on the Cross-River Structures and Channel Stability of the Tansui River." *Technical Report* No. 119, Hydraulic Research Lab., National Taiwan University, Taiwan (in Chinese).

Lehmer, D. H. (1951). "Mathematical Methods in Large-Scale Computing Units," *Ann. Comp. Lab.*, Harvard University Press, Cambridge, MA, **26**:141–146.

Li, S. T., and J. L. Hammond. (1975). "Generation of Psuedorandom Numbers with Specified Univariate Distributions and Covariance Matrix," *IEEE Transaction on Systems, Man, and Cybernatics*, 557–561.

Lu, Z., and Y. K. Tung. (2003). "Effects of Parameter Uncertainties on Forecast Accuracy of Xinanjiang Model," In *Proceedings of 1st International Yellow River Forum on River Basin Management*, Zhengzhou, China, 12–15 May.

MacLaren, M. D., and G. Marsaglia. (1965). "Uniform Random Number Generators," *Journal of the Association of Computing Machinery*, **12**:83–89.

Manache, G. (2001). "Sensitivity of a Continuous Water-Quality Simulation Model to Uncertain Model-Input Parameters," *Ph.D. Thesis, Chair of Hydrology and Hydraulics*, Vrije Universiteit Brussel, Brussels, Belgium.

Manache, G., and C. S. Melching. (2004). "Sensitivity Analysis of a Water-Quality Model Using Latin Hypercube Sampling," *Journal of Water Resources Planning and Management*, ASCE, **130**(3):232–242.

Marsagalia, G., and T. A. Bray. (1964). "A Convenient Method for Generating Normal Variables," *SIAM Review*, **6**:260–264.

McGrath, E. I. (1970). *Fundamentals of Operations Research*, West Coast University Press, San Francisco, CA.

McKay, M. D. (1988). "Chapter 4: Sensitivity and Uncertainty Analysis Using a Statistical Sample of Input Values," In: *Uncertainty Analysis*, Y. Ronen (ed.), CRC Press, Boca Raton, FL.

Melching, C. S. (1992). "An Improved First-Order Reliability Approach for Assessing Uncertainties in Hydrologic Modeling," *Journal of Hydrology*, **132**:157–177.

Melching, C. S. (1995). "Reliability Estimation," Chap. 3 in *Computer Models of Watershed Hydrology*, V. P. Singh (ed.), Water Resources Publications, Littleton, CO, pp. 69–118.

Melching, C. S. (2001). "Sensitivity Measures for Evaluating Key Sources of Modeling Uncertainty," In *Proceedings of International Symposium on Environmental Hydraulics, IAHR*, 5-8, December, Tempe, AZ.

Melching, C. S., and C. G. Yoon (1996). "Key Sources of Uncertainty in QUAL2E Model of Passaic River," *Journal of Water Resources Planning and Management*, ASCE, 122(2):105–113.

Melching, C. S., and W. Bauwens. (2001). "Uncertainty in Coupled Non-Point Source and Stream Water-Quality Models," *Journal of Water Resources Planning and Management*, ASCE, **127**(6): 403–413.

Miller, R. G. (1974). "The Jackknife—A Review," *Biometrika*, **6**(1):1–17.

Nataf, A. (1962). "D_termination des distributions de probabilit_s dont les marges sont donn_es," *Computes Rendus de l'Acad_mie des Sciences*, Paris, **255**:42–43.

Nguyen, V. U., and R. N. Chowdhury (1985). "Simulation for Risk Analysis with Correlated Variables," *Geotechnique*, **35**(1):47–58.

Nguyen, V. U., and A. J. Raudkivi (1983). "Analytical Solution for Transient Two-Dimensional Unconfined Groundwater Flow," *Hydrologic Sciences Journal*, **28**(2):209–219.

Olmstead, P. S. (1946). "Distribution of Sample Arrangements for Runs Up and Down," *Annals of Mathematical Statistics*, **17**:24–33.

Parrish, R. S. (1990). "Generating Random Deviates from Multivariate Pearson Distributions," *Computational Statistics & Data Analysis*, **9**:283–295.

Pebesma, E. J., and G. B. M. Heuvelink (1999). "Latin Hypercube Sampling of Gaussian Random Fields," *Technometrics*, **41**(4):303–312.

Press, W. H., B. P. Flannery, S. A. Teukolsky, and W. T. Vetterling (1989). *Numerical Recipes in Pascal: The Art of Scientific Computing*, Cambridge University Press, New York.

Press, W. H., B. P. Flannery, S. A. Teukolsky, and W. T. Vetterling (1992). *Numerical Recipes in FOR-TRAN: The Art of Scientific Computing*, Cambridge University Press, New York.

Press, W. H., B. P. Flannery, S. A. Teukolsky, and W. T. Vetterling (2002). *Numerical Recipes in C++: The Art of Scientific Computing*, Cambridge University Press, New York.

Quenouille, M. (1956). "Notes on Bias in Estimation," *Biometrika*, **43**:353–360.

Ronning, G. (1977). "A Simple Scheme for Generating Multivariate Gamma Distributions with Non-Negative Covariance Matrix," *Technometrics*, **19**(2):179–183.

Rosenblatt, M. (1952). "Remarks on Multivariate Transformation." *Annals of Mathematical Statistics*, **23**:470–472.

Rubinstein, R. Y. (1981). *Simulation and The Monte Carlo Method*, John Wiley and Sons, New York, p. 278.

Saltelli, A., and T. Homa. (1992). "Sensitivity Analysis for Model Output: Performance of Black Box Techniques on Three International Benchmark Exercises," *Computational Statistics and Data Analysis*, **13**(1):73–94.

Saltelli, A., T. H. Andres, and T. Homa. (1993). "Sensitivity Analysis of Model Output: An investigation of New Techniques," *Computational Statistics and Data Analysis*, **13**(1):211–238.

Sohrabi, T. M., A. Shirmohammadi, T. W. Chu, H. Montas, and A. P. Nejadhashemi. (2003). "Uncertainty Analysis of Hydrologic and Water Quality Predictions for a Small Watershed Using SWAT2000," *Environmental Forensics*, **4**(4):229–238.

Tung, Y. K., and L. W. Mays. (1981). "Reducing Hydrologic Parameter Uncertainty," *Journal of the Water Resources Planning and Management Division*, ASCE, **107**(WR1):245–262.

Tung, Y. K., and L. W. Mays. (1982). "Optimal Risk-Based Hydraulic Design of Bridges," *Journal of the Water Resources Planning and Management Division*, ASCE, **108**(WR2):191–203.

U.S. Army Corps of Engineers. (2001). *Hydrologic Modeling System, HEC-HMS: User's Manual*, Version 2.1, Hydrologic Engineering Center, Davis, CA.

Vale, C. D., and V. A. Maurelli. (1983) "Simulating Multivariate Nonnormal Distributions," *Psychometrika*, **48**(3):465–471.

von Neumann, J. (1951). "Various techniques used in connection with random digits," U.S. National Bureau of Standards, *Applied Mathematics Series*, **12**:36–38.

Yeh, K. C., and Y. K. Tung. (1993). "Uncertainty and sensitivity of a pit migration model," *Journal of Hydraulic Engineering*, ASCE, **119**(2):262–281.

Young, D. M., and R. T. Gregory (1973). *A Survey of Numerical Mathematics, Vol. II*, Dover Publications, New York.

Yu, P. S., T. C. Yang, and S. J. Chen. (2001). "Comparison of Uncertainty Analysis Methods for a Distributed Rainfall-Runoff Model," *Journal of Hydrology*, **244**:43–59.

Zhao, B., Y. K. Tung, K. C. Yeh, and J. C. Yang. (1997). "Storm Resampling for Uncertainty Analysis of a Multiple-Storm Unit Hydrograph," *Journal of Hydrology*, **194**:366–384.

ABOUT THE AUTHORS

YEOU-KOUNG TUNG, Ph.D., is a Professor of Civil Engineering at Hong Kong University of Science and Technology. The author of numerous technical papers on hydrology and risk analysis, he received his B.S. in Hydraulic Engineering from Tamkang University, Taiwan and his M.S. and Ph.D. in civil engineering from the University of Texas at Austin.

BEN-CHIE YEN, Ph.D., (deceased) was a Professor of Civil and Environmental Engineering at the University of Illinois at Urbana–Champaign. He held a B.S. in civil engineering from National Taiwan University and M.S. and Ph.D. degrees in civil engineering from the University of Iowa. He worked with surface water and urban hydrology problems, risk and reliability analysis, and open channel and river hydraulics for more than 30 years, and was the author of over 200 published technical papers and coauthor of eight books.

NOTES

NOTES

NOTES

NOTES

NOTES

NOTES

NOTES

NOTES

NOTES

NOTES

NOTES

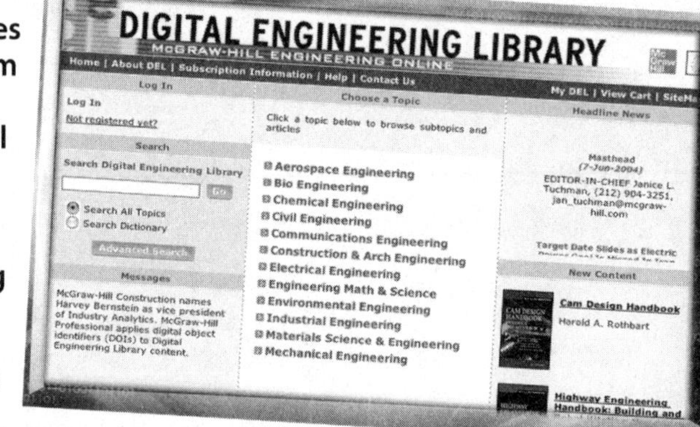